ECOSYSTEM DIVERSITY AND CARBON SEQUESTRATION

Climate Change Challenges and a Way Out for Ushering in a Sustainable Future

ECOSYSTEM DIVERSITY AND CARBON SEQUESTRATION

Climate Change Challenges and a Way Out for Ushering in a Sustainable Future

By
P.L. Gautam
Vir Singh
Uma Melkania

2021
Daya Publishing House®
A Division of
Astral International Pvt. Ltd.
New Delhi – 110 002

First Published, 2009
ISBN: 9789351240099 (Int. Edn.)

Published by : **Daya Publishing House®**
A Division of
Astral International Pvt. Ltd.
– ISO 9001:2015 Certified Company –
4736/23, Ansari Road, Darya Ganj
New Delhi-110 002
Ph. 011-43549197, 23278134
e-mail: info@astralint.com
Website: www.astralint.com

Digitally Printed at : **Replika Press Pvt. Ltd.**

Editorial

Revive our Tormented Planet: Ecosystem Diversity and Carbon Sequestration Hold the Key

Vir Singh, Uma Melkania and P.L. Gautam

Global temperatures are soaring. Planet Earth is warming up gradually and steadily. A formidable picture of the future is being projected: Earth will be 1.4–5.8^{0}C warmer by the end of the century, glacier melt, gloom on Antarctica, catastrophic rise in sea level, species extinction, danger to humanity, and the likes. Climate change is already proving a nightmare for the humanity. Earth's warming climate is estimated to contribute to more than 150,000 deaths and five million illnesses every year and, according to the World Health Organisation (WHO), this toll could double by 2030. Reports pouring in suggest that the ominous climate change is driving up rates of malaria, malnutrition and diarrhoea throughout the world. Worst outbreak of dengue fever in South Asia is also attributable to warmer temperatures.

What will happen to our Earth, the only known living planet in the cosmos? What will be the fate of the whole cosmos if the planet ceases to be living one? There is no answer in black and white to these questions. But we know the current state of our planet. It is definitely in a pathological state. One thing that is also certain is that it has not yet assumed the state of no-return. It can be revived. It can be revived because its pathogenesis is neither a cosmic phenomenon nor a global one, but is triggered by anthropogenic activities. It is not that humanity is not capable to effectively control and reverse the global problem. We are capable. Our governing systems are capable. Our institutional machinery is capable. Since the problem is evidently anthropogenic, its solution also lies in our own control and command. We can and we should revive our tormented planet. We have to do that, for the revival of the planet is linked with our destiny which is bound with the welfare of its life systems.

Evolution has been transforming the planet Earth for 3.5 billion years or so, and in this period it has brought in life and is subsequently maintaining it. Earth's evolution as a living planet was possible thanks to appropriate environmental conditions. Appropriate temperature range was a decisive factor for the environmental conditions conducive to perpetuation of life. It is thanks to the appropriate temperature range on account of which life flowered in myriad forms–numerous ecosystems, millions of species and their genotypes. Biodiversity evolved and multiplied thanks to appropriate environmental conditions largely set by temperature factor. Biodiversity, in turn, also contributed to maintain appropriate environmental conditions including maintenance of planet's average temperature.

Increasing global temperature, or the global warming, is leading to climate change which is one of the greatest challenges humanity has ever faced. But, neither global warming nor its consequence–the climate change–is the basic issue. These are merely a consequence of increasingly accumulating carbon dioxide and other greenhouse gases in the atmosphere. Rate of accumulation of these gases exceeds the capacity of natural ecosystems and biological processes to trap them into plant biomass and soils. Thus, reduced rate of the removal of carbon from the atmosphere and maintaining greenhouse gases at a minimum level in the atmosphere are the real issues that need be tackled and resolved. In other words, carbon sequestration is the real issue that needs to be addressed. Since carbon sequestration is inevitably linked with the ecosystems of the Earth, state of biodiversity at ecosystem level is the another issue that needs to be addressed.

An Agenda for the New World

Our world has undergone sea changes. The changes have occurred only during last few years of the last century. One of the most significant changes that without doubt has become irreversible is this that we now no longer live in or are a part of natural ecosystems, but live in and are the part of anthropogenic ecosystems. These ecosystems are not absolutely the solar-powered ecosystems. The process of further expansion, intensification and modification of the human-dominated ecosystems is gradually and gradually pushing the natural and last-remaining virgin ecosystems towards periphery.

Human species becoming dominant in the bio-community operates the anthropogenic ecosystems at the great expense of fossil fuels. Anthropogenic dimensions rest on exhaustion of enormous fossil fuel energy, manipulate the natural ecosystems in short-term favours of the human community, pollute the environment and release greenhouse gases in the atmosphere.

The processes of privatization, liberalization and globalization are bound to rampantly affect the whole biosphere. Economic growth virtually has inverse relationships with ecosystem resources and quality of the environment. Ecosystem resources diminish and environmental quality deteriorates proportionately with increase in economic growth. The globalization we are witnessing is economy-centred and gives way to rampant exploitation of ecosystem and energy resources to the further shrinkage of nature and environmental pollution. Can the international protocols relating to carbon emission reduction and biodiversity conservation, millennium goal, etc. be of any relevance in our economically globalised, culturally shattered and ecologically fissured world?

This issue has been attempted in the first Section of the book–Agenda for the Revival of Our Tormented Planet: Issues Facing the Ecologically Shattered and Economically Globalised World. The first chapter by Vir Singh and PL Gautam presents an overview of ecosystem diversity and carbon sequestration and discusses some contemporary issues confronting humanity. The second chapter by SP Singh goes deeper into global climate change which is posing a challenge before humanity. The third chapter by JS Bali discusses forest management for livelihoods and carbon sequestration and calls for harmony within natural elements and regards this call a mantra for human happiness. In chapter fourth, Vishal Mahajan and Kamal Kishor Sood present a vision on carbon sequestration, while in chapter fifth BS Tewari and Ajay project carbon as a material for the 21st century, highlighting its prospects and promises.

Ecosystems' Well-being for Sustenance and Posterity of Life

Carbon sequestration is one of the most indispensable phenomena for the sustenance and enhancement of life on Earth. Ecosystems are the organs of living body of the Earth. Their well-being is of absolute importance for the well-being of the living planet. Ecosystem diversity is an indicator of ecosystem health and well-being. Different ecosystems have different capacities for trapping carbon dioxide in plants and soils. The issues of ecosystem diversity and carbon sequestration are inseparable. Hence, we have attempted to integrate the twin issues into the main title of the book: *Ecosystem Diversity and Carbon Sequestration.*

Carbon sequestration in nature is of critical value for resolving vital issues of our times, namely the state of ecological paucity, natural resource management, global warming, climate change and sustainable development. It is the free carbon in nature, particularly in the form of CO_2 that is responsible for most of the ills of our environment and that makes future of life on Earth bleak and unsustainable. Carbon sequestration in living ecosystems, therefore, is the major challenge our contemporary society faces at this juncture of time. Time always goes like an arrow, its pointed head always pointing towards future. Future of time is inevitable even without humanity. The sustainable future is one that is to be witnessed by blossoming and flourishing humanity. Hence, the subtitle of the book: *Climate Change Challenges and a Way out for Ushering in a Sustainable Future.*

Living planet combats cosmic entropy and helps put the whole cosmos in a balance. Unfortunately, our living planet is tormented thanks to callous anthropogenic activities. Reviving our tormented Earth is not only necessary but is an imperative of our destiny. This book, in essence, calls for the revival of our tormented Earth, which is not just possible, but is also not a difficult task should we have little restraint on our ways of living.

Second Section of the book is devoted to Ecosystem diversity. India can boast of her unique diversity in ecosystems. Scenarios, anthropogenic dimensions and management relating to ecosystems have been elaborated in five chapters in this

Section. JB Lal presents a panorama of forest ecosystems in India and elaborates their role in carbon sequestration. Sandeep Tripathi and VRS Rawat have looked into operationalisation of CDM afforestation and reforestation projects in India and have analysed the barriers at national and international level. Microbial diversity is an indicator of soil fertility status. Saurindra Narayan Goswamy and Soneswar Sharma ponder over this aspect and put forward useful tips for redevelopment of humid subtropical perturbed ecosystems. B Mohan Kumar attempts to link ecosystem diversity with sustainability. Anubhav, Kundan Singh, Akanksha Rastogi and Vir Singh present a panorama of the somewhat neglected sacred groves in India which are vital instruments of nature to keep the carbon dioxide levels at bay.

Life on Edge

Global warming and consequent climate change thanks particularly to anthropogenic activities including reduced rate of carbon sequestration are being in long-term set to bring life to a point of total disaster. Alarming trends gradually culminating into species extinction is an evidence that life on Earth is on edge. There are conspicuous examples of species extinction from Costa Rica, Europe (particularly United Kingdom), Central America, Antarctica, and many evidences from oceanic life (Ragupathy Kannan, Chapter 11). Climate change is also set to trigger negative trends in agriculture and create a condition of food insecurity (BS Mahapatra, AP Singh, AK Chaubey and DK Shukla, Chapter 12). Effect of climate change would lead to significant decline in overall agricultural productivity which would also have phenomenal repercussions on our food security. It therefore requires necessary adjustments in agriculture (SK Saini, Yogendra Pal and Amit Bhatnagar, Chapter 13). Global annual production of methane from the livestock is as high as 80 Tg from enteric fermentation and 25 Tg from their excreta. India's contribution to this production is around 11 and 1 per cent, respectively, with about 15.2 per cent of the domesticated ruminants of the world. Although, India is responsible for producing less methane per head of livestock, but we must look into the methods/ techniques to reduce methanogenesis as we are bound by Kyoto Protocol to reduce methane emission by 2012 to the level of methane produced in 1990, which may not be an easy task (DN Kamra and Someshwar Zadbuke, Chpater 14).

A Search for Viable Options

Bio fertilizers have a crucial role to play in the management of soil fertility in hill agro-ecosystems (Susheela Negi, GK Dwivedi and RV Singh, Chapter 15). The Green Revolution bowl of Western Uttar Pradesh reels under catastrophic industrial pollution. Sugarcane cultivation has led to proliferation of sugarcane industries. Effluent of sugar industries has phenomenal effect on the germination of seeds and hence on the agriculture of the area (Neelam and Ila Prakash, Chapter 16). Soil carbon sequestration is affected by the type of vegetation. How the hybrid *Eucalyptus* plantation affects this natural phenomenon is discussed on the basis of the information derived from a research experiment (Asha Upadhyay and Uma Melkania, Chapter 17). Gassification technology could be applicable for solving energy crises in rural areas and this would also be critical to the processes of sustainable development (Raj

Narayan Pateriya and Sadachari Singh Tomar, Chapter 18). Arbuscular micorrhizal fungi could be of potential implications for carbon sequestration (Rashmi Srivastava, Shruti Chaturvedi, Preeti Chaturvedi and AK Sharma, Chapter 19). Molecular means for promoting organic farming are not very familiar to people. Role of plant transcription factor–dof (dna binding with one finger only) contributes to enhance nitrogen use efficiency, which would be vital for the development of organic farming (Dinesh Yadav, Nidhi Gupta, Anil Kumar, Pushpa Lohani, Munna Singh and US Singh, Chapter 20). Our textile industry can also contribute to enhance carbon sequestration. There are several plants which yield fibre for textile industry while playing their natural role of trapping carbon dioxide of the atmosphere and in this way we can bank on ecological attributes of economic plants (Sapna Gautam and Uma Melkania, Chapter 21).

Rejuvenating the Fragile Mountain Ecosystems

Himalayan mountains are the youngest, the highest and the most fragile ecosystems on Earth. Their inherent characteristics or specificities are altogether distinguishable from those of other ecosystems on Earth. The fragility-, marginality- and inaccessibility-ridden ecosystems embrace very high degree of diversity and ecological niches and require altogether different approaches to their rejuvenation. Rejuvenated, these ecosystems would have enormous bearing on carbon sequestration and, consequently, will substantially contribute to contain temperature rising in a big way. A Section in the book, therefore, has been devoted to the Himalayan Mountains. It contains three crucial chapters. Isabelle Providoli, Sanjeev Bhuchar, Keshar Man Sthapit, Madhav Dhakal and Eklabya Sharma have elaborated on the potential effects of sustainable land management options on carbon sequestration in upland soils in the Himalayas. This piece is the product of their multidisciplinary research at the International Centre for Integrated Mountain Develepoment (ICIMOD) based in Kathmandu, Nepal (Chapter 22). Rangelands occupy the largest chunk of geographical area in the Himalayan mountains. An effective management of these ecosystems would, therefore, have phenomenal impact on carbon sequestration. Vir Singh and RD Gaur gave brought to the fore interesting panorama of the fodder species occurring in the Himalayan mountains. One of the objectives of the range management in the fragile mountain areas should be to enhance carbon sequestration (Chapter 23). ML Dewan emphasizes on Himalayan conservation and development, for that would help the Earth reverse the trend of global warming (Chapter 24).

A Life-building, Life-saving, Life-enhancing and Life-sustaining Phenomenon

Life on Earth is attributable to the vital phenomenon of photosynthesis. With photosynthesis is associated carbon sequestration. How and why carbon sequestration is a life-building, life-saving, life-enhancing and a life-sustaining phenomenon has focused through four articles in a separate section. Munna Singh elaborates on global warming mitigation through improved carbon economy linked with photosynthesis (Chapter 25). B Mishra and KP Raverkar ponder over the state of carbon sequestration on agricultural lands (Chapter 26). Soil is a potential source of carbon sequestration and can help mitigate climate changes. JS Chauhan, Bineet Singh and JPN Rai have discussed on such issues (Chapter 27). Shiwani Bhatnagar and AK Karnatak ponder over some startegies of enhancing the phenomenon of carbon sequestration (Chapter 28).

Potential Mechanisms to Trap Carbon

We need to operationalise carbon sequestration enhancement in a number of ways. There are potential technological and institutional mechanisms including carbon trading and policies which might arrest climate change and reverse the trends of global warming. This section collates articles on such issues. Alok Mukherjee presents beautiful account of the coastal wetland ecosystems in sequestering carbon directly by geologic repositories and phytoplankton fertilization, discussing valuable strategies for maintaining ecological integrity of such ecosystems (Chapter 29). Analysing the various aspects relating to carbon sequestration, Vikram Singh Rathore brings out philosophical and practical aspects of mitigating environmental and socio-economic impacts of global warming (Chapter 30). Govind Singh Kushwaha discusses carbon mitigation and social issues relating to forest management (Chapter 31). AK Singh and Virendra Singh bring to the fore some economic issues relating to the enhancement of carbon sequestration in India (Chapter 32). Tirthankar Banerjee, Jyotsana Pathak and RK Srivastava ponder over climate change and Kyoto Protocol (Chapter 33). Rinku Verma analyses the state of global climate and the laws meant to control climate (Chapter 34).

Towards a Sustainable Future

This section devotes to eco-ethics, eco-philosophy and psychology which are central elements pivoting conservation-oriented transcendental development. Vanmathy. A and Abha Ahuja attempt to synthesise eco-ethics of biodiversity conservations for sustainable development (Chapter 35) and socio-cultural values contributing to promote nature's conservation (Chapter 38). A debate on environmental services or ecosystem services appears to have sparkled these days. Vir Singh tries to resolve the dilemma by suggesting a framework for valuing environmental services against the backdrop of eco-philosophy and by advocating the goal of global happiness (Chapter 36). Subaran Singh presents some tips on conservation of ecosystems for carbon sequestration and also tries to equip common man with the workable knowledge of common terms and notions (Chapter 37). In the last Chapter, Govind Singh Kushwaha and Vir Singh discuss environmental psychology in landscaping which serves as an important tool in the operationalisation of sustainability (Chapter 39).

Acknowledgements

This work would have not been accomplished without the support of the Department of Science and Technology and the GB Pant University of Agriculture and Technology. We would like to acknowledge the help rendered by Drs. BRK Gupta, Dean College of Basic Sciences and Humanities, RK Gupta, JPN Rai, RK

Srivastava, Babita Bohra, Govind Singh Kushwaha, AK Gaur, Munna Singh, Alok Shukla, SK Guru, Vikram S Rathore, JB Lal, Alok Mukherjee, DN Kamra, Susheela Negi, BM Kumar, Reeta Goel, Neelendra Kumar Joshi and Isabelle Providoli.

Post-script

The views and interpretations in the book are those of the author(s) and are not necessarily attributable to the editors or their respective organizations. We earnestly hope that the book would serve its purpose and will, in a big way, be usable for the cause of the Earth.

Contents

ECOSYSTEM DIVERSITY IN INDIA
Scenarios, Anthropogenic Dimensions and Management

LIFE ON EDGE
Climate Change, Critical Environmental Problems, Alarming Trends, Species Extinction and the Likes

ENVIRONMENTAL MANAGEMENT
A Search for Viable Options

HIMALAYAN MOUNTAINS
Rejuvenated Fragile Ecosystems can Give Appropriate Response to Global Warming

CARBON SEQUESTRATION
A Life-building, Life-sustaining and Life-enhancing Phenomenon on Earth

ENHANCING CARBON SEQUESTRATION IN NATURE
Potential Technological and Institutional Mechanisms, Carbon Trading and Policies

USHERING IN A SUSTAINABLE FUTURE
Eco-ethics, Eco-philosophy and Psychology as Core Elements Pivoting Conservation-oriented Transcendental Development

List of Contributors

Abha Ahuja Department of Human Development and Family Studies, College of Home Science, G.B. Pant University of Agriculture and Technology, Pantnagar–263 145, Uttarakhand, India

Ajay Department of Physics, G.B. Pant University of Agriculture and Technology, Pantnagar–263 145, Uttarakhand, India

AK Chaubey Department of Agronomy, College of Agriculture, G.B. Pant University of Agriculture and Technology, Pantnagar–263 145, Uttarakhand, India

AK Karnatak Department of Entomology, G.B. Pant University of Agriculture and Technology, Pantnagar–263 145, Uttarakhand, India

AK Sharma Department of Biological Sciences, CBSH, G.B. Pant University of Agriculture and Technology, Pantnagar–263 145, Uttarakhand, India

AK Singh Department of Agricultural Economics, College of Agriculture, G.B. Pant University of Agriculture and Technology, Pantnagar–263 145, Uttarakhand, India

Akanksha Rastogi Department of Environmental Science, College of Basic Sciences and Humanities, G.B. Pant University of Agriculture and Technology, Pantnagar–263 145, Uttarakhand, India

Alok Mukherjee SASCOM, Centre on Global Change, RASD/TEC Building, NPL, Dr. K.S. Krishnan Marg, New Delhi-110012

Amit Bhatnagar Department of Agronomy, College of Agriculture, G.B. Pant University of Agriculture and Technology, Pantnagar–263 145, Uttarakhand, India

Anil Kumar Department of Molecular Biology and Genetic Engineering, College of Basic Sciences and Humanities, G.B. Pant University of Agriculture and Technology, Pantnagar–263 145, Uttarakhand, India

Anubhav Department of Environmental Science, College of Basic Sciences and Humanities, G.B. Pant University of Agriculture and Technology, Pantnagar–263 145, Uttarakhand, India

AP Singh Department of Soil Science, College of Agriculture, G.B. Pant University of Agriculture and Technology, Pantnagar–263 145, Uttarakhand, India

Asha Upadhyay Department of Environmental Science, College of Basic Sciences and Humanities, G.B. Pant University of Agriculture and Technology, Pantnagar–263 145, Uttarakhand, India
(E-mail: asha_upadhyayenv@yahoo.co.in)

B Mishra Department of Soil Science, College of Agriculture, G.B. Pant University of Agriculture and Technology, Pantnagar–263 145, Uttarakhand, India

B Mohan Kumar Department of Social Sciences and Humanities, College of Basic Sciences and Humanities, G.B. Pant University of Agriculture and Technology, Pantnagar–263 145, Uttarakhand, India

Babita Bohra International Centre for Integrated Mountain Development (ICIMOD), Khumaltar, Lalitpur, GPO Box 3226, Kathmandu, Nepal

Bineet Singh Ecotechnology Laboratory, Department of Environmental Science, G.B. Pant University of Agriculture and Technology, Pantnagar–263 145, Uttarakhand, India

BS Mahapatra Department of Agronomy, College of Agriculture, G.B. Pant University of Agriculture and Technology, Pantnagar–263 145, Uttarakhand, India

BS Tewari Department of Physics, G.B. Pant University of Agriculture and Technology, Pantnagar–263 145, Uttarakhand, India

Dinesh Yadav Department of Molecular Biology and Genetic Engineering, College of Basic Sciences and Humanities, G.B. Pant University of Agriculture and Technology, Pantnagar–263 145, Uttarakhand, India

DK Shukla Department of Agronomy, College of Agriculture, G.B. Pant University of Agriculture and Technology, Pantnagar–263 145, Uttarakhand, India

DN Kamra Centre of Advanced Studies in Animal Nutrition, Indian Veterinary Research Institute, Izatnagar–243 122, India

Eklabya Sharma International Centre for Integrated Mountain Development (ICIMOD), Khumaltar, Lalitpur, GPO Box 3226, Kathmandu, Nepal

GK Dwivedi Hill Campus, GB Pant University of Agriculture and Technology, Ranichauri–249199, Tehri Garhwal, Uttarakhand

Govind Singh Kushwaha Department of Social Sciences and Humanities, College of Basic Sciences and Humanities, G.B. Pant University of Agriculture and Technology, Pantnagar–263 145, Uttarakhand, India
(E-mail: drgovindsingh1973@rediffmail.com)

Ila Prakash Departmant of Botany, D. N. College, Meerut–250 002, (U.P.) India

Isabelle Providoli International Centre for Integrated Mountain Development (ICIMOD), Khumaltar, Lalitpur, GPO Box 3226, Kathmandu, Nepal

JB Lal E-16, Shankar Nagar, Bhopal–462016, M.P., India (E-mail: balij_dehradun@vsnl.net)

JPN Rai Ecotechnology Laboratory, Department of Environmental Science, G.B. Pant University of Agriculture and Technology, Pantnagar–263 145, Uttarakhand, India

JS Bali 157, Madhuban, New Delhi–110 092, India

JS Chauhan Ecotechnology Laboratory, Department of Environmental Science, G.B. Pant University of Agriculture and Technology, Pantnagar–263 145, Uttarakhand, India

Jyotsana Pathak Department of Environmental Sciences, College of Basic Sciences and Humanities, G.B. Pant University of Agriculture and Technology, Pantnagar–263 145, Uttarakhand, India

Kamal Kishor Sood Division of Agroforestry, Shere-Kashmir University of Agricultural Sciences and Technology, Faculty of Agriculture, Chatha, Jammu (J and K) India

Keshar Man Sthapit International Centre for Integrated Mountain Development (ICIMOD), Khumaltar, Lalitpur, GPO Box 3226, Kathmandu, Nepal

KP Raverkar Department of Soil Science, College of Agriculture, G.B. Pant University of Agriculture and Technology, Pantnagar–263 145, Uttarakhand, India

Kundan Singh Department of Environmental Science, College of Basic Sciences and Humanities, G.B. Pant University of Agriculture and Technology, Pantnagar–263 145, Uttarakhand, India

Madhav Dhakal International Centre for Integrated Mountain Development (ICIMOD), Khumaltar, Lalitpur, GPO Box 3226, Kathmandu, Nepal

ML Dewan N-44, Panchshila Park, New Delhi–110 017

Munna Singh Department of Plant Physiology, College of Basic Sciences and Humanities, G.B. Pant University of Agriculture and Technology, Pantnagar–263 145, Uttarakhand, India (e-mail: drmunnasingh@yahoo.com)

Neelam Departmant of Botany, D. N. College, Meerut–250 002, U.P., India

Nidhi Gupta Department of Molecular Biology and Genetic Engineering, College of Basic Sciences and Humanities, G.B. Pant University of Agriculture and Technology, Pantnagar–263 145, Uttarakhand, India

PL Gautam Indian Council of Agricultural Research (ICAR), Krishi Bhawan, New Delhi–110 001, India

Preeti Chaturvedi Department of Biological Sciences, CBSH, G.B. Pant University of Agriculture and Technology, Pantnagar–263 145, Uttarakhand, India

Pushpa Lohani Department of Molecular Biology and Genetic Engineering, College of Basic Sciences and Humanities, G.B. Pant University of Agriculture and Technology, Pantnagar–263 145, Uttarakhand, India

Ragupathy Kannan University of Arkansas–Fort Smith, USA

Raj Narayan Pateriya Deptt. of Farm Machinery and Power Engineering, College of Technology, G.B. Pant University of Agriculture and Technology, Pantnagar–263 145, Uttarakhand, India

Rashmi Srivastava Department of Biological Sciences, CBSH, G.B. Pant University of Agriculture and Technology, Pantnagar–263 145, Uttarakhand, India

RD Gaur Department of Botany, HNB Garhwal University, Srinagar–246 174 (Uttarakhand) India

Rinku Verma Karnataka Environment Research Foundation, Bangalore, India

RK Srivastava Department of Environmental Sciences, College of Basic Sciences and Humanities, G.B. Pant University of Agriculture and Technology, Pantnagar–263 145, Uttarakhand, India

RV Singh Hill Campus, G.B. Pant University of Agriculture and Technology, Ranichauri–249 199, Tehri Garhwal, Uttarakhand

Sadachari Singh Tomar Division of Agricultural Engineering, Indian Agricultural Research Institute (IARI), Pusa, New Delhi–110 012

Sandeep Tripathi Biodiversity and Climate Change Division, Indian Council of Forestry Research and Education, P.O. New Forest, Dehradun–248 006

Sanjeev Bhuchar International Centre for Integrated Mountain Development (ICIMOD), Khumaltar, Lalitpur, GPO Box 3226, Kathmandu, Nepal

Sapna Gautam College of home Science, CSK University of Agriculture and Technology, Palampur, Himachal Pradesh

Saurindra Nr Goswami Ecorestoration, Dimoria College, Khetri, Assam–782 403

Shiwani Bhatnagar Department of Entomology, G.B. Pant University of Agriculture and Technology, Pantnagar–263 145, Uttarakhand, India

Shruti Chaturvedi Department of Biological Sciences, CBSH, G.B. Pant University of Agriculture and Technology, Pantnagar–263 145, Uttarakhand, India

SK Saini Department of Agronomy, College of Agriculture, G.B. Pant University of Agriculture and Technology, Pantnagar–263 145, Uttarakhand, India

Someshwar S Zadbuke Centre of Advanced Studies in Animal Nutrition, Indian Veterinary Research Institute, Izatnagar–243 122, India

Soneswar Sarma Department of Biotechnology, Gauhati University, Guwahati–14, Assam

SP Singh HNB Garhwal University, Srinagar Garhwal, Uttarakhand, India (E-mail: surps@yahoo.com)

Subaran Singh Department of Molecular Biology and Genetic Engineering, College of Basic Sciences and Humanities, G.B. Pant University of Agriculture and Technology, Pantnagar–263 145, Uttarakhand, India

Susheela Negi Department of Science and Technology, Government of India, New Delhi, India

Tirthankar Banerjee Department of Environmental Sciences, College of Basic Sciences and Humanities, G.B. Pant University of Agriculture and Technology, Pantnagar–263 145, Uttarakhand, India

Uma Melkania Department of Environmental Science, College of Basic Sciences and Humanities, G.B. Pant University of Agriculture and Technology, Pantnagar–263 145, Uttarakhand, India (E-mail: umamelkania@yahoo.co.in)

US Singh Department of Plant Pathology, College of Agriculture, G.B. Pant University of Agriculture and Technology, Pantnagar–263 145, Uttarakhand, India

Vanmathy A Department of Human Development and Family Studies, College of Home Science, G.B. Pant University of Agriculture and Technology, Pantnagar–263 145, Uttarakhand, India

Vikram S Rathore Flat 803/IC Whispering Palms, Kandivili (E), Mumbai–400 101, India (E-mail: vic_rathore@yahoo.com)

Vir Singh Department of Environmental Science, College of Basic Sciences and Humanities, G.B. Pant University of Agriculture and Technology, Pantnagar–263 145, Uttarakhand, India (E-mail: drvirsingh@rediffmail.com)

Virendra Singh Department of Agricultural Economics, College of Agriculture, G.B. Pant University of Agriculture and Technology, Pantnagar–263 145, Uttarakhand, India

Vishal Mahajan KVK, Poonch, Shere-Kashmir University of Agricultural Sciences and Technology, Jammu (J and K) India

VRS Rawat Indian Council of Forestry Research and Education, P.O. New Forest, Dehradun–248 006

Yogendra Pal Department of Soil Science, G.B. Pant University of Agriculture and Technology, Krishi Vigyan Kendra, Matela (Kosi) Almora–263 651, Uttarakhand, India

AGENDA FOR THE REVIVAL OF OUR TORMENTED PLANET

Issues Facing the Ecologically Shattered and Economically Globalised World

Ecosystem Diversity and Carbon Sequestration: Climate Change Challenges and a Way Out for Ushering in a Sustainable Future *Pages 3–9*
Editors: P.L. Gautam, Vir Singh and Uma Melkania
Published by: DAYA PUBLISHING HOUSE, 2009

Chapter 1

Ecosystem Diversity and Carbon Sequestration: Some Issues Confronting Humanity

Vir Singh and PL Gautam

Terrestrial ecosystems are in perilous state. Aquatic ecosystems too are turning inhospitable. Industry and transport sectors are turning the environment increasingly less hospitable. Agriculture sector too is not environment-friendly and contributing in one way or the other to the worsening on our environment. Balance in photosynthesis by green plants and breathing by animals is at stake with accumulation of carbon dioxide (CO_2) to a critical level, and further set to cause greater concerns about the state of our biosphere, of our planet and of our lives.

Not only does the increased concentration of free CO_2 happen to be a matter of concern, methane (CH_4), nitrous gas (N_2O), CFC-12 (CCl_2F_2), HCFC-22 ($CHClF_2$), perfluoromethane (CF_4), sulphur hexafluoride (SF_6) are the other greenhouse gases which are increasing at an alarming rate.

Capacities of the Earth's ecosystems to sequester CO_2 are squeezing rapidly. Carbon sequestration, of course, offers potential solution to the vital issues of global warming and climate change confronting the humanity today.

The Ecosystem Diversity

Ecosystem diversity is crumbling at an alarming rate. A glimpse of deforestation would reveal it. We are losing about 11 million to 17 million hectare of forests in the whole world (http://www.kidsforfuture/ourfuture/000032.html 2006), *i.e.*, little less than one hectare to about five hectare per second. We thus can imagine the rate of CO_2 concentrations released into the atmosphere thanks to depletion of the ecosystem capability to sequester CO_2 as a result of callous anthropogenic activities. Again, "dead forests" are also being increasingly witnessed. These are the forests which are getting destroyed even without being cut down. They are dying due to acid rain. And, should the temperatures continue to rise thanks to global warming, more ecosystems are likely to vanish. If these trends continue, days are not far when this green planet of ours would turn brown.

In June 1992, the world community gave birth to the Convention on Biodiversity. One hundred and sixty-seven countries of the UNO signed the Convention. The Convention on Biodiversity came into force on December 29, 1993. As the best evidence of this crucial Convention's comprehensiveness is found in Article 1, which outlines the Convention's objectives, *viz.* Crattiger *et al.* (1994):

- ✰ The conservation of biodiversity at the genetic, species and ecosystem levels;
- ✰ The sustainable use of its components; and
- ✰ The fair and equitable sharing of benefits derived from the use of genetic resources.

These objectives embody the realization that biodiversity is a cross-sectoral issues with biological, ecological and socio-economic considerations. The Convention on Biodiversity should be regarded as one of the landmark events in the history from the viewpoint that the world community put its fingers on the pulse of life, intactness of life systems and, of course, the very principle of sustainability of biosphere and of the living planet. The Convention has been of phenomenal importance on account of its emphasis placed on approaching biodiversity conservation from a development perspective. Although virtually no, or only little values seem to have been attached with the obvious linkages between biodiversity conservation and carbon sequestration enhancement. A further emphasis on this important phenomenon of life vis-à-vis biodiversity conservation would be in favour.

India is one of the fewer mega diversity centres and has natural credit of two hot spots. We are endowed not only with very rich species and genetic diversity but also with unique ecosystem diversity. This ecosystem diversity in India ranges from tropical rain forests to wetlands to deserts, from alpine meadows to temperate forests to subtropical forests, from glacier-fed rivers to forest-fed rivers, from glacial lakes to vast open oceans, and the likes.

A spectacular body of literature on the diversity of ecosystems is available. However, their potential capabilities of carbon sequestration remain a less addressed issue. An assessment of the carbon sequestration capability of different types of ecosystems would impart a new dimension to the ecosystem/natural resource/forest management. This strategy would greatly help mitigating climate change or global warming with carbon-sequestration-oriented ecosystem management as the indispensable strategy.

The Greenhouse Gases

Carbon dioxide (CO_2), methane (CH_4), nitrous gas (N_2O), CFC-12 (CCl_2F2), HCFC-22 ($CHClF_2$), perfluoromethane (CF_4), sulphur hexafluoride (SF_6) are the greenhouse gases that are playing havoc. While CFC-12 (CCl_2F_2), HCFC-22 ($CHClF_2$), perfluoromethane (CF_4), sulphur hexafluoride (SF_6) existed at zero level in the pre-industrial age, the carbon dioxide, methane and nitrous oxide gas had only as low as 278000, 700 and 275 ppbv industrial concentrations in that age. Fossil combustion, land use conversion and cement production were attributable to exceedingly high release of carbon dioxide. Fossil fuels, rice cultivation, waste dumps and livestock have been the major contributing factors of methane increase. Fertilizer, industrial processes and combustion have led to substantial increase in the concentration of nitrous gas. Liquid coolants and foam are responsible for CFC. Liquid coolants have also contributed to HCFC. Production of aluminum is largely responsible for perfluoromethane. Dielectric fluid gave rise to sulphur hexafluoride.

Amongst all the greenhouse gases, CO_2 is the most responsible one for the warming of Earth's environment, which can be blamed for more than 60 percent of the cause of greenhouse effect. Methane is at second place that is responsible for about 20 per cent of this effect. Chlorofluorocarbons, the third most damaging agents, cause about 10 per cent of the greenhouse effect.

India's greenhouse gases inventory suggests that Energy and Transformation Industries contribute nearly 29 per cent to the greenhouse gases, other industries 12 per cent, transport about 7 per cent, industrial processes 8 per cent and about 22 per cent from agriculture (15 per cent from enteric fermentation and 7 per cent from rice cultivation). Considerably large CO_2 emissions are generated from households. In addition to breathing, we contribute to CO_2 production (and other greenhouse gases) while watching television, turning on lights and refrigerators, air-conditioners, cooking, etc. Our elevated living standards and expensive lifestyles and production of greenhouse gases and pollutants virtually go hand-in-hand.

The Greenhouse Effect and Climate Change

The earth's climate is driven by a continuous flow of energy from the sun. The earth must send this energy back to sun in the form of infra red radiation. Greenhouse gases in the atmosphere, however, block this radiation from escaping directly from the surface of the Earth. Concentration of the greenhouse gases is increasing thanks to anthropogenic activities. As a result, since 1800, levels of CO_2 have risen by over 30 per cent. This increase over last 200 years is three times larger than that occurred in the earlier 10,000 years. Climate models predict that the global temperature will rise by about 1.4–5.8°C by the year 2100.

Carbon dioxide gas makes up only about 0.03 per cent of the air, but it acts as the "earth's sweater". And this "sweater" is becoming thicker due to which the earth is becoming warmer.

When there is global warming, there are numerous problems not simply that air is too hot. The world is experiencing more typhoons, hurricanes and storms because of global warming (<http: / /wwww.kdsforfuture.net/ourfuture/000022.html 2006>). If this trend continues, this much rise in the global temperature would be catastrophic for life systems. The possible consequences of global warming and climate change would be effect on glaciers and stream flows, very adverse effects forest ecosystems, flora and fauna, agriculture, and on society as a whole. Decline in winter rain fall and spell of drought during summer and increase in monsoon rainfall, etc. might also be experienced.

An overall picture of the global climate change can be summed up as follows (IPCC 2001):

ΔT over the 20th Century	+0.6±0.2°C
Rate of ΔT increase since 1950	+17°C/decade
Sea level rise over 20th Century	+0.1-0.2 m
Change in precipitation	+0.5-1 per cent/decade
Extreme events	+2-4 per cent

Climate change is already proving a cataclysm for the humanity. Earth's warming climate is estimate to contribute to more than 150,000 deaths and five million illnesses every year and, according to the World Health Organisation (WHO), this toll could double by 2030. Climate change is driving up rates of malaria, malnutrition and diarrhea throughout the world. Worst outbreak of dengue fever in South Asia is also attributable to warmer temperatures. WHO data also reveal that rising temperatures disproportionately affect poor countries that have done little to create the problem. The regions most at risk from climate change include the Asian and South American coasts along the Pacific Ocean, as well as the Indian Ocean coast and sub-Saharan Africa. Climate-sensitive diseases are more prevalent there because those regions are most vulnerable to abrupt shift to climate. Large cities also are likely to experience more severe health problems because they produce what scientists refer to as the urban "heat island" effect (Eilperin 2005).

What is Carbon Sequestration?

Earth's gradually but steadily becoming warmer is one of the grimmest and gravest issues humanity on earth has ever faced in the recorded history. An environment of optimism nevertheless prevails. CO_2 and other greenhouse gases cause havoc only when they are freed in the atmosphere. Carbon can be sequestered into ecosystems and excess of its can remain sunk where it should be. It can be fixed into plant biomass or in soil.

Carbon sequestration is the mechanism through which global warming and climate change could be controlled to a great extend. What is then carbon sequestration, the buzzing word of the day?

It can be defined as the capture and secure storage of carbon that would otherwise be emitted to or remain in the atmosphere, or to keep carbon emissions produced by human activities from reaching the atmosphere, by capturing and diverting them to secure storage. It is through this process that agriculture and forestry practices remove CO_2 from the atmosphere.

Scenario 1: Carbon Sequestration in Indian Soils

With a large land area and diverse eco-regions, there is a considerable potential of terrestrial/soil carbon sequestration in India. A study by Rattan Lal, Director of the Carbon Management and Sequestration Centre at the Ohio State University (2004) would reveal that of the total land area of 329 million hectares (Mha), 297 Mha is the land area comprising 162 Mha of arable land, 69 Mha of forest and woodland, 11 Mha of permanent pasture, 8 Mha of permanent crops and 58 Mha is other land uses. The soil organic carbon (SOC) pool is estimated at 21 Pg (petagram = Pg = 1×10^{15} g = billion ton) to 30-cm depth and 63 Pg to 150 cm depth.

The soil inorganic carbon (SIC) pool, according to Lal's (2004) estimates, is 196 Pg to 1-m depth. The SOC concentration in most cultivated soils is less than 5 g/kg compared with 15 to 20 g/kg in uncultivated soils. Low SOC concentration is attributed to plowing, removal of crop residue and other bio solids, and mining of soil fertility. Accelerated soil erosion by water leads to emission of 6 Tg C/y.

Lal (2004) further relates that important strategies of soil C sequestration include restoration of degraded soils, and adoption of recommended management practices (RMPs) of agricultural and forestry soils. Potential of soil C sequestration in India is estimated at 7 to 10 Tg C/y for restoration of degraded soils and ecosystems. Potential of soil C sequestration in India, according to Lal (2204), is estimated at 7 to 10 Tg C/y for restoration of degraded soils and ecosystems, 5 to 7 Tg C/y for erosion control, 6 to 7 kg C/y for adoption of RMPs on agricultural soils, and 22 to 26 Tg C/y for secondary carbonates. Total potential of soil C sequestration, thus, is 39 to 49 (44±5) Tg C/y.

Scenario 2: Carbon Sequestration Potential of Indian Forests

The forests are the most important terrestrial ecosystems from the point of view of carbon sequestration. The forestry sector not only sustains its carbon but also has the potential to absorb carbon from the atmosphere.

M. Lal and Roma Singh of the Centre for Atmospheric Sciences, Indian Institute of Technology, New Delhi have attempted to estimate carbon sequestration potential of Indian forests. India has maintained approximately 64 Mha of forest cover for the last decade. Lal and Singh (2004) maintain that rate of afforestation in India is one of the highest among the tropical countries, currently estimated to be 2 Mha per annum.

The annual productivity has increased from 0.7 m^3 per hectare in 1985 to 1.37 m^3 per hectare in 1995. Increase in annual productivity, according to Lal and Singh (2004), directly indicates an increase in forest biomass and hence higher carbon sequestration potential.

The carbon pool for the Indian forests is estimated to be 2026.72 Mt for the year 1995, as per the estimates of this case study. Estimates of annual carbon uptake increment suggest that our forests and plantations have been able to remove at least 0.125 Gt of CO_2 from the atmosphere in the year 1995. Lal and Singh (2004) further assume that that the present forest cover in India will sustain itself with a marginal annual increase by 0.5Mha in area of plantations and that we can expect our forests to continue to act as a net carbon sink in future.

Global Response

Carbon Trading

Let us rewind to Kyoto Protocol. Kyoto Protocol was signed in 1997; came into force since 16 Feb 2005. It was ratified by more than 130 countries. All countries were required to reduce their greenhouse gas emissions by 5 per cent–from 1990 levels–by 2012 or pay price to those that do. The idea was to make developed countries pay for their wild ways with emissions while at the same time monetarily rewarding countries with good behaviour in this regard (http://www.goodnewsindia.com/index.php/Supplement/article/320/2006). It is also a mechanism to enable Annex I parties to "purchase" GHG emission reduction from projects in developing nations. Annex I parties include the industrialised countries who have specific commitments to reduce greenhouse gas emissions under the 1992 United Nations Framework Convention on Climate Change (UNFCCC) and the Kyoto Protocol. The only exceptions are Turkey and Belarus which are in Annex I but do not have reduction commitments under the Kyoto Protocol.

A commodity Certified Emission Reduction (CER), equivalent to 1 ton of CO_2 is proposed. When a project activity in Annex 1 does better than "Business as Usual" scenario the emission reduction over that level is translated into CER.

The World Bank has built itself a role in this market as a referee, broker and macro-manager of international fund flows. The scheme has been entitled clean development mechanism (CDM) in 2000, which is a flexible mechanism that creates a "Market Based Instrument", which 'commoditizes' environmental performance over business-as-usual. CERs accrue each year after the project performance is "verified" through a pre approved methodology (http://www.goodnewsindia.com/index.php/Supplement/article/320/2006).

The purpose of the clean development mechanism (CDM) is defined in Article 12 of the Kyoto Protocol to the United Nations Framework Convention on Climate Change. The CDM has a two-fold purpose: a) to assist developing country Parties in achieving sustainable development, thereby contributing to the ultimate objective of the Convention; and b) to assist developed country Parties in achieving compliance with part of their quantified emission limitation and reduction commitments under Article 3. Purpose of clean development mechanism: Emission reduction, Application of geo-engineering, Forestry options, Carbon storage in wood products, Introduction of carbon credits.

Sinks/Sequestration

"Sinks" and "sequestration" are the generic names given to projects that soak up and store carbon in biomass such as trees, formally known as Land Use Change and Forestry projects (LUCF). In general, afforestation is the planting of trees on land that historically has not had tree cover, while reforestation is the planting of trees on land that was cleared *before* 1990, the base year from which most emission reductions under the Kyoto Protocol are measured.

The amount of credits from CDM sinks projects that can be used by an industrialised country to meet their Kyoto target is limited to 1 per cent of their total emissions in their *base year* (usually 1990) multiplied by 5 (1 per cent for each year of the first commitment period which is 2008-2012).

From Kyoto to Bali

The Thirteenth Conference of the Parties to the UN Framework Convention on Climate Change (UNFCCC) held in Bali, Indonesia towards the end of 2007 is another milestone in the slowly but steadily consolidating global battle against climate change. Representatives from 190 countries made persistent efforts for over two weeks to come to an agreement on negotiating for an international treaty that would replace the 1997 Kyoto Protocol by the end of 2009.

After heated debate over quantitative targets for reduction of emissions, a compromise was reached with the Bali Action Plan recognizing "deep cuts" in global emissions. There was also consensus that forest protection would earn credits and that the adaptation fund would finance clean technology transfer to developing countries.

References

Burhenne-Guilmin, F. and Glowka, L., 2004. An introduction to the Convention on Biological Diversity. In: *Widening Perspectives on Biodiversity*, (Eds.) Krattiger, A.F., Mc Neely, J.A., Lesser, W.H., Miller, K.R., Hill, Y.S. and R. Senanayake. IUCN, Gland, Switzerland and Cambridge, U.K., and International Academy of the Environment, Geneva, Switzerland, p.15-18.

Eilperin, J., 2005. WHO's climate warning. *The Wall Street journal*. November, 18-20, 2005.

http://www.goodnewsindia.com/index.php/Supplement/article/320/. 2006.

http://wwww.kdsforfuture.net/ourfuture/000032.html. 2006. It's getting hot.

http://wwww.kdsforfuture.net/ourfuture/000032.html. 2006. The Earth is turning brown.

IPCC, 2001. *Climate Change 2001*. WMO and UNEP.

Lal, M. and Singh, R., 2004. Carbon sequestration potential of Indian forests. *Environment Monitoring and Assessment*, 60(3): 315–327.

Lal, R., 2004. Soil carbon sequestration in India. *Climatic Change*, 65(93): 277–296.

Ecosystem Diversity and Carbon Sequestration: Climate Change Challenges and a Way Out for Ushering in a Sustainable Future *Pages 10–25*
Editors: **P.L. Gautam, Vir Singh and Uma Melkania**
Published by: **DAYA PUBLISHING HOUSE, 2009**

Chapter 2

Global Climate Change: A Challenge before Humanity

S.P. Singh

The world at present is warmer than any other time during last 10,000 years. This warming has taken place because of the rise in the concentration of greenhouse gases (GHGs) (Table 2.1). These gases differ widely in lifetime in atmosphere, from 5 to 200 yrs, the difference being most within different forms of CO_2 (Table 2.1). Obviously, once they accumulate they will continue to raise temperature long after their emission stops. The GHGs trap sun energy and slow the escape of long wave radiation back to outer space. The phenomenon of trapping and radiating heat by GHGs in the atmosphere is called greenhouse affect. These radiatively active gases differ in their ability to trap heat in the atmosphere. A molecule of methane is 21 times more effective than of carbon dioxide. Similarly, nitrous oxide is 210 times more effective and chlorofluro-carbons are 1200-1500 more effective than CO_2. Carbon dioxide has the lowest radiative forcing per molecule, but because of its higher concentration it contributes most (53 per cent) to global warming (Table 2.2).

Table 2.1: Greenhouse Gases Concentrations in Atmosphere

Greenhouse gases	*Change in Concentration*		*Atmospheric Life Time (yrs)*
	Pre-1750	*2003*	
Carbon dioxide, CO_2 (ppm)	280	375	5-200
Nitrous oxide, N_2O (ppb)	270	319	114
Methane, CH_4 (ppb)	730	1,852	12
Tropospheric ozone, O_3 (ppb)	25	34	-
Carbon tetrachloride, CCl_4 (ppt)	0	94	84

ppm: Parts per million; ppb: Parts per billion; ppt: Parts per trillion.

Table 2.2: Contribution of Different Gases to Radiative Forcing Since the Industrial Revolution (Post 1750)

Greenhouse Gas	*% Contribution*
Carbon dioxide	53
Methane	17
Ozone	13
Halocarbons	12
Nitrous oxide	5

Globally, a total of 24,7900 million tones CO_2 equivalents were emitted in year 2000. Nearly two-thirds of CO_2 and other GHGs emissions occurred because of the burning of fossil fuels (oil, coal and natural gas) primarily for electricity, transportation, heating and cooling, and industrial processes. Man made sources of other GHGs are shown in Table 2.3.

Table 2.3: Sources of Major Greenhouse Gases

GHG	*Man-made Sources*
Carbon dioxide (CO_2)	Land use changes, fossil fuels burning, industrial processes, electricity manufacturing and transportation.
Methane (CH_4)	Rice cultivation, coal mining, energy production and rearing of livestock, landfills.
Nitrous oxide (N_2O)	Nitrogen-based fertilizers, manures.

The doubling in concentration of these atmospheric greenhouse gases is predicted to cause 1.5^0–4.5^0C rise in global temperature, most likely rise being 3^0C (Kerr 2004). Already, the world has warmed by 0.74^0C as the updated 100^{-yr} linear trend during 1906 to 2005 indicates. The warming rate of last 50 years is nearly twice that of the last 100 years (IPCC 2007).

Global Carbon Cycle and Carbon Sinks

Since CO_2 is the principal greenhouse gas, and its emission still keeps on rising we take a closer look at the global carbon cycle. Carbon is taken from the atmosphere in two ways, but is released back into the atmosphere in many different ways. The uptake of carbon from the atmosphere occurs (*i*) as a result of photosynthesis in which CO_2 is converted into carbohydrates and oxygen is released, and (*ii*) as CO_2 dissolves in water at the surface of oceans near poles, when water becomes cooler (Figure 2.1).

The release of carbon back to the atmosphere can occur thro ugh: (*i*) respiration of plants and animals involving the break down of glucose or other organic molecules into CO_2 and water (an exothermic reaction); (*ii*) decomposition of plant and animal matter, releasing CO_2 (if oxygen is present) or methane (if oxygen is absent); (*iii*) combustion of organic material, producing CO_2 and other things like smokes (*e.g.*, burning of fossil fuels, such as coal and petrol, stored in the geosphere for

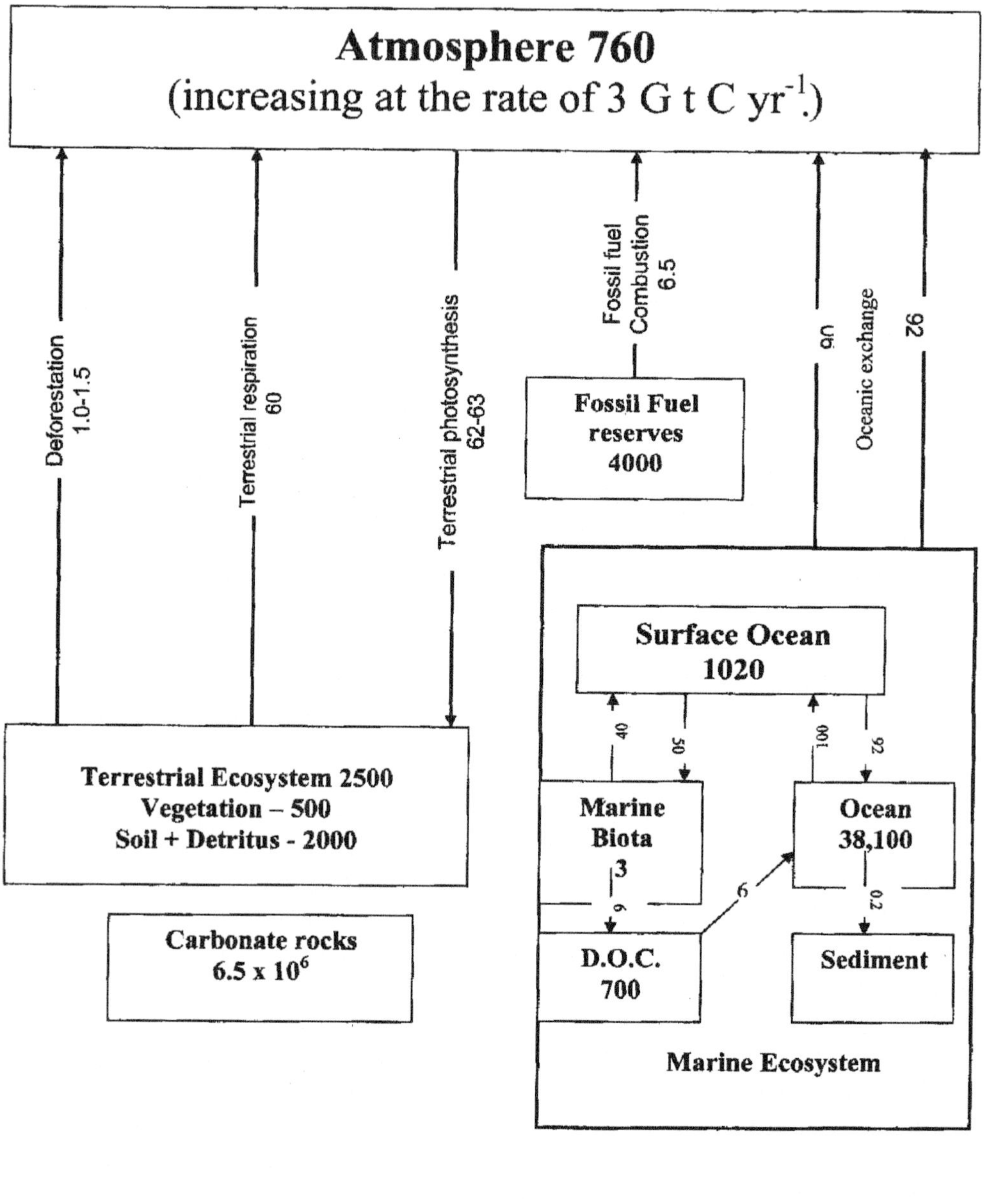

Figure 2.1: A Representation of Global Carbon Cycle During 1990s.

The carbon stocks are in billion tonnes of C or Gt C. The carbon fluxes, shown with labeled arrows, are in Gt C yr^{-1}. Doc=dissolved organic carbon both terrestrial and marine ecosystems are net absorbers of CO_2, yet atmospheric stock is increasing approximately at the rat of 3 Gt yr^{-1} (shown in parenthesis within the atmospheric box) because of the fossil fuels combustion (and also cement production) and deforestation (adapted from Grace *et al.*, 2000).

millions of years; (*iv*) erosion by water of calcium carbonate-rich rocks *viz.*, limestone, marble and chalk (breakdown products include CO_2 and carbonic acid) and production of cement and lime by heating limestone; (v) the release of dissolved CO_2 from the surface water of oceans as a consequence of warming; and (vi) volcanic eruptions.

The total carbon emission from human activities, *viz.*, fossil fuels burning, cement production and deforestation on an average is about 8 Gt yr^{-1} (Figure 2.1). This is more than the average rate at which CO_2 is increasing in atmosphere, 3 Gt yr^{-1} (Schimel 1995). Obviously, there are sinks on the earth surface, and they account for this difference. It is agreed that the global carbon sink is equally divided between the ocean and terrestrial ecosystems (Press *et al.*, 2000). Understanding the variation in the strength of the carbon sinks and their locations are major challenges and important for managing carbon cycle in the biosphere. The current average rate at which oceans of the world are absorbing CO_2 in about 2Gt C yr^{-1}, with strong sinks in north Atlantic and Pacific. The mid-pacific, on the other hand is a source of CO_2 (Peng *et al.*, 1998).

Compared to oceans terrestrial communities generally have much larger standing crops, and capacity to take up CO_2 per unit area. Many forest ecosystems, as example, are known to have more than 200 t C ha^{-1} in biomass, compared to generally less than 10 t C ha^{-1} in oceans. Soils of forests located in cold climates, such as boreal forests have large capacities to store carbon (Table 2.4).

Table 2.4: Carbon Stocks and Increments in Major Forest Types of the World (from Malhi *et al.*, 1999)

	Area (10^6 km^2)	*Carbon Stock (t ha^{-1})*		*Annual Increment in Stock of Biomass (t ha^{-1} yr^{-1})*
		Biomass Total	*Soil Organic Matter*	
Tropical forests	17.6	285	162	2.30
Temperate forest	10.4	1258	56	4.19
Boreal forest	13.7	67	390	1.40
Indian Himalayan forest (means)	0.23	148	120	2.59

* Values of Indian Himalaya are only gross approximations.

Some of the conclusions of the measurements of net uptake of carbon using sensors mounted above vegetation on towers distributed in different parts of world are as following:

- ☆ Most of boreal and temperate forests are reasonably good C sinks, sequestering between 0.5-8 t C ha^{-1} yr^{-1} (see in Press *et al.*, 2000).
- ☆ Contrary to general perception old growth Amazonian forests are also strong carbon sinks, sequestering C in range of 0.5-6 t C ha^{-1} yr^{-1} (Malhi *et al.*, 1998).
- ☆ Of the respiratory fluxes of carbon, fluxes from soil far exceed the fluxes from the above ground parts of plants.

- ☆ Potential for soils to sequester carbon is considerable. For example, forests in Finland show that soil carbon content stabilizes only after 2000 years (Liski *et al.*, 1998). Even agricultural soil has considerable ability to store carbon.

Evidences of Climate Change

Weight of evidences in support of global warming is considerable, the commonly cited evidences include as following:

- ☆ Since the instrumental recording of global surface temperature began in 1850, 11 out of the lat 12 years during 1995 to 2006 rank among the 12 warmest years.
- ☆ The average leafing time in trees has advanced by 6 days and leaf shedding in them delayed by similar number of days.
- ☆ Species have begun to march pole-ward on plane surface and upward in mountains. In India's Valley of Flowers woody species can be seen marching upward (Kimothy's information based on remote sensed data).
- ☆ There are several instances of well established ice and snow-melts. The period of ice cover in the Arctic sea has shortened, reducing the hunting period of polar bears. The shortage of food is affecting their health and reproduction.
- ☆ The retreat of glaciers is being witnessed in Andes, Himalayas and Alps. Lake formation has been seen in Bhutan.
- ☆ In Siberia temperature has risen by 3°C during last 40 years, resulting in snow-melt and formation of lakes.
- ☆ In Alaska roads and buildings are subsiding as a consequence of the melt of permafrosts.
- ☆ Larsen B ice-shelf, with an area of 3,250 square kilometer has broken away from the Antarctic Peninsula. The surface melt of Greenland ice cap has increased substantially from 1992 to 2005 (Figure 2.2).
- ☆ All glaciers surveyed in Himalaya are in retreat; the Meren glacier of Java disappeared in 2000; in New Zealand nearly 34 glaciers are in retreat; Alaska is retreating and thinning; in N Andes, Peru the largest glacier is retreating by 30 m/yr; snow capped peaks of Kilimanjaro and mount Kenya are predicted to disappear by 2050; and in the Rokies tree stumps now get exposed because of glaciers are receding.
- ☆ In Indian subcontinent temperatures rose above 50°C at many places in 2003, causing 1,500 fatalities. In Europe about 35,000 deaths occurred because of unusual heat waves in 2005, indicating that even advanced societies are vulnerable to climate change. At some places temperatures were 10°C higher than normal temperatures. In Australia weather was unusually dry during 2002-2005, forcing farmers to sell their livestock. Some of the recent winters in Spain were the driest of the recorded history.

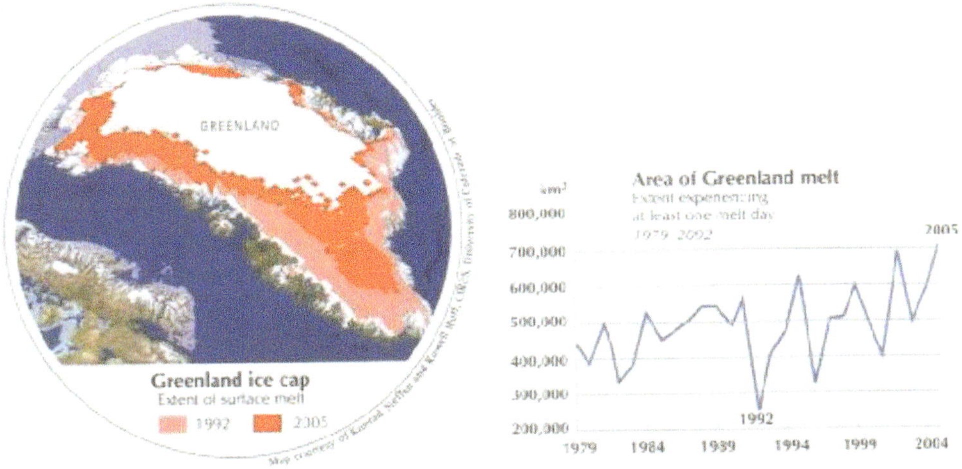

Figure 2.2: Extent of Surface Melts in Greenland Ice Cap

- ✰ In the flood of 2004 in Bangladesh 1,800 people were left dead and millions stranded. In the same year 150,000 people had to be evacuated in Argentina. In 2003 flood in Brazil 160 persons were killed.
- ✰ In 1998 in Kumuan region of Himalaya predawn tree water potential of oaks declined up to –6 MPa, compared to –2 MPa in normal years. A particular population of the oak had to wait over a long period until the commencement of rainy season to produce new leaves. Normally they produced new leaves about 3 moths before the monsoon set in. Floods are becoming more intense.

Consequences of Climate Change

Global warming is the product of failure of humans to achieve sustainability, and it would be at a great risk with growing GHGs concentrations. As time progresses more and more dramatic effects of the accumulation of GHGs in the atmosphere are to be seen. Poor, who directly depend on natural resources, are going to be most affected. It is important that the temperature rise remains modest, dealing with the problems of 3°C or more rises would be almost impossible. Some of the easily predictable consequences of global warming include extinction of species, change in water availability, and increase in diseases, disturbances associated with rise of oceans and changes in them.

Mass Extinction of Species

The area of Central American forests is predicted to experience a drastic reduction in rainfall, which may threaten the survival of many of the endemic mammals, amphibians, birds and 1700 plant species which live there. Species would find difficult to disperse to climatically suitable areas because of fragmentation of vegetation. In many areas species would have no corridors or stepping stones to escape from a place rendered unsuitable due to the global warming. Many tropical species are growing at or near the upper side of their temperature tolerances at present therefore further increases in temperature are likely to eliminate sensitive ones. Ecosystem near mountain tops would have no land surface available to move up. In Himalaya Alpine grasslands, timberline trees like birches (*Betula utilis*), rhododendrons and animals like pikas would be the first to vanish from the Himalayan regions.

Table 2.4: A Broad Outline of Possible Impact of Global Warming

Temperature rise

- ☆ Ice melting in polar and high mountain areas.
- ☆ The rise of sea level and flooding of coastal cities.
- ☆ Increased methane release from permafrost as it melts.

Wider fluctuations in weather

- ☆ More incidents of heavy rainfall
- ☆ More incidents of severe droughts

Ecosystem disruption

- ☆ Stress and death of vegetation
- ☆ Species migration and extinction, poleward as well as toward high altitudes

Changes in phenology of plants and animals

- ☆ Changes in ecosystem processes

Human health

- ☆ Heat stress
- ☆ Migration of disease vectors
- ☆ Increase in incidents of diseases like malaria

Change in land availability and economy

- ☆ Tropical countries are to lose habitable area.
- ☆ Cold countries like Russia to get more land.

Species primarily living in nature parks (*e.g.*, Spanish imperial eagle) may find hard to locate unspoilt areas with suitable temperature.

Likely Changes in the Distribution of Species and Ecosystems in India and Himalaya: A Few Examples

With increasing warming species and ecosystems are likely to shift from lower latitudes and altitudes to higher ones. Generally, temperature decrease by altitude at the rate of 5-10°C/km across various mountains of the world. The species would need to migrate upward to survive. However, the upward movement of alpine species occurring near the mountain peaks is likely to be restricted by the lack of space and soil. Since mountain tops are smaller than bases, the species near tops would get smaller and smaller area to occupy with the warming. They may be severely affected by the smallness of populations. Some of the important alpine species of Himalaya that may face immediate extinction include the oak, *Quercus semecarpifolia* (Singh *et al.*, 1997), birch (*Betula utilis*) and some rhododendrons, several herbs of medicinal values, and mammals like pikas, brown bear and snow leopards.

According to a projection for the year 2085, depending upon the rise in CO_2, 68–77 per cent of the forested grids in India are likely to experience change in forest types. The shift generally would be towards wetter types in the north-eastern parts, and drier types in the north-western parts (Ravindranath *et al.*, 2006). The study also indicates doubling of net primary productivity by 2085.

Tropical evergreen forest type is predicted to expand extraordinarily, from 3 per cent of the grids at present to 21.5 per cent and 35 per cent, under low and high increases in the atmospheric CO_2 concentration, respectively. In contrast, the impact of tropical dry deciduous forest in likely to be negligible.

Changes in mountains are expected to be much sharper. It is estimated hat snowline will rise by about 150 m for each degree Celsius increase in temperature. The warming would affect vegetation also by reducing snow pack duration, its amount, and water availability from snowmelt. Plant growth in alpine meadows where many medicinal plants occur can be severely affected by early snowmelt. Many species may be exposed to severe frosts with thinning of the snow cover. Species requiring winter cooling for regeneration are likely to be most vulnerable to warming. In Himalaya, junipers for which perhaps winter chilling is not necessary may survive warming.

At the species level there are three likely adaptational responses (Huntley 1991): (*i*) replacement of dominant species by a more heat-loving species; (*ii*) replacement of climax species by pioneer species having adaptation capacity for wider ranges of environment; and (*iii*) better expression of the less important species of the same community.

Changes in Phenology

A number of studies carried out in different parts of world indicate that global warming of last three decades has advanced a few days several spring-time activities, such as leaf production and flowering in plants, breeding in birds, and arrival time of migrant birds. Similarly, there are indications for delayed colouration of leaves during autumn. In Himalaya there are many dominant forest tree species (*e.g., Shorea robusta, Quercus floribunda and Q. semecarpifolia*) in which seed maturation is synchronized with the commencement of monsoon, and their seed vitality is unusually short, one to two weeks. Early maturation of seeds due to the warming or stress may break this synchronization, and thereby suppress regeneration of such species (Singh and Tewari, unpubl.).

Spread of Tropical Diseases

The warming is likely to cause the outbreak of at least some diseases. For example, warm water of oceans may cause the outbreak of diseases, such as one that killed thousands of dolphins in 1990.

The occurrence of many vector-borne diseases until now was not seen in cold latitudes and altitudes. At elevations above 1500m in Himalaya and other subtropical and tropical mountains, the *Anopheles* mosquito can neither breed nor survive (Craig *et al.*, 1999). The warming is likely to lead to new distributions of vector-borne diseases. The malarial transmission is predicted to increase in a warmer and wetter climate. Predications are that even Himalayan states like Himachal Pradesh, Arunachal Pradesh, Nagaland, Manipur and Mizoram are likely to be prone to malaria (Battacharya *et al.*, 2006). Condition in Nepal is not going to be any different.

Lyme disease (carried by deer ticks) is spreading in the USA and Europe. The disease can cause a long-term disability. In any case, in the absence of the subzero temperatures which kill insects, rodents, bacteria, etc., they are going to widen their habitat ranges.

More Droughts and Floods

Many centres of great civilizations have connections to snowmelt water of mountains. This connection could be lost with global warming, as snow cover depletes and glaciers disappear. The age-old civilizations of the Gangetic Plains, Nile, Euphrates and Tigris, and other are going to have hard time. Water scarcity is going to be a serious problem as many countries already extract more than 40 per cent of renewable water (Table 2.5).

Table 2.5: Categorization of the Countries from the Stand Point of Water Extraction as the Percentage of Renewable Water

Level of Water Extraction	*Countries by Examples*
94 per cent and above	Saudi Arabia, Egypt, Libya
40-71 per cent	Yemen, Morocco, Bulgaria, Pakistan Afghanistan, Iran, Belgium
20-39 per cent	India, Algeria, Kazakhstan, Poland
Below 20 per cent	China, USA, Canada, Russia, Australia, Brazil

On the other hand, warming by enhancing hydrological cycle can, cause a problem of too much of water in certain areas during a part of a year. For example, in India the eastern Himalaya may get wet. In certain rain-shadow areas of Himalaya rains are being seen as new events. People there had not reckoned with this situation while constructing their houses, consequently they are now faced with the problems of the roof leakage.

Consequences of Sea Level Rise

The sea level is reported to have risen by 15-20 cm during 20th century because of the snow-melt and expansion of warm water. The warm water holds less CO_2. The complete melt of the Greenland ice cap alone may raise the global sea level by 7 m. Even 1m sea level rise which is likely to occur in the coming century, is enough to destroy precious agriculture land and cities of many areas. For example, 15 per cent of the habitable land of Egypt is predicted to get inundated by such a rise. Windstorms are predicted to frequently inundate low land areas along sea coasts. The loss of land can be a serious problem for the economy. In fact, it is being suggested that climate change would have varying effects across the countries of world. Tropical countries which have just begun to make sizeable progress economically are most of suffer on account of the loss of habitable land. On the other hand the warming may release large areas from snow and, permafrost in Siberia for cultivating crops and other economic activities. Similarly, the US may gain in case land is released from snowmelt in Alaska.

Coral Reef Bleaching

Coral reefs and complex systems, basically involving anthozoan corals and their symbiotic endozoan dinoflagellate and coralline algae. Though occurring in nutrient poor tropical oceans, they support a high diversity of colourful organisms. There are indications that coral bleaching (reduction in the density of dinoflagellate algae and their pigments) and warmest years coincide.

The above description only gives an outline of the impact of global climatic change. Since ecosystems function in a complex way and have capacity to modify the course of warming, many changes under the influence of global warming could be different from what are being predicted. The CO_2 enrichment is likely to have many direct influences on biota and soil component, and thus modify the path of global change.

Responses to Climate Change

There can be two strategies of addressing the crisis of global climatic change: (1) *Mitigation*, which pertains to reducing the increase of GHGs by sequestering carbon and developing technologies with lower emissions (*e.g.*, efficient energy use) and substitution of fossil fuels. (2) *Adaptation* which means taking measures to deal with the problem arising out of global warming. For example, suitable habitats could be provided for species to migrate poleward in plains and upward in mountains. Farmers in the Gangetic plains of India may need to replace wheat with rice in some areas for winters may not be long enough to allow wheat to grow well.

Responsibility of Developed Countries

Countries differ widely in their energy uses. For example, CO_2 emission/person/yr from burning fossil fuels (oil, natural gas and coal) is 20 t for the USA, about 11 t for Russia 9.5 t for Japan, 2.7 t China, and 1 t for India. The USA and Europe alone account for about half of the global CO_2 emission from fossil fuels. As for nitrous oxide (N_2O) emission, Asia (31 per cent) is the greatest contributor.

Among the major sources, while deforestation accounts for 18 per cent of the emission, house heating contributes 25 per cent, and transportation 12 per cent.

Though the per capita emissions of India and China are low, because of their large populations they are already major contributors to GHGs. Given the its rapid economic growth, China is going to exceed the USA's level of emission soon. Here, the important question relating (particularly China) to negotiation is that whether responsibilities should be assigned at per capita level or country level. Individuals in developing countries who have just shrugged off their age-old poverty would justify their increases in energy use. While Americans may raise the issue of uncontrolled population growth in developing countries like India and Nepal, and argue that their country-level emissions are going to be soon comparable to theirs. Obviously, Indians and Chinese are not going to stop increasing their per capita consumption. If a country is to be the unit of managing climate change and frame rules, large countries would be in disadvantage because of the larger populations irrespective of their population densities and per capita emission. Therefore, density of population along with per capita emission should be considered to develop protocols.

Another way of looking at the emission issue is how countries differ in carbon intensity that is measured as tonnes of CO_2 emitted per $1,000 GDP. This value is lower in developed countries, and very high in oil-rich countries (Table 6). Among developing countries like China and India it is intermediate, but these countries show far more inefficient energy use than developed countries. This is one of areas where technology can play a significant role in improving efficiency, and China and India need to improve their carbon budget by addressing this.

Kyoto Protocol and Beyond It

It took several years to achieve Kyoto Protocol and progress is still limited. The international cooperation to address the crisis began taking shape with the formation of United Nations Framework Convention on Climate Change (UFCCC). Its goal is to "stabilize the emission of GHGs at a level that would prevent dangerous anthropogenic interference with the Climate system". It has to be achieved within a time frame. Agreed in Rio 1992 the convention came in force in March 1994. Kyoto Protocol, the first addition to the treaty was adopted in 1997 and came into force in February 2005. The industrialised countries (USA, Europe, Japan etc.) and 12 countries of transition economy (Croatia, Bulgaria, Hungary, Romania, Poland, and Russia) would reduce their combined CO_2 emission (all emissions of all greenhouse gases expressed as CO_2 equivalents, CO_2e) by 5.2 per cent of their levels in 1990, by a target date between 2008-2012.

The carbon trade encourages energy efficiency, use of renewable energy, carbon sequestration and other means. Reponses of nations to Kyoto Protocol vary and the world is divisible as following: Non-Signatories countries, *e.g.* Iraq, Taiwan, Western Sahara. Countries signed the UNFCC but not adopted the Kyoto Protocol- *e.g.*, USA, Australia, Kazakhstan, Afghanistan, Libya, Chad, Gabon, and Zimbabwe. Signatories–most others.

Under the Kyoto Protocol an industrialized and countries with economy in transition are entitled to purchase "carbon credits" from developing countries and from other industrialized countries whose emissions are below the required level (*e.g.*, Germany and Sweden). In this emissions of all GHGs are included, and are expressed as CO_2 equivalents (CO_2e). The trade is expected to encourage investment in energy efficiency, development of renewable energy and other means for reducing emissions. Much of this trade occurs through commercial organizations. Markets are divisible into project-based and allowance-based types. The former markets promote investment in companies or schemes committed to decrease GHGs emissions. Among the major sellers of carbon are India, Brazil and China, while Africa is way behind. In allowance-based markets main players are large companies, such as energy producers. They purchase emission allowances under schemes administered by international organizations like EU Emissions Trading Scheme (ETS).

Kyoto protocol only considers reforestation (plantations at sites which had forest in the past but were bare on 31 December 1989) and afforestation (plantations raised on sites which never had forest). It does not give value to avoiding deforestation. A young plantation may sequester C at higher rates than do existing mature forests, but it may take several decades to accumulate carbon equal to that in a mature forest. For addressing the problem of global warming next 10-20 years are crucial. Avoiding deforestation is also important because it leads to soil loss, and soil has more carbon than biomass *plus* atmosphere. However, one knows little about the fate of carbon associated with soil loss. How much of it enters the atmosphere, and how much is deposited elsewhere including deep beds of ocean are unknown.

Table 2.6: Carbon Intensity (Tones of CO_2 per $1000 GDP in 2003)

Country	*Carbon Intensity*
Austria	0.37
Azerbaijan	5.09
Belarus	6.00
Belgium	0.60
Brazil	0.55
Bangladesh	0.68
Canada	0.80
China	2.58
Croatia	1.06
Cuba	4.58
Ethiopia	0.61
Mongolia	7.91
India	1.86
Indonesia	1.89
Iran	4.39
Israel	0.59
Japan	0.25
Kyrgyzstan	3.25
Nepal	0.50
Russia	5.24
South Africa	2.97
United States	0.56

The above countries are separable into several negotiating groups.

Carbon markets under Kyoto Protocol account for less than 0.5 per cent of GHGs annual emissions. Environmentalists find fault with them because they think that projects taken such as single-species plantations and large hydroelectric dams are unsustainable. Then, getting a forestry project accepted under CDM is very different. Just one afforestation/reforestation CDM project-in China- has been accepted. However, the recent success of voluntary carbon markets has generated new hopes (see Ricardo Bayon, who along with Al Gore and several other authors, has published a book called, Voluntary Carbon Markets). The voluntary carbon offsets market which amounted to only $ 10-20 million in 2006 has doubled during last three months. The voluntary market is supported by companies in two broad ways: (1) To neutralize the CO_2 emissions of air travel of company workers. (2) To neutralize the carbon entailed in the manufacture of the company products. Thus carbon cost is embedded in their products. These companies are doing it also out of a sense of decency, and to appeal to their customers and shareholders sense of decency. Some of the big companies (located in the USA and Japan) have given substantial importance to forestry. In the past the contributions of forestry projects to the carbon offsets purchased by them was 58 per cent. However, in future energy projects are going to be more important. Voluntary market is now also considering accepting the protection of forests (*e.g.*, Trees for Travel). These companies are willing to consider only tree projects that produce 1000 t CO_2 equivalent per year.

There are certain other encouraging signs. Markets have been set up in the USA and Australia, the countries which have not ratified the Kyoto Protocol. The protocol encouraged investment in research to reduce GHGs emissions. Worldwide voluntary market is developing rapidly. People purchase "carbon credits" to channel their money into projects leading to a reduction in emissions. A number of websites have been set up to assist organizations to evolve plans to address the issues of climate change and cut down GHGs emissions.

Several adaptive measures are being suggested, *e.g.*, protecting flood hazard zones, developing building designs which could help reducing heat stress and conserving water resources. Poor are the suffer the most in the warming world (Figures 2.3 and 2.4), as they greatly depend on wild plants and animals, and ecosystem service flows for their day to day living.

Figure 2.3

Figure 2.4

It is important to have a better understanding of problems of different countries, their limitations and strengths, and to be considerate to the problems of those who are less capable to deal with the ravages of climatic change to effectively deal with the crisis of global change (Figure 2.5).

High exposure			
Coastal and high mountain zones	Most Andean countries, Bangladesh, Indonesia, Nepal, many small islands, etc.	Brazil, China, India, Malaysia, Thailand, S. Africa, etc. Russia	Germany, Switzerland, UK, Etc.
Arctic nations			US, Canada, Norway, Finland, Sweden, Iceland,
Mediterranean nations	Albania, Egypt, Morocco en.	Croatia, Turkey, Tunisia, en.	France, Greece, Norway, Sweden, Iceland
Low exposure	Low GHGs emitters Low ability to pay →		High emitters High ability to pay

Figure 2.5: Coalitions of Countries those are Likely to Emerge in View of Risks of Climate Changes and their GHGs Emission and Ability to Pay (from Morlet *et al.*, 2005)

References

Bhattacharya, S., Sharma, C., Dhiman, R.C. and Mitra, A.P., 2006. Climate change and malaria in India. *Current Science*, 90: 369–375.

Craig, M.H., Snow, R.W. and Le Sueur, D., 1999. A Climate-based distribution model of malaria transmission in Africa. *Parasitology Today*, 15: 105–111.

Huntley, B., 1991. How Plants Respond to Climate Change: Migration Rates, Individualism and the Consequences for Plant Communities'. *Ann. Botany.*, 67: 15–22.

Liski, J., Tlevesniemi, H., Makela, A. and Starr, M., 1998. Model analysis of the effect of soil age, fires and harvesting on the carbon storage of boreal forest soils. *European Journal of Soil Science.*

Malhi, Y., Nobre, A.D., Grace, J. and Others, 1998. Carbon dioxide transfer over a central Amazonian rain forest. *Journal of Geophysical Research*, 103: 593–631.

Morlet, J.C., Smith, J. Agrawal, S. and Frank, T., 2005. Long-term goals and post-2012 communities: Where do we go from here with climate policy? In: *Climate Policy Options, Post-2012*, (Eds.) Metz, B. and Hulme. Earthscan, London, pp. 251–272.

Peng, T.H., Wanninkho, F.R., Bullister, J.L., Feely, R.R. and Takahashi, T., 1998. Quantification of decadal anthropogenic CO_2 uptake in the ocean based on dissolved inorganic carbon measurements. *Nature*, 396: 560–563.

Press, M.C., Huntley, N.J. and Levin, S., 2000. *Ecology: Achievement and Challenge.* Blackwell Science, Oxford.

Ravindranath, N.H., Joshi, N.V., Sukumar, R. and Saxena, A., 2006. Impact of climate change on forests in India. *Current Science*, 90: 354–361.

Singh, S.P., Rawat Y.S. and Garkoti, S.C., 1997. Failure of brown oak (*Quercus Semecarpifolia*) to regenerate in Central Himalaya: A case of environmental semi-surprise. *Current Science*, 73: 371–374.

Schimell, D.S., 1995. Terrestrial ecosystems and the global carbon cycle. *Global Change Biology*, 1: 77–91.

Ecosystem Diversity and Carbon Sequestration: Climate Change Challenges and a Way Out for Ushering in a Sustainable Future *Pages 26–33*
Editors: P.L. Gautam, Vir Singh and Uma Melkania
Published by: DAYA PUBLISHING HOUSE, 2009

Chapter 3

Management of Ecosystems for Livelihoods and Carbon Sequestration in India: Harmony within Natural Elements a Mantra for Human Happiness

J.S. Bali

Too many people striving for too high a standard of living and pursuing the so called "rates of growth and "aping the unsustainable life styles" of the West are destroying the natural regulatory powers of the Earth. Fossil fuels used for energy for home heating and transport in individual cars have taken out too much carbon stored safely by the earth and can today be held to be 70 per cent responsible for greenhouse effect and consequent climate warming. How many people can live on this earth without hazard to the ecosystems depends upon our choice of sustainable lifestyle and society. Development of women and care for their health and education would lead to lowering of their fertility rate by their own choice. Coercive methods would fail. Ecosystems management in India has to take care of both, the concerns of people's livelihoods and carbon sequestration, in order to be successful in our country. The model of Bioindustrial Watershed Management is suggested. It will address issues of four P's: Protection of Environment, Ecology and Biodiversity; Production of food and bio–fuels from plants and animals in a sustainable manner; Processing of the bio-produce in people-owned plants in the watershed itself and an enabling favourable Policy regime. When people's incomes are thus enhanced and farming is made profitable, there will be less pressure on the forests and village common property resources. More technology and capital will flow to the rural areas. Forests and grasslands would be rejuvenated. Lastly, there is no hope for mankind unless dependence on fossil fuels is reduced and atomic energy takes the central place.

Introduction

Global warming and climate change is occurring today due to increasing input of solar radiant energy and decreasing self-regulatory cooling capacity of the Earth. Both these aspects have been explored in the following pages. Increase of population, faulty meaning of development, land clearance for agriculture and consequent decrease of forest cover have resulted in climate warming. Burning of fossil fuels in too large quantities in the modern transport and energy systems has played a major role in increasing carbon dioxide accumulation and greenhouse effect. Alternate energy systems have to be developed and brought within affordable zone. Ultimately, atomic energy mixed with renewable energy sources may provide an answer. Reforestation and increase of vegetative cover have to take care of people's livelihoods, especially in the developing countries. The principle: polluter *pays*, will have to be accepted for an equitable glonal solution.

Glaciers are receding, threatening water supplies. Some experts like V. Raina think they are not R.K. Pachauri, Chairman Intergovernmental Panel on Climate Change and Jagdish Bahadur think they are Satellite data confirm glacier receding in India at least for the 25 out of 9,575 glaciers which are under study.

The West pollutes the most. Will they come forward to pay? Will the West voluntarily reduce its energy wasting lifestyle? Will this world ever become equitable before it is too late? Let us examine the various aspects of this gigantic global warming and carbon sequestration problem in the context of India and see what is practically possible to manage ecosystems consistent with livelihoods of our people, specially the rural people.

Population and the Pursuit of Unsustainable Lifestyles

First the facts: On September 29, 2006, India's population touched 1,098, 949,509, which was 16.81 per cent of the world population of 6,537,357,028. Annual exponential growth rate of India is 1.38 against the world average of 1.14. Number of children per woman in India is 2.73, against the world average of 2.59.

Table 3.1: Rates of Increase in Population

	India	*World*
Per year	1,55,31,000	7,77,60,000
Per month	12,73,033	66,80,000
Per day	42,434	21,16,000
Per hour	768	9,000
Per minute	29	150

Source: Medindia.net.

Today, the 20 per cent population of the Western countries is consuming 80 per cent of the world's resources. If the Western lifestyle is to be made the goal and endless economic rate of growth the path, we shall need five Earths. Only one planet Earth cannot fulfill such ambitions of every nation. In an attempt to follow such unsustainable paths, we shall be promoting further carbon dioxide production and global warming.

As a consequence, food shortages and large scale inundation of the coastal areas would occur. If major glaciers of the world were to recede, sea levels would rise 14 metres. Wheat crop will vanish from Punjab. World population would die of starvation. We have still to find practical methods of carbon sequestration. Population stabilization, rather reduction to half the present level, should be adopted as the goal, if global warming has to be checked.

If action is not taken early and seriously, it would be too late. It would be a case of leaving smoking when lung cancer has already spread. The way to reduce population is quick development. When women are developed and educated, they themselves take effective action to reduce their fecundity. This has been proven by the population data. Legal coercion has been tried in India in the Emergency days in 1975 and failed. Better woman and child health facilities, spreading awareness and eradication of poverty are the real remedies for checking population growth.

Understanding the Solar Source of Heat and Earth's Regulatory System

Sun is like a huge fusion reactor. Its fuel, hydrogen, under pressures created by a temperature of 150 million degrees Celsius, will produce the inert gas, helium, with the release of tremendous amounts of energy. The radiant energy is absorbed by the earth and converted into the larger wavelength heat energy. While the atmosphere is transparent to the light waves, its carbon dioxide and other gaseous layers are opaque to the heat waves which get trapped, thus causing global warming. Incidentally, gases like natural gas which we in India use as a *clean* fuel is 24 times as damaging for trapping heat as carbon dioxide. Can we remove this carbon dioxide and sequester it safely somewhere? That is the problem facing humanity today. But let us first examine the ideas, though looking wild, on reducing solar energy reaching the earth

The sun is a medium size star. By the very nature of the fusion process, sun is 0.5 per cent hotter today than it was 55 million years ago. This is equivalent to about 0.5 degrees C in global temperature. It supplies 1.35 kilowatts of energy for every square metre of the earth that is in direct sunlight. Inside the sun, hydrogen–to–helium conversion will go on rising and in a billion years the energy received on earth will become 2 kilowatt per square metre. The sun will ultimately explode and reach its terminal state as a shining small object. Can we shade the earth to lessen the radiant energy reaching the earth?

Lowell wood and Ken Caldiera presented a courageous answer and suggested building of a sunshade in space between the Earth and the Sun. It would be a disc 11 kilometres in diameter, weigh about 100 tons and would be assembled and spun out in space. John Lotham suggested production of marine stratus clouds near the surface by turning sea water into aerosols of tiny particles. Ocean algae help in the process. But acidity produced by increased carbon dioxide will reduce the growth of algae.

These ideas are not yet practical. We must reduce the production of carbon dioxide. The present annual rate of carbon dioxide production will provide enough gas which if cooled, frozen and heaped up, would form a mountain 1.6 km high and 19 km in diameter. Klaus Lackner suggests sequestering the carbon dioxide produced in the form of magnesium carbonate (by reacting it with the natural rock serpentine) which can be used as building material (James Lovelock, 2006).

James Lovelock says: climatology was only one part of global change. Just as important as curbing emissions is the need to recognize that the Earth's natural ecosystems regulate the climate and the chemistry of the Earth and are not there merely to supply us with food and raw materials. Our attempts to replace these ecosystems with farmland or forestry plantations have led in recent years, in Indonesia and elsewhere in the tropics, to disaster both regional and global in scale.

In order, therefore, to allow the self-regulatory Earth to resume its regulatory role in full, we have to stop putting more land under agriculture, stop deforesting the Amazon Valley of its primeval rainforests, and fully afforest the less dense forestland of India and other countries. This is a tall order. But it must be done in order the climate change to reverse itself. Otherwise there is little hope in the direction of carbon sequestration.

As already stated the burning of the fossil fuels in heating the homes of the better to do Western nations and in the faulty transport system of the world again led by the West, lie at the root of the globally growing greenhouse conditions. The present dirty internal combustion engines based upon fossil fuels must be replaced by the hybrid engines comprising hydrogen cell and back up of straight hydrogen fuel. Such engines have been built. But there is resistance to change from the present car industry of the West and Japan who have vested interests in petroleum and the present jigs and fixtures capital investment of the present manufacturing system. Cleaner Energy Systems must be brought in quickly.

Examining the Clean Energy Systems

Clean energy systems provide the only hope for mankind for effective carbon sequestration. All other solutions would only touch the fringe of the problem. Clean Energy Systems would include the so-called alternate energy systems and atomic energy or a judicious mixture of these, which may differ from country to country.

Hydrogen

Supply of hydrogen gas on a large scale for all energy needs would require huge investments and would also pose a threat to safety. But hydrogen use in the fuel cell is a feasible proposition. Fuel cells are reversible in action. You can produce electricity between the electrodes of hydrogen and oxygen by supplying hydrogen and also produce hydrogen and oxygen gases from water when electricity is passed between them during the night or other periods when the electricity demand is less than the supply. Prototypes of fuel cells are already in use in cars and other power systems.

Solar Energy

Sun is the primary source of energy. Our food from the plants is dependent on the energy of the sun. Men and animals and all living creatures are powered by the sun. It makes the wind blow and the waves to rise. Even coal and petroleum represent stored solar energy. We receive 1.35 kilowatts of solar energy on every square metre of the earth's surface. It can be focused to produce heat and therefore electricity. It can also be directly converted into electricity in the photoelectric cell panels. Unfortunately, however, we have not yet been able to produce solar electricity at economical rates and in large quantities. Its large- scale use has to await further research which may take decades. But use of direct solar heat for water heating is practical at present and is being adopted as a roof top facility in many homes and institutions. Some provisions for it have also been made in the Delhi Master Plan 2021.

Wind Energy

Wind is driven by the sun. Man started using this source of energy centuries back in the form of sailing ships. It has a great promise. Wind mills have been in use in Europe for centuries. Large scale wind farms have already been set up in the Indian desert region where winds are strong. Electricity from these wind farms is directly fed into the regional electricity grids. It is metred and duly paid for. But the quantity needed by mankind is so large that this source alone cannot provide it. Hundred metre high one megawatt wind turbines will be required. There are other practical difficulties. Wind is not uniformly available. Storage poses a problem. Large scale storage in the form of pumped up water to higher elevations and its use for electric production at peak hours requires high ground nearby which is not always available. Transport from distant deserts is also not economical. Its use would, therefore, remain limited.

Tidal Energy

This form of energy is really the stored gravitational energy of the earth, moon and the sun. It can supply energy in the coastal areas. It can also supplement of nuclear energy sources. Its larger use has, however, to await further development which may take 20 to 40 years.

Hydroelectricity

Water power is India's oldest energy source. In Punjab, local water power use systems are called *gharats*. *Gharats* are made by village carpenters. The set-up comprises a wooden turbine driven by a fast water flow. The fast flow is created by utilizing the topography. A part of a natural stream is diverted by an earthen or stone or brushwood temporary barrage. This barrage is washed away during the rainy season and rebuilt immediately in a day. The water flow is brought in a contour channel and dropped steeply at the *gharat* site. The fast flowing water drives the vanes of the wooden turbine at the lower end. The top of the wooden turbine is connected with a moving stone over another stationary stone. The contraption is used for grinding of wheat into wheat flour.

About 25 per cent of electricity in India comes from hydro-electricity produced at the large dams. Smaller and even micro hydel projects have also come up especially in the hills. Micro hydel projects, however, are not economical to set up every where. Canada, Norway and Sweden use hydroelectricity to meet half their needs.

Large dams for hydro-electricity need natural sites which are only a gift from nature and are not available to all nations in the numbers required. They have problems of submergence of land and rehabilitation. Their life is also limited due to sedimentation All dam sites, however, must be exploited to the full. Hydro electrity output fluctuates with river flow. They need other power stations to make the output even for the consumers.

Bio-fuels

Large scale plantations of *Jatropha curcas* have been created in Chhattisgarh, where the Forest Department has already planted 82,000 hectares. But they have not done it in the bare or sparsely vegetated forestland under their complete legal control. Instead, the plantations have been done on the village grasslands. They have deprived the poorest of the poor of the villagers who are landless. They subsist on the common property resources of the village, the chief among them being the grazing lands or *gowchars*. Incidentally this writer has not noticed a single scheme for the development of the grasslands and fodder resources. Is it because the very poor people have no time and energy to protest or even to know how they are deprived.. Pongamia and other species also produce oil which can be mixed with diesel. The current plans are just to mix diesel to the tune of 5 per cent. Brazil uses sugarcane plantations for alcohol to mix with petrol to the tune of 50 per cent or more. Burning of the organic matter would send carbon back to the atmosphere and there will be no net gain for carbon sequestration. Poisonous plantations near villages pose problems not only for the animals but also for children. Land in India is limited. Forest land should also ideally be put back to dense forest trees rather than use it for plantations of bio-fuels. Forests best aid the regulatory system of the earth. On the balance, there is little long term hope from this source. Immediately, however, it looks attractive and may save us some dollars.

Atomic Energy

Atomic energy will be the future source of energy for the mankind. At present the atomic energy comes from fission of uranium, thorium and plutonium. India does not have much uranium, but thorium is plenty. India is, therefore, foremost in the world today for the thorium based nuclear technology. The latest design of nuclear reactors not only consume the fissionable materials but also create them. Danger from nuclear wastes has been exaggerated. Nuclear waste management technology is quite advanced now. We already live in the nuclear universe. Our bodies are radioactive. But we are alive and the average life span is on the rise. Ultimately, fusion energy which powers the stars has to be the answer. It has already been produced and sustained for a few seconds creating temperature existing in the sun–100 million degrees Celsius. Hydrogen isotopes–Deuterium and Tritium–when heated to such temperatures, convert into helium releasing energy. This energy was quantified by Einstein's well–known equation: $E = m.c^2$, where, c is the velocity of light (300,000 kilometres per second).

In fact, even chemical reactions are atomic in nature where electrons are involved. Any flame is an atomic process in which electrons flow. But the flame produces a voltage of only 0.82 volts compared to 21 million volts of the atomic energy process.

Ultimate Answer: *Mixing of Energy Sources*

Different countries could mix atomic energy, and alternate source energy according to their own conditions. Even fossil fuels have a place in the energy system if the affluent gases can be fully sequestered. Carbon can also be sequestered as magnesium carbonate. The ultimate planning for energy research and use should proceed on creating physically and economically optimum mixtures for different sites. But the reduction in fossil fuel use must start *now*. We must see more of the fuel cell cars or hybrid cars on the street.

Human Livelihoods and Carbon Concerns

Cultivated agriculture on the land cleared from the natural jungle territory, was the original "sin" of mankind. The sin of the livelihood compulsion still goes on. Human livelihood activities compel combustion of the buried fossil fuels and hence become the cause of carbon abundance in the atmosphere and climate changes. Food production and livelihoods are to be the priorities. What is not a priority is the unnecessary production of non-essentials in the name of "standards of living"

The West exploited the natural resources and now insists on maintaining a standard of living based upon the fossil fuels. That is the major source of the whole carbon sequestration problem. The developing nations want to ape the West. If man abandons this model of living based upon exploitation and adopts a system of living in harmony with nature, all problems would be solved. Of course the present standards of living might come down but the quality of living would improve. Mankind has to work together as one community dependent upon the mother Earth. There is no sign today that the "haves" are willing for the sacrifice involved in the needed new paradigm.

A healthier system of *ecosystem management* has to be evolved. The new system of resource management must include concerns of the poor of the world. Agriculture is vital for food production. There are some scientists or futurists who say that food could all be produced chemically and micro-biologically. Leaving the land free for a dense cover of forests which would sequester the excess quantities of carbon dioxide in the atmosphere. But these are just dreams yet. Practical solutions are required immediately.

Agriculture has to be made more productive. Horizontal expansion is not possible. Sustainable ecosystem management has to be developed. Here is proposed such a system which is called *Bioindustrial Watershed Management.* For the developing nations there is no alternative but mix industry with agriculture. This system would address the concerns of ecology, environment, economics, employment and equity. The unit of development planning should be a watershed. Top lands should go to well-conserved forests, and not the present forests which are full of conflict, corruption and crime.

The lower lands should go to sustainable practices of agricultural production based upon soil and water conservation. The practices would include contour cultivation, conservation cropping systems and organic farming. It will include grassland development, water harvesting and maintenance of rural common property resources. Land holdings should be consolidated. The marginal holdings should be coalesced into cooperative economic units.

The middle lands should go to animal husbandry where the animals of the landless would graze or stall-fed from the grasses and fodder growth. Fodder trees should be promoted. Ponds should be built in millions for water and promotion of fodder resources.

The produce will be marketed only after some packaging, processing and value addition. This will not only fetch more profits but also give more rural employment. Rural Computer Centres should connect farmers with the Universities on a day to day basis. Private companies should have contract deals with the farmers in which private companies will provide extension, inputs and plant protection. The company would buy back the produce at pre-announced prices. Ideally, farmers' cooperatives themselves should also own the processing industry. Storage and marketing, local, national and international will be developed.

This Bioindustrial model shall need infrastructure of power, roads, transport and communications. It would also need an enabling policy umbrella with financial concessions, credit facilities and ensuring that *farming becomes productive and profitable.* At present even the cost of cultivation is not returned to the farmers. Middle men become rich. Farmers are made to subsidise the urban consumers in the present system of farm marketing. Direct marketing of farmers to consumers will have to be developed (Bali 2005).

Conclusion

Aiming at achieving American living standards has created the present difficulties for the mankind. Such pre-occupation with growth rates has to be abandoned. A proper mix of energy sources has to be planned. Forests have to be protected and fully stocked. Agricultural land must not expand horizontally, but aim at only vertical increases in productivity. Pollution has to be strictly controlled. This would include the sea and ocean pollution. Polluters should be made to pay. Kyoto Convention should be signed by the biggest consumer of energy and creator of carbon problem–USA. Humanity has to be made aware of the perils of tinkering with the Earth's natural regulatory system. These are some of the steps which mankind, acting together, must take for their survival.

Human happiness depends upon harmony between the natural elements. The Hindus discovered this mantra centuries back, as evidenced by the Shanti Mantra which evokes *harmony on the Earth, harmony in water, harmony in our body, and harmony in the Cosmos.*

References

Bali, J.S., 2005. *Bioindustrial Watershed Management*. Soil Cons. Soc. of India.

Clark, Ronald W., 1978. *Einstein: The Life and Times.*

James, Lovelock, 2006. *Revenge of Gaia.*

Hawking, Stephen W., 1988. *A Brief History of Time.*

Jagdish Bahadur, *A History of Almost Everything Glaciology.*

Ecosystem Diversity and Carbon Sequestration: Climate Change Challenges and a Way Out for Ushering in a Sustainable Future *Pages 34–42*
Editors: **P.L. Gautam, Vir Singh and Uma Melkania**
Published by: **DAYA PUBLISHING HOUSE, 2009**

Chapter 4

Carbon Sequestration: A Vision

Vishal Mahajan and Kamal Kishor Sood

With iincreasing human population the natural resources are being exploited extensively world-wide. This has resulted in many concomitant environmental hazards. Climatic change including global warming with its adverse impact on the life on the earth is one of them. Increase in atmospheric CO_2 is one of the causes for such a change. The increase in CO_2 concentration results in climate changes which are directly or indirectly attributed to human activity that changes the composition of global atmosphere adversely. Scientific models and observations over the past 1000 years provide evidences that global warming is due to anthropogenic increase in Green House Gases (GHG's) including that of CO_2 (Pandey 2002). Thus there has emphasis to reduce this concentration through the process know as carbon sequestration. With the advent of the Kyoto Protocol and its recognition of the use of forestry activities and carbon sinks as acceptable tools for recouping the build-up of atmospheric carbon, the potential role of planted forests as a tool for carbon sequestration has taken on a new significance. Growing global concern on increasing carbon emissions in the atmosphere has led to the search for various mitigating options. Atmospheric concentrations of carbon dioxide can be lowered either by reducing emissions or by taking carbon dioxide out of the atmosphere and storing in it terrestrial, oceanic, or freshwater aquatic ecosystems. Carbon sequestration is squeezing of carbon out of the atmosphere, and its absorption and storage in a terrestrial or aquatic body. Capturing and storing carbon in biomass and soils in the agriculture and forest sector has now gained widespread acceptance as one potential greenhouse gas mitigation strategies.

The long-term conversion of grassland and forestland to cropland (and grazing lands) has resulted in historic losses of soil carbon worldwide but there is a major potential for increasing soil carbon through restoration of degraded soils and widespread adoption of soil conservation practices. The two major sources of rising CO_2 concentration are fossil fuel consumption, and global deforestation and is perhaps the major cause of green house effect with implications on Global Climate Change (IPCC 2002). Current understanding of the global carbon cycle suggests that managing forests and agricultural lands to increase the sequestration of greenhouse gases (GHG) would provide credible policy options (Brown 1996, Brown *et al.*, 1996). The annual exchange of carbon between forests and the atmosphere, and the amounts of carbon stored in forests, varies widely with the nature of forest cover. Sustainable management of forests can significantly increase the long-term sequestration of atmospheric CO_2. Notwithstanding this the tropical forests are faced with deforestation and destruction.

Characteristics of Carbon Sequestration

The idea of sequestering carbon by growing trees has attracted considerable interest in the climate change policy of community since physicist Freeman Dyson first proposed the idea about 20 years ago. The characteristics of CO_2 sequestration are delineated as:

1. Carbon sequestration is compatible with the continued large-scale use of fossil fuels, as well as greatly reduced emissions of CO_2 in the atmosphere. Carbon sequestration could be a major tool for reducing carbon emissions arising from use of fossil fuels.
2. Given the magnitude of carbon emission reductions needed to stabilize the atmospheric CO_2 concentration, multiple approaches to carbon management will be needed. Carbon sequestration should be researched in parallel with increased energy efficiency and decarbonization of fuel.
3. The natural carbon cycle is balanced over the long term but dynamic over the short term; historically, acceleration of natural processes that emit CO_2 is eventually balanced by an acceleration of processes that sequester carbon and vice-versa. The current increase in atmospheric carbon is the result of anthropogenic mining and burning of fossil carbon, resulting in carbon emissions into the atmosphere, are unopposed by anthropogenic sequestration.

The gases with special optical properties are responsible for global warming include: carbon dioxide (CO_2), water vapours (H_2O), methane (CH_4), nitrous oxide (N_2O), oxides of nitrogen (NO_x), stratospheric ozone (O_3), carbon monoxide (CO) and chlorofluorocarbons (CFCs). The gases are collectively called as Green House Gases (GHGs). Among all the sources, CO_2 plays a vital role. It contributes to 50 per cent of total green house effect which is followed by CH_4 (20 per cent) and nitrous oxide (6 per cent). The concentration of atmospheric CO_2 has now increased upto 370 ppm and is increasing at the rate of 1.5 μ L yr^{-1} giving rise to alarming situation (Lee and Dodson 1996). Carbon is accumulating in the atmosphere at the rate of 3.5 GT (10^{15}g) yr^{-1} as a result of fossil fuel combustion and tropical deforestation (Rosenberg 1999).

India's Contribution to Climate Change

Developing countries like India still continue to use technologies which are cheaper but may be outdated and harmful for the environment. Inefficient use of fossil fuels and other resources is from both a lack of state-of-art technology and proportionally higher use of coal and biomass, which emit more of GHG's per unit of energy than do petroleum products and natural gas. This further increases their vulnerability to global as well as local environmental and climatic changes. Though the developed countries contributes major chunk to the global warming, still they are not as vulnerable as that of developing countries.

The total carbon-dioxide equivalent emissions from India are estimated to be 1,001,352 Gg (1000 Mt), which is about three percent of the total global carbon dioxide-equivalent emissions. The per capita carbon dioxide-equivalent emissions for 1990 are estimated to be 1.194 tonnes or 325 kg of carbon (Singh and Lal 2000) compared to 2400 and 5400 kg of carbon in Japan and the US respectively (Singh and Lal 2000). This implies that India's CO_2 emissions are far below than that of developed countries. Notwithstanding this India will be one of the nations which could possibly be seriously affected by global warming and the resulting climate change, partly due to its high population density particularly along its long coastline.

Role of Forests in Co_2 Sequestration and Kyoto Protocol

Forests are both a source and a sink of carbon. Thus managing forests for carbon storage will help to absorb atmospheric carbon dioxide (IPCC 2001). Agricultural crops, forests sequester carbon from the atmosphere due to photosynthesis. Trees can play an important role in CO_2 sequestration because they have a much longer lifespan; they act as long-term reservoirs that lock up the carbon for decades, even centuries, in the form of cellulose and lignin.

Forests and Kyoto Protocol (KP)

Forests have assumed greater significance after the inclusion of afforestation and reforestation activities under the Kyoto Protocol of the United Nations Framework Convention on Climate Change (UNFCCC). UNFCCC is an international environmental treaty formulated at the United Nations Conference on Environment and Development (UNCED), informally known as the Earth Summit, held in Rio de Janeiro in 1992. The treaty aimed at reducing emissions of greenhouses gases. The treaty as originally framed set no mandatory limits on greenhouse gas emissions for individual nations and contained no enforcement provisions; it is, therefore, considered legally non-binding.

The Kyoto Protocol (KP) was adopted in December 1997 and came into effect in February 2005. The objective of this agreement is to stabilize greenhouse gas concentrations in the atmosphere at a level that would prevent dangerous anthropogenic interference with the climate system. Countries, which ratify this protocol, commit to reducing their emissions of carbon dioxide and five other greenhouse gases, or engage in emissions trading if they maintain or increase emissions of these gases. The industrialized countries are required to reduce their collective emissions by 5.2 per cent compared to the year 1990. In addition to this the industrialized countries are also required to consider ways to minimize adverse affects of their greenhouse gas emissions on developing countries. The developing countries are not bound by the Kyoto Protocol to reduce their emissions, as the per-capita emission rates of these countries are a tiny fraction of those in the developed world. However, the protocol emphasized strengthening the capacity of the developing countries to enhance their contribution to fighting climate change. Capacity building and technology transfer needs to be a central part of the global approach to combating climate change.

Carbon mitigation in developing countries remains an area of lower priority as these countries have various other immediate problems to deal with and developmental needs to meet with. Many developing countries are of the view that taking up any obligation regarding carbon sequestration would reduce them to carbon colonies, and that they would be asked to compromise with their development to comply with the agreed terms of such projects. At the present stage of development, reducing energy consumption would mean compromising with the developmental potential of the country. What can be done at this stage is to ensure that the increased CO_2 emission due to continuous use of energy sources is mitigated continuously. This can be achieved by afforestation and reforestation activities. So the implementation of carbon mitigation projects must be integrated within the general national economic policies (IPCC 2001).

The Kyoto Protocol also includes three market-based instruments known as the Kyoto Mechanisms that allow countries to earn or buy credits outside their borders. They are Clean Development Mechanism (CDM), Joint Implementation (JI), and International Emissions Trading (IET). These mechanisms enable Parties to access cost-effective opportunities to reduce emissions or to remove carbon from the atmosphere in other countries. While the cost of limiting emissions varies considerably from region to region, the benefit for the atmosphere is the same, wherever the action is taken. The developing countries can participate in two of these mechanisms, directly in CDM activities and indirectly in IET through CDM activities. The CDM is a way to earn credits by investing in emission reduction projects in developing countries. International Emissions Trading (IET) will permit developed countries that have taken on a Kyoto target to buy and sell credits among themselves. Additionally, the emergence of tradable emission permits and now tradable carbon offsets provides a vehicle for financially capturing the benefits of carbon emission reductions and carbon offsetting activities.

Forests and agricultural soils can remove and store carbon dioxide from the atmosphere. These carbon sequestration activities are called 'sinks'. Sinks can be enhanced through sustainable management practices in forestry and on farms. The Kyoto Protocol provides for inclusion of sinks as part of a country's strategy to meet its obligations with respect to climatic change. India signed and ratified the Protocol in August 2002. Since India is exempted from the framework of the Treaty, it is expected to gain from the protocol in terms of transfer of technology and related foreign investments.

The best possible forestry scenario among the various options available needs to be identified such that the demand of wood and wood products is met optimally, besides mitigating carbon (IPCC 2001). After the energy sector, the focus of the Climate Convention and scientific literature is largely on the forestry sector for undertaking various projects under Land-Use, Land-Use Change and Forestry (LULUCF) (Hitz and Smith 2004). Absorbing CO_2 from air and injecting it into the biomass is the only practical way of removing large volumes of GHG's from the atmosphere.

The idea of carbon sinks based on growing trees rests on an understanding of the carbon cycle. Enormous amounts of carbon are naturally stored in trees. As part of photosynthesis trees absorb carbon dioxide from the atmosphere and store it as carbon while oxygen is released back into the atmosphere. Rapidly growing trees absorb a larger amount of carbon dioxide. Mature trees grow less rapidly and thus have a lower intake of carbon dioxide. Mature forests, made up of a mixtures of various aged trees as well as dead and decaying matter, may be carbon neutral above ground. In the soil, however, the gradual buildup of slowly decaying organic material will continue to accumulate carbon, but at a slower rate than an immature forest. Trees are about 50 per cent carbon by weight. While all trees die and rot, releasing most of the stored carbon back to the atmosphere, the forest as a whole continues to store carbon as dying or harvested trees are replaced by natural regeneration. In effect, forests are carbon dioxide stores, and the sink effect exists only when they grow in size: it is naturally limited. It seems clear that the use of forests to curb climate change can only be a temporary measure. Even optimistic estimates come to the conclusion that the planting of new forests is not enough to counter-balance the current level of greenhouse gas emissions, as stipulated in the Kyoto Protocol. Although a forest is a net CO_2 sink over time, the plantation of new forests may also initially be a source of carbon dioxide emission when carbon from the soil is released into the atmosphere.

Other studies indicate that the cooling effect of removing carbon by forest growth is counteracted by the effects of the forest on reflection of sunlight, or albedo. Mid-to-high latitude forests have a much lower albedo during snow seasons than flat ground, and this contributes to warming. To prevent the stored carbon from being released into the atmosphere when the trees die, there have been suggestions of sinking the trees into the ocean. Such suggestions raise serious questions about feasibility. The dead trees, plants, and moss in peat bogs undergo slow anaerobic decomposition below the surface of the bog. This process is slow enough that in many cases the bog grows faster and fixes more carbon from the atmosphere than is released. Over time, the peat grows deeper. Peat bogs inter approximately one-quarter of the carbon in land plants and soils.

Forests are important in the global carbon cycle because they store more than 55 per cent of the global carbon stored in vegetation and about 45 per cent of that stored in soils, exchange carbon with the atmosphere through photosynthesis and respiration, are sources of atmospheric carbon when they are disturbed by human or natural causes and become atmospheric carbon sinks during re-growth after disturbance. Forests can influence climate change by affecting the level of CO_2 in the atmosphere; through the production of other greenhouse gases such as carbon monoxide, ozone, and nitrous oxide; and through changes in albedo of land as forests are converted to other land cover types. Globally, forest vegetation and soils contain approximately 359 and 787 Pg of C, respectively (Pg = 1,000 million tonnes). Some other projections ranged from 953 to 1400 Pg of global C. The allocation of carbon between vegetation and soils differs by altitude, with a large part of the vegetation (25 per cent) and soil (59 per cent) carbon pools located in the high latitude forests. Mid-latitude forests account for a small portion of the global carbon pool (16 and 13 per cent of the vegetation and soil, respectively). Low altitude tropical forests are relatively heterogeneous and contain 59 and 27 per cent of global forest vegetation and soil C, respectively (Dixon *et al.*, 1994).

Because there are provisions in the KP that acknowledge benefits from terrestrial carbon sequestration, scientists in a variety of fields are engaged in applying their expertise to estimate quantities of carbon that are stored in forest and agricultural ecosystems. Periodically, the Intergovernmental Panel convenes meetings of groups of these scientists from around the globe on climate change (IPCC) to develop consensus documents to establish a scientific basis for policy surrounding the KP. In the years since the KP was signed, the sequestration provisions of the protocol have been debated widely in scientific, policy, and public arenas. Issues range from economic impacts of sequestration efforts, to equity among developed and developing nations, to concerns about the certainty with which we can estimate carbon storage, to the advantages and disadvantages of substituting sequestration efforts for emissions reductions. Thus, the KP offers a case study of an environmental policy issue that has drawn widespread public attention in the US, scientific disagreement, and international contention.

Techniques of Carbon Sequestration

Carbon dioxide sequestration is a main long term method of mitigating the effects of fossil fuel consumption. Before carbon can be sequestered, it will have to be collected in the form of gaseous CO_2. CO_2 capture is estimated to amount to around 75 per cent of total sequestration costs (including capture, storage, transport, and sequestration). Although separation and capture systems are fairly well developed, as part of industrial processes such as synthetic ammonia production, hydrogen production, and cement production, they are not developed enough for large-scale sequestration. The important techniques in carbon sequestration are:

A. Industrial Techniques to Sequester Carbon

- ☆ Pre-combustion capture, in which the carbon content of the fuel is captured before the final combustion (*i.e.* not in the flue gas). This approach is used in coal gasification plant, which generates a syngas made mostly of CO and H_2. The CO and H_2 are then separated and reacted in a controlled environment releasing almost pure CO_2.
- ☆ Post-combustion capture, in which the CO_2 is captured from the flue gas. With this approach various amine-based processes already exist, and retrofit to existing plant is possible. Presently, these techniques are cost expensive.
- ☆ Oxyfuel technologies: These are an alternative approach, involving the use of enriched oxygen as feedstock for the combustion process. This, together with recycling of combustion products, results in CO_2 and H_2O products, which are readily separated.

B. Increasing Vegetal/Biomass Cover to Sequester Carbon

The vegetal cover extracts CO_2 from the atmosphere and thereby provides the opportunity to sequester carbon in the biomass as long as the biomass is not harvested or if harvested is stored in long-lived industrial products. This prospect provides the basis for carbon management strategies in which CO_2 emissions from fossil fuel burning in one part of the world might be offset by carbon sequestration realized by growing trees in another part of the world.

Although the plant-and-store approach can provide carbon sequestration benefits a variety of other public benefits such as watershed protection (water supply and soil stabilization), maintenance of biodiversity, creation of wildlife sanctuaries, it offers few private benefits (except possibly enhanced tourism). Widespread public acceptance of tree-growing strategies will probably require that a significant part of the overall sequestration effort be devoted to perpetual-rotation strategies, in which the local populations can earn income from the sale of timber, pulpwood, fuel wood, etc., products generated from the harvested biomass and can often use many of these products in serving local needs (Parikh and Reddy 1997; Sathaye *et al.*, 1995; Xu 1995). Xu (1995) indicated that sequestration carried out in China using the sustainable rotation approach on 130 million hectares of land under Forests that is available for afforestation plus 70 million hectares of dry croplands under agriculture that is suitable for agroforestry development could sequester 9.7 billion tonnes of carbon (49 tonnes C per hectare) for an initial cost (of $19.3 billion (@ $2 per tonne of C). This sequestration potential is equivalent to 11 years of China's total CO_2-equivalent GHG emissions at the 1995 emission rate.

Considering the local benefits and the importance of minimizing opportunity costs that could be very significant if trees were planted on prime agricultural lands has also led to emphasis in recent studies on carbon sequestration projects that might be carried out on degraded lands (Parikh and Reddy 1997; Ravindrath and Somashekhar 1995; Xu 1995), which could simultaneously generate local benefits by providing watershed protection, maintaining biodiversity, providing biomass products to serve local needs, and generating rural employment.

Large degraded land areas, a common feature of the landscape in many countries, are suitable for carbon sequestration. In many instances these lands were forested until recently, and they retain adequate soil fertility to support biomass growth (Sathaye *et al.*, 1995). In India, for example, some 53 million hectares of degraded non-crop and non-forest land (16 per cent of the total land area of India) is potentially available for reforestation projects (Gautam 1997; Ravindranath and Somashekhar 1995). Another estimate is that 950 million hectares may be technically available for revegetation (Dixon *et al.*, 1994). Moreover, Trexler *et al.* (1989) estimate that it is typically socio-economically feasible to utilize about 70 per cent of technically available land.

Conclusion

Indiscriminate and inefficient use of fossil fuels, unsustainable use of natural resources and deforestation and forest degradation has lead to increase in CO_2 of the atmosphere resulting in the global warming and many concomitant environmental hazards. Reduction in atmospheric CO_2 concentration has now gained wide importance in light of Kyoto Protocol. CO_2 sequestration is one of the means to achieve this goal which be achieved through industrial processes as well as by regenerating the forests and afforestating the degraded lands. The latter being cost effective and have wide-applicability than the industrial processes. Therefore, regeneration and afforestation is encouraged at local level as mass movement which will ameliorate the climate not only at local level but also at regional and global level.

References

Brown, S., 1996. Tropical forests and the global carbon cycle: Estimating state and change in biomass density. In: *The Role of Forest Ecosystems and Forest Management in the Global Carbon Cycle*, (Eds.) M. Apps and D. Price. NATO ASI Series No. 140, Springer-Verlag, NY, pp. 135–144.

Brown, S., Sathaye, J., Cannell, M.G.R. and Kauppi, P., 1996. Management of forests for mitigation of greenhouse gas emissions. In: *Climate Change 1995. Impacts, Adaptations and Mitigation of Climate Change.* (Eds.) R.T. Watson, M.C. Zinyowera, R.H. Moss, and D.J. Dokken. Working Group II, Second Assessment Report of the IPCC. Cambridge University Press, p. 773–797.

Dixon, R.K., Brown, S., Houghton, R.A., Solomon, A.M., Trexler, M.C. and Wisniewski, J., 1994. Carbon pools and flux of global forest ecosystems. *Science*, 263:185–190.

Gautam, N.C., 1997. Extent of wastelands in India estimated by remote sensing. In: *Sustainable Regeneration of Degraded Lands*, (Eds.) J.K. Parikh and B.S. Reddy. Tata McGraw-Hill, New Delhi, India, pp. 43–55.

Hitz S. and Smith, J., 2004. Estimating global impacts from climate change. *Global Environmental Change*, 14: 201–218.

http://unfccc.int/files/essential_background/application/pdf/beginner_02_en.pdf

IPCC, 2001. Climate Change 2001: Working Group II: Impacts, Adaptation and Vulnerability. http://www.grida.no/climate/ipcc_tar/wg2/

Lee, Jeffrey J. and Dodson, Rusty, 1996. Potential carbon sequestration by afforestation of pasture in the South-Central United States. *Agronomy Journal*, 88: 381–384.

Pandey, D.N., 2002. Carbon sequestration in agroforestry systems. *Climate Policy*, 2: 367–377.

Parikh, J.K. and Reddy, B.S., 1997. *Revisiting Carbon Sequestration for Joint Implementation*. Indira Gandhi Institute of Development Research, Mumbai, India.

Ravindranath, N.H, and Somashekhar, B.S., 1995. Potential and economics of forestry options for carbon sequestration in India. In: *Forestry and Climate Change: A Special Issue of Biomass and Bioenergy*, (Eds.) J. Sathaye and W.R. Makundi, 8(5): 323–336.

Sathaye, J., Makundi, W. and Andrasko, K., 1995. A comprehensive mitigation assessment process (COMAP) for the evaluation of forestry mitigation options.

In: *Forestry and Climate Change: A Special Issue of Biomass and Bioenergy*, (Eds.) J. Sathaye and W.R. Makundi, 8(5): 345–356.

Singh, R. and Lal, M., 2000. Sustainable forestry in India for carbon mitigation. *Current Science*, 78: 563–567.

Trexler, M., Faeth, P. and Kramer, J., 1989. *Forestry as a Response to Global Warming: An Analysis of the Guatemala Agroforestry and Carbon Sequestration Project*, World Resources Institute, Washington, DC.

UNEP and UNFCCC, 2002. Understanding Climate Change: A beginner's guide to the UN Framework Convention and its Kyoto Protocol. United Nations Environment Program Information Unit for Conventions, Geneva, Switzerland, United Nations Framework Convention on Climate Change, Bonn, Germany.

Winjum, J.K., Dixon, R.K. and Schroeder, P.E., 1993. Forest management and carbon storage: analysis of 12 key forest nations. *Water, Air and Soil Pollution*, 70: 239–257.

Xu, D., 1995. The potential for reducing atmospheric carbon by large-scale afforestation in China and related cost/benefit analysis. In: *Forestry and Climate Change: A Special Issue of Biomass and Bioenergy*, (Eds.) J. Sathaye and W.R. Makundi, 8(5): 337–344.

Ecosystem Diversity and Carbon Sequestration: Climate Change Challenges and a Way Out for Ushering in a Sustainable Future *Pages 43–47*

Editors: P.L. Gautam, Vir Singh and Uma Melkania

Published by: DAYA PUBLISHING HOUSE, 2009

Chapter 5

Carbon–A Material for the Twenty First Century: Prospects and Promises

B.S. Tewari and Ajay

Recently, one of the important challenges before the environmental scientists is to check the content of carbon, especially carbon dioxide, in our environment for sustainable and safe future. But if one looks at carbon from a material scientist's point of view, it is a very promising electronic material and may play an important role in the development of society to which we owe our existence. The present paper reviews the potential application of carbon as a next generation material, most suited for fabrication of electronic devices. In nature carbon exists in two allotropic (crystalline) form sp^3 bonded Diamond and sp^2 bonded Graphite. Further, carbon is the most flexible atom when it comes to bonding. This flexibility of carbon makes it a backbone of bio-molecules and fabrication at nano scale. Recently, due to ultra fabrication technologies and characterization techniques material scientists harnessed carbon into clusters like Buckyballs (C_{60}), also atom by atom fabricated sheets/ribbon of carbon atoms called Graphene are now possible. These clustered and atom by atom fabricated structures of carbon have shown amazing electronic and optical properties to make these materials a candidate for carbon based electronic material to replace Silicon in the electronic industries in near future. Therefore, the present paper deals with the review of the recent developments of free carbon that exists in nature as an established electronic material with future prospects.

Introduction

The name carbon comes from Latin word *carbo*, means charcoal. Carbon is the fourth most abundant chemical element in the universe by mass, after oxygen, hydrogen and helium. Carbon is abundant in the Sun, Stars and Comets and in the atmosphere of most planets. Further, carbon is essential to all known living systems, and without it one can not find the existence of life. As a free element it forms allotropes from different kinds of carbon-carbon bonds, such as in graphite and diamond. Coal is the main source of carbon in the mineral form and contains nearly 95 per cent of carbon in anthracite (Reed, 1993).

There was a balance between the different elements in the environment of our earth, but today, due to industrialization, frequently used motor vehicles and deforestation are making this balance unstable and increasing the level of carbon dioxide in our atmosphere and now it becomes necessary to check carbon percentage in our environment. A lot of research is being carried out to find the amicable solution of this problem by environmental scientists. Here, we are looking carbon from material scientist point of view that carbon may be a next generation material due to its availability in different forms and hence various applications come true and remaining applications are in the way to be explored (http://www.research.ibm.com/nanoscience/nanotube) and may out perform silicon, the established semi conducting material, available for electronic devices.

Carbon is the fourth group element of the periodic table that has atomic number 6. An abundant nonmetallic, tetravalent element, carbon has several allotropic forms. Its position in periodic table made it as a more flexible atom when comes to bonding. In nature, carbon exists in two basic forms: the sp^2 bounded diamond (insulator) and sp^3 bonded graphite (semimetal). Beside these two basic forms, now atom by atom fabrication of carbon atom as a single atom ribbon is possible (known as Graphene) due to ultra fabrication and characterization.

The diamond is the hardest known natural material. In diamond, each atom is bonded tetrahedrally to four others, making a 3 dimensional network of puckered six-membered rings of atoms. It is an excellent abrasive because it can be scratched only by other diamonds. It is a type of perfect insulator having the band gap of the order of 5eV and not suitable for electronic purposes at room temperature but after doping with some impurity like Boron one may make it semiconductor and electronic conduction may occur. The semiconductor devices of diamond can bear 20-50 time higher electric current as compared to Si and Ge based electronic devices. At very low temperature and ambient pressure it shows superconducting phase. The important thing is that diamond sheet posses a high quality interface and could be attracted to biomolecules for biomolecular electronics.

Nanostructured Carbon: Carbon Clusters, Sheets and Tubes

Carbon is essentially graphite in its amorphous form but not held in a crystalline macrostructure. It is, rather, present as a powder which is the main constituent of substance such as charcoal, lampblack and activated carbon. In graphite, each atom is bonded trigonally to three other atoms, making a 2 dimensional network of flat six-membered rings; the flat sheets are loosely bonded. It is the softest material in comparison to diamond. The graphite is good conductor of electricity. Graphite is combined with clay to form the 'lead' used in pencils, it is also used as a lubricant and pigment. The graphite is the stable form of carbon due to change in climate, surface temperature and pressure, diamond in the earth crust is changing to graphite.

In fact, all diamond at or near the surface of the earth are currently undergoing a transformation into graphite. This reaction, fortunately, is extremely slow otherwise we would have nearly vanished the diamond.

The other important form of carbon is cluster form known as fullerene, *e.g.* C_{60} and C_{70} (the cage of carbon atom). Fullerene is made up of large molecule formed completely of carbon bonded trigonally, forming spheroids of which the best known and simplest is the buckminster fullerene or buckyball, because of its soccer ball-shaped structure (http://www.nanoscript.com/nanotubes/newsletter; Guangyu *et al.*, 2005). This form of carbon is attracting material scientists now-a-days as they posses fascinating electronic properties. They show superconductivity at around 20K. The Nobel Prize in chemistry was awarded for the discovery of Fullerene in the year 1996 to R.F. Curl, H.W. Kroto and R.E. Smalley (Rice and Sussex Universities).

Recently, thin single atom thick carbon sheets are now possible to grow in laboratory and may serve new generation electronic material. In March 2007, A.K. Geim and K.S. Novoselov group at University of Manchestor Oxford (U.K.) claimed world's thinnest material in the form of one atom carbon sheets called grapheme. In grapheme, the electrons are moving with a very high speed and called Dirac (relativistic) electrons. Transistors are being built with this highly conducting graphene sheets and relativistic Hall Effect is also been predicted in these thin sheets (Meyer, *et al.*, 2007). Once a graphene sheet is available, recent material fabrication advancement made it easy to roll these graphene sheets to form single wall nanotube and if one has multi-graphene sheets then multiwall nanotube are be developed. The diameter of these is tube is of the order of nanometer and can be few microns long. Further, these are carbon tubes out of defects and can be converted into new geometry by capping up a half of buckyball to ends. The geometrical and electronic arrangement of these carbon nanotubes leads to fascinating electronic properties like semiconductor to metallic transition if one go from bulk carbon to nanostructure and other thermal, electrical and magnetic properties are also depends on surface to volume ratio.

Electronic Properties of Carbon Based Nanostructures

The special nature of carbon combines with the molecular perfection of single-wall carbon nano tubes (CNTs) to endow them with exceptional material properties, such as very high electrical and thermal conductivity, strength, stiffness, and toughness. No other element in the periodic table bonds to itself in an extended network with the strength of the carbon-carbon bond. The delocalized pi-electron donated by each atom is free to move about the entire structure, rather than remain with its donor atom, giving rise to the first known molecule with metallic-type electrical conductivity (free electrons). Furthermore, the high-frequency carbon-carbon bond vibrations provide an intrinsic high thermal conductivity (Stankovich *et al.*, 2006; Hill *et al.*, 2006).

In most bulk materials, however, the actual observed material properties–strength, electrical conductivity, etc. are degraded very substantially by the occurrence of defects in their structure. For example, high-strength steel typically fails at only about 1 per cent of its theoretical breaking strength. CNTs, however, achieve values very close to their theoretical limits because of their defects free molecular structures.

This aspect is part of the unique story of CNTs. CNTs are an example of true nanotechnology: they are only about a nanometer in diameter, but are molecules that can be manipulated chemically and physically in very useful ways. They open an incredible range of applications in materials science, electronics, chemical processing, energy management, and many other fields (Sofo *et al.*).

Applications: Electronic Devices and Fabrications

The exploration of Bucky tubes in biomedical applications is just underway, and has significant potential in drug delivery system. Cells have been shown to grow on Bucky tubes, so they appear to have no toxic effect if interfaced with human organs. The cells also do not adhere to the Bucky tubes, potentially giving rise to applications such as coatings for prosthetics and anti-fouling coatings for ships. The ability to chemically modifying the sidewalls of Bucky tubes also leads to biomedical applications such as vascular stents, and growth of neuron and regeneration as a cutting edge technologies having biomedical applications.

There is a wealth of other potential applications for CNTs, such as solar collection; nonporous filters; catalyst supports; and coatings of all sorts. There are almost certainly many unanticipated applications for this remarkable material that will come to light in the years ahead, and which may prove to be the most important and valuable ones of all. Many researchers are looking into conductive and or water proof paper made with CNTs. CNTs have also been shown to absorb Infrared light and may have applications in the I/R Optics Industry (Nilsson *et al.*; Gein and Novoselav, 2007) in near futures.

Conclusions

Although carbon is relatively safe due to low toxicity and resistance to most chemical attacks (including fire) at normal temperature, inhalation of fine soot in large quantities can be dangerous. Diamond dust can do harm as an abrasive if ingested or inhaled. Carbon may also spawn flames at very high temperatures and burn vigorously and brightly. The great variety of carbon compounds includes such lethal poisons as cyanide (CN^-), and carbon monoxide (CO). The increasing concentration of carbon in carbon dioxide form is the primary concern of environmental scientists for sustained and safe future. We will conclude with the remark that carbon has its various structure like clusters, nanotubes, grapheme sheets that possess remarkable electronic properties at nano scale and may play a pivotal role as established electronic material and after a glimpse of what is possible if the silicon, the back bone of electronic industry, become unavailable. Hence, challenge before material scientists is to use the sequestrate carbon that exists in our surrounding as an electronic material for fabrication of devices and that out-performs silicon for development of our society.

References

Gein, A.K. and Novoselav, K.S., 2007. *Nature*, 6: 183.

Guangyu, Z. *et al.*, 2005. *Science* 102: 45.

Hill, E.W. *et al.*, 2006. *IEEE Trans. Magn*, 2: 2694.

http://www.research.ibm.com/nanoscience/nanotube.

http://www.nanoscript.com/nanotubes/newsletter.

Meyer, J.C. *et al.*, 2007. *Nature.*

Nilsson, J. *et al.*, Cond-mat/0606589.

Reed, M.A., 1993. *Scientific American*, 98, January.

Schonenberger, C., 2000. *Physics World*, June.

Sofo, O. *et al.*, Cond-mat/0606704.

Stankovich S. *et al.*, 2006. *Nature*, 442, 282.

Yakobson, B. and Smalley, Richard E., 1999. *American Scientist.*

ECOSYSTEM DIVERSITY IN INDIA
Scenarios, Anthropogenic Dimensions and Management

Ecosystem Diversity and Carbon Sequestration: Climate Change Challenges and a Way Out for Ushering in a Sustainable Future *Pages 51–56*
Editors: P.L. Gautam, Vir Singh and Uma Melkania
Published by: DAYA PUBLISHING HOUSE, 2009

Chapter 6

Forest Ecosystems and Carbon Sequestration in India: Keeping the Greenhouse Gas at Bay

J.B. Lal

Introduction

Forest is both a source and a sink of carbon dioxide. However, the relationship between forest systems and carbon sequestration on one hand, and deforestation and carbon emission on the other, is not only complex but variable. In India where geographical variability of forest systems is immense to find the precise equations to describe the relationships is much more difficult. The forest ecosystems in India vary from moist alpine to wet evergreen. There are as many as 16 major type groups of forests, 3 in alpine zone, 3 in temperate, 3 in sub-tropical, and 7 in tropical. These 16 groups are further divided in 221 ecologically stable vegetation types (Lal 1989). A forest type is defined as "a unit of vegetation which possesses broad characteristics in physiognomy and structure sufficiently pronounced to permit its differentiation from other such units". The definition certifies that a forest type qualifies as a distinct ecosystem.

The Two Basic Reactions

The two basic reactions that link the atmospheric carbon dioxide with living plants are photosynthesis, *i.e.*,

$$6CO_2 + 6\,H_2O + \xrightarrow{\text{Solar Energy}} = C_6H_{12}O_6 + 6O_2$$

and, its opposite, respiration, or oxidative metabolism, *i.e.*,

$$6O_2 + C_6H_{12}O_6 \rightarrow 6CO_2 + 6H_2O + \text{Heat}$$

Photosynthesis and respiration are vital activities not only in plant ecology, but also in human ecology. Photosynthesis fixes the carbon energy, and respiration uses most of that energy, and the balance is the Net Primary Productivity (NPP) of a vegetational unit. The primary production from forests is 49 per cent of the total production from land (Lal, 1995). Whittaker and Likens (1973) estimated the NPP from various forests as follows:

Table 6.1

Forest Ecosystem Type	*NPP (Mean) g/Square Meter/Year*
Tropical evergreen	2200
Tropical deciduous	1600
Temperate evergreen	1300
Temperate deciduous	1200
Shrubland	700
Alpine	140
Desert and semi-desert	90

Forest and Climate Change

Forests affect the climate and in turn are influenced by the climate. Forests are crucial element of climate change in their capacity as carbon emitting source through deforestation and degradation, and a carbon sink through reforestation and afforestation. Simulated models suggest that about 30 per cent of global forest will experience big changes in composition and structure if carbon dioxide level doubles from pre-industrial level. The impact of climate change will be much more on temperate forests than on tropical ones.

NPP of Forests and Atmospheric CO_2

NPP of various forests is important to the humans not only for matter and energy, but also, as the recent observations and researches prove, for its role in stabilizing carbon dioxide concentration in the atmosphere and checking the consequent global warming. The carbon dioxide concentration in the atmosphere has been steadily rising from about 290 ppm (part per million) in the latter half of the nineteenth century to over 358 in 1994, to over 370 ppm at the beginning of 21st century. The steep rise in the carbon dioxide concentration in the atmosphere leading to global warming, dangerous not only to humans, but to all kinds of life, is attributable mainly to two factors: first, exponential increase in burning of fossil fuel, and secondly, rapid decrease in earth's forest cover.

The addition of carbon-dioxide to the atmosphere on account of fossil fuel burning is estimated at 5x 10^{15}g C/yr (Lal, 2005). The additional input of carbon dioxide on account o f deforestation, has not been estimated with any degree of certainty. As a matter of fact, in establishing a correlation between deforestation and carbon emission so many variables have to be taken into accounts that precise estimates are, indeed, very difficult. Various evaluations, however, suggest a figure of 2.8×10^{15} g C/yr for carbon emission on account of loss of forest cover. Addition of the two figures, gives the figure for total carbon emission as 7.8×10^{15} g C/yr.

If working out of emission figures is difficult, the exercise in relation to 'carbon sink' is virtually confounding. Of the estimated carbon emission in the world, only 3.8×10^{15} g C/yr accounts for the rise in atmospheric carbon dioxide. What are the sinks for the balance of 4.7×10^{15} g C/yr. Estimates vary enormously. If one estimate puts oceans as top most sink, the other puts them at the bottom. Whatever may be the situation, the role of terrestrial ecosystems, especially of forests, is significant, and their carbon assimilating capacity needs to be thoroughly researched upon.

Forestry Situation in India

FAO defines forests as ecosystems with a minimum of 10 per cent crown cover of trees and/or bamboos, generally associated with wild flora, fauna, and natural soil conditions, and not subject to agricultural practices.

Deforestation in the Indian as well as in the international context refers to a change of land use with the depletion of tree cover to less than 10 per cent. The change in a forest crop on account of the lowering of the crown density with the consequent lowering of the productivity is termed degradation. Unfortunately, degradation is mostly not assessed.

In India, the biennial report brought out by Forest Survey of India (FSI) since 1987 give the figures of forest cover in the country in various years based on the interpretation of satellite imagery. FSI has adopted the same definition of forests as FAO. It classifies forest with crown density of 40 per cent or more as 'dense', with less than 40 per cent as 'open'.

Table 6.2: Forest Cover in India

Category	*Forest Area in Million Hectares*			
	1987	*1993*	*1997*	*2003*
Dense	36.14	38.56	36.72	39.06
Open	27.66	25.03	26.13	28.77
Mangrove	0.40	0.42	0.48	Not assessed separately
Total	64.20	64.01	63.33	67.83

It would appear from the Table 6.2 given above that the forest cover in country was nearly static in nineteen-nineties, and has increased with the turn of the century.

This statement is only technically true. True in the sense that the area of forest lands with crown density of 10 per cent or more has stayed close to or over 64 million hectare. These figures, however, do not reflect degradation. Even if the crown density of a forest reduces from 90 per cent to 60 per cent, and growing stock to two-third, the assessment would indicate only status quo. That degradation is taking place and on account of, first, consumption of firewood in excess of sustainable level of production, grazing in forest beyond carrying capacity, and recurring fires. The annual consumption of firewood in the country was estimated at 235 million cubic meter by the author in1987 (Lal, 1989), and at 270 million cubic meter by FAO in 1994. The gap in consumption and sustainable production of firewood was estimated at 152 million cubic meters by the author (Lal, 1995), and at 147 million cubic meters by FAO.

The current growing stock of wood in forests is estimated by FSI (2003) at 4,781 million cubic meters. FSI also estimated the growing stock of trees outside forests, and the figure for this stock is 1632 million cubic meters.

Carbon Emission in India

Even assuming that on account of a little more efficiency in use, the firewood consumption has not increased in the country since FAO made the estimate, the country is emitting 85 million tons of carbon by burning wood. Since each ton of carbon emitted into the air results in 3.7 tons of carbon dioxide, the country has been adding about 315 million tons of carbon dioxide annually.

The average of various estimates of carbon emission on account of fossil fuel burning works out at about 450 million tons. Thus the total of carbon dioxide released in the country annually on account of fossil fuel burning and firewood consumption is about 765 million tons. The country may plan to sequester carbon in as near quantity to this figure as possible.

Carbon Sequestration in Forest Biomass

We stated earlier that the growing stock (woody biomass above the ground) in the forests of the country is estimated at about 4781 million cubic meters equivalent to 1484 million tons of carbon. This figure, however, represents only a fraction of living phytomass; this does not include foliage, shrubs, herbs, and other ground flora and the root. The annual increment on this quantity, *i.e.*, potential stem wood productivity is worked out by the author at 92 million cubic meters, corresponding to 29 million tons of carbon.

Table 6.3: Potential Stem-wood Productivity of India's Forests

Sl.No.	*Region*	*Forest Cover (million cu.m.)*	*Potential Productivity (cu.m./ha/yr)*	*Total Potential Productivity (million cu.m./yr.)*
1.	Western Himalaya	5.38	2.21	11.89
2.	Eastern Himalaya	6.46	2.03	14.85
3.	North-East	7.33	1.66	12.17
4.	Western Coast and Andaman and Nicobar Islands	1.95	3.85	7.51
5.	Deccan	15.90	1.35	21.46
6.	Central India	17.49	1.05	18.36
7.	Gangetic Plain	4.75	0.80	3.80
8.	Indus Plain	1.44	0.41	0.59
	Total	60.70		90.63

Alpine, semi-desert forests and mangroves are not included in the calculations. If they are included, total potential productivity would be about 92 million cubic meters.

If we use the mean NPP figures worked out by Whittaker and Likens (1973) and multiply by forest cover figures estimated by FSI (2003), the net primary production of India's forests works out at about 1010 million cubic tons of dry matter (DM) per annum corresponding to 460 million tons of carbon.

The NPP thus worked out is many times higher than stem-wood productivity worked out by the author. This is because of the fact that 50 per cent of total NPP is lost in saprobe respiration, and 10 per cent in animal respiration (Lal, 1989). Only 40 per cent forms the accumulated dry matter (DM). And of accumulated DM all is not living phytomass, *i.e.*, wood and leaves. It also includes litter and carbon accumulated in soil. In tropical forests, on average, the break-up of phytomass is:

Wood stem	57 per cent
Wood root	20 per cent
Leaf	5 per cent
Soil carbon	16 per cent

On the basis of the break-up indicated above, the estimate of portion of NPP accumulated in stem-wood in India's forests works out at 104 million tons. If we reduce it by 10 per cent to account for shrub wood, the figure comes down to 93.6 million tons. That this figure is more than 3 times higher than the figure estimated by the author indicates that country's forests suffer from immense biotic pressure, and are in far from ideal ecological situation. Nevertheless, it can be safely assumed that afforestation or reforestation made over an hectare of land would, on average, sequester 1.5 tons of carbon.

The Strategy

Before we suggest a strategy, we should take into account the following:

- ☆ Mature communities are more stable, but immature communities, *i.e.*, seral stage forests and plantations, are more productive, *i.e.*, store more carbon.
- ☆ As said before India can be divided in as many as 8 physiographic zones. In a more refined division (as done by FSI in its 2003 Report) it can be divided in as many as 14 physiographic zones.
- ☆ India has many as 221 forest types. Each type plays quantitatively a different role in carbon sequestration.

Considering these statements, and taking into account facts mentioned earlier, the following strategy with 6 components is suggested to check the rise of carbon concentration in the atmosphere:

First, to use firewood more efficiently, and increase the present level of thermal efficiency which is only 8 per cent to a minimum of 25 per cent. This would reduce our firewood consumption to one-third of the present quantity, and we would emit 60 million ton less carbon in the air.

Second, the protected area in the country be increased from 14 to 20 million ha. The measure is likely to increase annual carbon sequestration by 4 million tons.

Third, we afforest a minimum of one million hectare in a year to secure additional sequestration of 15 million tons of carbon.

Fourth, we have virtually ignored the scientific management of fourth bio-productive system, *viz.* grassland (the other 3 systems are cropland, forest, and fisheries). If we manage 12 million hectares officially recorded as pastures, scientifically, we would secure additional sequestering of 6 million tons of carbon.

Fifth, wherever possible, integrated land-use systems such as agri-silviculture, silvi-pastoral, and agri-silvi-pastoral be adopted to make more efficient use of land resource.

Sixth, forests be protected more earnestly from excessive grazing, and recurring fires. If the strategy succeeds, a net carbon fixation of 30 million tons by forests and allied systems could be achieved.

References

FAO, 1994. *Year Book of Forest Production*. FAO, Rome.

FSI. *Forest Survey Reports, 1987, 1991, 1997, 2003*. FSI, Dehradun.

Lal, J.B., 1989. *India's Forests: Myth and Reality*. Natraj Publishers, Dehradun.

Lal, J.B., 1995. Forest Biomass and Carbon Sequestration in India. *Journal of Tropical Biodiversity*, Vol. 2, No. 1.

Lal, J.B., 2005. *Economic Tools and Institutional Framework for Forest Development Planning*. IBD, Dehradun.

Whittaker, R.H. and Likens, G.E., 1973. *Carbon in the Biota*. US Dept. of Commerce.

Ecosystem Diversity and Carbon Sequestration: Climate Change Challenges and a Way Out for Ushering in a Sustainable Future *Pages 57–70*
Editors: P.L. Gautam, Vir Singh and Uma Melkania
Published by: DAYA PUBLISHING HOUSE, 2009

Chapter 7

Operationalizing CDM Afforestation and Reforestation Projects in India: Analysis of Barriers at National and International Level

Sandeep Tripathi and V.R.S. Rawat

The Clean Development Mechanism (CDM) of Kyoto Protocol is one of the flexible mechanisms for project-based emission reduction activities in developing countries. The Land use, Land use Change and Forestry (LULUCF) sector can provide relatively low cost opportunities to combat Climate Change, either by increasing the removal of GHGs from the atmosphere through carbon sinks (*e.g.* by planting trees) However, the development of forestry Projects under CDM has not been a great success in India so far. The CDM project submission process to CDM EB is very cumbersome. The approval rate of methodologies is very low. Poor understanding of CDM A/R modalities and procedures has been one of the primary reasons for high rate of rejection of these projects. There is a potential of 59.6 million ha of land for afforestation and reforestation activities in the country. A sizable amount of this potential can be undertaken for CDM A/R projects. India has a potential of 175 million tonnes of carbon to be sequestered annually through LULUCF activities. So far only 4 CDM A/R Projects have been developed in India and only two of them could make up to CDM EB.

In the present paper an analysis of Barriers to CDM A/R projects has been analysed. Some of the recommendation suggested are: (*i*) Delineation of Potential CDM A/R area, (*ii*) Inclusion of CDM Component in Forest Working Plans, (*iii*) Opening of a CDM Cells in research and extension wings of SFDs, (*iv*) Incorporation of CDM component in JFM Programs, (*iv*) Promotion of forestry based CDM Projects other than CDM A/R, such as biomass based renewable energy projects and fuel switching from *Jatropha* oil etc. (*v*) Finance for CDM Projects, and (vi) Institutional Strengthening and capacity building for stakeholders and creating market for CDM A/R Projects.

Introduction

Few global issues have received more attention than climate change in recent years. Amidst all the discussion about the greenhouse effect one fact is certain that atmospheric CO_2 levels are rising. Climate change is recognized as one of the greatest environmental threat facing the world today. It threatens to have major adverse impact on natural world and human society as well. The inter governmental panel on climate change (IPCC) representing the vast majority of scientists internationally predicts that if no action were taken to limit greenhouse gases emissions, temperature would rise in the range of 1.8-4.0°C by the year 2100 (IPCC 2007). This would be a faster rate of warming at any time since the end of the last ice age, 10,000 years ago. Most of the observed increase in the globally averaged temperature since the mid-20th century is very likely due to the observed increase in anthropogenic greenhouse gas concentrations. The primary source of the increased atmospheric concentration of carbon dioxide since the pre-industrial period results from fossil fuel use, with land-use change providing another significant but smaller contribution.

In the year 1992 most countries joined an international treaty–the United Nations Framework Convention on Climate Change (UNFCCC)–to begin to consider the actions to reduce global warming and to cope with whatever temperature increases are inevitable. The Kyoto Protocol of UNFCCC legally binds Annex 1 (developed countries) countries that have ratified the protocol to reduce their green house gas emissions by 5.2 per cent before 2012 with targets set for each nation based on their 1990 level emissions. The Clean Development Mechanism (CDM) is one of the flexible mechanisms established by Article 12 of the Kyoto Protocol for project-based emission reduction activities in developing countries. The CDM allows an Annex-1 party to implement a bilateral project that reduces green house gas emissions or removes green house gases by carbon sequestration in the geographical territory of a non-Annex 1 (developing countries) party.

Approaches to Carbon Management

Predictions of global energy use in the next century suggest a continued increase in carbon emissions and rising concentrations of CO_2 in the atmosphere unless major changes are made in the way we produce and use energy, in particular, how we manage carbon. Greenhouse gas emissions arise from numerous human activities and natural processes. Opportunities for mitigating atmospheric accumulation of these gases will therefore arise from a number of different sectors within society. Current thinking about climate change mitigation options look at following three board categories.

1. Avoiding production or release of CO_2 in the first place: this option to manage carbon includes improving energy production and conversion

efficiency and end use so that fewer units of primary fossil energy are required to provide the same energy service.

2. A second approach is to substitute lower-carbon or carbon-free energy sources for our current sources. For example, this strategy might involve substituting lower-carbon fossil fuels such as natural gas for coal or oil; using renewable energy supplies such as solar, wind, or biomass; or increasing the use of nuclear power.
3. Carbon sequestration could represent a third approach in addition to efficiency improvement and evolution toward low-carbon fuels.

Forestry Based Mitigation Options

There is a close relationship between climate and forests. Forests act as reservoirs by storing carbon in biomass and soils and as carbon sinks when their area or productivity is increased, which results in greater uptake of atmospheric carbon dioxide (CO_2), the most important greenhouse gas. Various forestry practices have a significant role in helping to slow down the accumulation of CO_2 in the atmosphere. Conversely, forests are a source of greenhouse gases when biomass burns or decays. Some practices involved in plantation development, forest management and agroforestry, such as tilling and use of natural fertilizers, can release greenhouse gases. The primary kind of land use based measures that specifically has been discussed in the context of helping mitigate climate change are:

- *Slowing or stopping the loss of existing forests, thus preserving current carbon reservoir.* Project Opportunities include forest protection, forest management, harvest management, and off site agroforestry.
- *Adding to planet's vegetative cover thus enlarging living terrestrial carbon reservoirs.* Carbon offset opportunities include reforestation, afforestation, natural and assisted regeneration, fertilization, farm forestry and soil management practices.
- *Increasing the carbon stores in the carbon reservoirs such as agricultural soils and wood products.* Offset opportunities include paper recycling, substitution of wood for steel and concrete and changes in tillage and fertilization practices.
- *Substituting sustainable biomass energy sources for fossil fuel consumption.* Offset opportunities include transfer of biomass technologies, efficiency improvement in existing biomass utilization and subsidization of biomass energy development.

Forestry and Kyoto Protocol

The Kyoto Protocol (KP), adopted in 1997, complements the UNFCCC by providing an enforceable agreement with quantitative targets for reducing GHG emissions. For fulfilling their emission-limitation commitments under the KP, industrialized countries (listed in the KP's Annex I)) can use land-based activities, such as reducing deforestation, establishing new forests (afforestation and reforestation) and other vegetation types, managing agricultural and forestlands in a way that the 'carbon sink' is maximized.

Annex I countries may also claim credit for carbon sequestration in developing countries by afforestation and reforestation (AR) through the Clean Development Mechanism (CDM), one of the 'Kyoto Mechanisms' that allow countries to achieve reductions where it is economically efficient to do so. For the period 2008-12, forestry activities under the CDM have been restricted to afforestation and reforestation on areas that were not forested in 1990. In addition, CDM projects must lead to emission reductions or net carbon uptake additional to what would have occurred without the CDM funding.

Forests Under Climate Change Negotiations

Next to energy sector the focus of attention of the climate convention and scientific literature is largely on forestry sector because it provides large and relatively low cost mitigation opportunities to address climate change. The various issues related to forestry sector are referred to as Land Use, Land Use Change and Forestry (LULUCF) in the Kyoto Protocol. The protocol imposed a cap on CER generate through Forestry Projects: *"...for the first commitment period, the total of additions to and subtractions from the assigned amount of a Party resulting from eligible LULUCF activities under Articles 12, shall not exceed 1 per cent of base year emissions of that Party..."* As a result of Kyoto Protocol carbon has become a tradable commodity. One ton of CO_2 reduced through a CDM project when certified by a Designate Operational entity (DOE) is known as a CER which can be traded.

The A and R activities and areas eligible for such activities are governed by A and R definitions as provided in decision 19/CP.9 of the UNFCC. The activities may be large or small scale, single or multiple species, pure forestry or farm forestry systems such as:

- ☆ Establishment of woodlot on communal lands. Reforestation of marginal areas with native species, *e.g.* riverine areas, steep slopes, around and between existing forest fragments (through planting and natural regeneration).
- ☆ New large-scale industrial plantations.
- ☆ Establishment of biomass plantations for energy production replacing fossil fuels.
- ☆ Small-scale plantations by land owners, especially low income communities.
- ☆ Introduction of trees into existing agricultural systems (agroforestry, if it fulfils the definition of forest).
- ☆ Rehabilitation of degraded areas through tree planting or assisted natural regeneration.

India has a vast potential of CDM forestry mitigation projects but still such projects are striving for a good start as far as their operationalization at the local level is concerned. The implementation of such Projects has not been a great success in India so far. There are several barriers that not only a Project developer but also all the stakeholders involved in the development of CDM forestry Project do face from time to time. Taking this into consideration an attempt has been made to identify the barriers for these projects. The root cause of the problems along with the possible solutions needs to be found out so as to make the forestry mitigation projects successful and a profitable venture for the nation as a whole.

CDM Afforestation and Reforestation related issues at International level.

1. AR CDM Methodology Submission and Approval Process

It has been more than four years ago when the Marrakech Accords (UNFCCC, 2001) drafted the basic set of rules and modalities for CDM but it do not provide detailed methodological process for developing baseline methodology for CDM project activities rather it is left to project developers to submit a new methodology together with their Project Design Document (PDD) unless an approved methodology already exists for that project type (UNFCCC, 2003) and the project developer choose to adopt that approved methodology.

Executive Board (EB 21) in September 2005 (Meeting of 21st CDM Executive Board) has issued fresh guidelines for "Procedure for submission and consideration of proposed new methodology for afforestation and reforestation project activities under CDM". EB in its 26the meeting revised these procedures as contained in Annex 18 of its report. The COP12/MOP-2 held in November 2006 in Nairobi has requested for further submission on it. The proposed new baseline and monitoring methodology complete in most recent version along with draft PDD is submitted to CDM EB (UNFCCC Secretariat) through a DOE/AE. The secretariat forwards the document to one member of ARWG (Afforestation and Reforestation Working Group) to assess the quality of submission. The AR WG taking into consideration of public comments and recommendations of the desk reviewers prepare its preliminary recommendations regarding the approval of proposed new A/R methodology to the EB. The final recommendation is then forwarded to EB. After the receipt of the recommendations EB decision is made publicly available as 'A' for approval, 'B' for resubmission and 'C' for rejection. A 'B' case methodology with required changes can be resubmitted to EB only once and on its resubmission a 'B' case is reconsidered by AR WG without further desk review.

2. Success Rate of CDM A/R Methodologies

Due to late finalization of Modalities and Procedures (M and P) for CDM Afforestation and Reforestation (AR) project activities at COP 9 (UNFCCC, 2003: CDM A/R M and P), the first forestry project (and methodology) has been submitted only by October, 2004. The approval rate of methodologies is very low. At the time of writing of this report there has been 9 round of baseline methodology submission and total of 25 methodologies have been submitted so far. Of these 25 submitted methodologies 5 are in 'public comments' phase (submitted in 9th round). Only three methodologies have been approved by CDM-EB. Michaelowa and Rawat (2007) has analysed in details the methodologies so far submitted to CDM EB and the reasons for their rejection or approval. As on date there has been 13 round of submissions and 5 methodologies approved. Only one forestry project (ARAM 000) popularly known as China Project has been registered as CDM project by CDM EB.

The submission process takes a minimum 4-5 months from its submission to final recommendations of EB. The first approved CDM A/R methodology (AR AM0001) took 5 months form it submission to final approval. The two other methodologies (AR AM0002 and AR AM0003) were also approved within 4-5 months of their submission which is quite satisfactory as CDM M and P specifies 4 months times to approve or reject a methodology. The average time for approving the methodologies (other than CDM AR) has been very long, *i.e.* 9 to 10 months (ECON 2005).

3. Lessons Learnt from CDM A/R Methodologies

There are continuously high rate of rejection of methodologies especially in energy efficiency, transport and forestry sector (Michaelova 2005). The submission of new CDM A/R methodologies has been a learning process for all involved in the CDM forestry issues. During the first year of methodology submission, the primary issues that caused new methodologies to be rejected included improper or lacking explanation regarding: additionality, methods for determining the project boundary; description of the baseline approach, justification for this approach and land-use scenario determination, consideration and selection of carbon and non-CO_2 greenhouse gas pools, methods for determining net anthropogenic greenhouse gas removals by sinks, as well as inadequacy in making recommended changes if the new methodology was being submitted for a second time. Secondary issues that also caused new methodologies to fail included improper or lacking explanation regarding: methods for creating a baseline of net greenhouse gas removals by sinks, methods for estimating actual net greenhouse gas removals by sinks, systems for addressing leakage, methods for compiling project emissions, improper or inadequate description of models, formulas, algorithms and data sources used, methods for addressing uncertainties, as well as the overall quality, drafting and language (Schlamadinger, 2005; Pearson, 2006).

4. Transaction Costs: Consultancy, Validation and Verification Services

Development of a CDM project right from inception to the issuance of CERs is an expensive affair. Only recently, transaction costs became an issue in the discussion on the Kyoto mechanisms leading to the Marrakech Accords clause according special treatment to small scale CDM projects. Transaction costs of the project-based mechanisms like Clean Development Mechanism (CDM) and Joint Implementation (JI) are likely to entail considerable costs of baseline development, verification and certification. Michaelowa (2005) estimated that the development of a standard PDD for a straightforward project type with an approved methodology costs around 15,000 €; if a new methodology has to be developed and fought through, costs can easily triple. Validation costs are between 7,000 to 15,000 €. Verification costs are still not clear but seem to be around 5,000-10,000 € per turn, probably declining over time. Michaelowa *et al.* (2003) estimated that projects below 20,000 annual CERs would not be able to cover their transaction costs. Michaelowa *et al.* (2003) suggested following measures to reduce the transaction costs of a CDM Project that include (*i*) Bundle projects to jointly undertake each step of the project cycle; (*ii*) Verification and certification not annually but at long intervals; (*iii*) Exempt projects from one or more steps of the project cycle; this however endangers environmental credibility and could lead to moral hazard; (*iv*) Streamline the information needs on each step of the project cycle; (*v*) Undertake unilateral CDM projects that reduce search and negotiation costs (*vi*) The CDM Executive Board/JI supervisory committee set a fee schedule proportional to the size of the project for the steps of registration and certification and thus cross-subsidize the smaller projects.

Concerning consultancy, several companies from developing countries have been able to establish themselves in the market, often as supporters of unilateral projects. MGM International, which is based in Argentina provides PDD consultancy all over Latin America. Brazilian Econergy and Ecoinvest also supported number of PDDs. In India, subsidiaries from multinational companies support high numbers of PDDs while the small domestic consultants have also supported some PDDs. If one analyses all projects, the overall share of host country consultants in India reaches 49 per cent. In a big contrast to the consultancy market, so far, only 0.5 per cent of projects are validated by a company from a Non-Annex B country. Although these companies are from Korea which is an OECD member. As the validation market is characterised by a strong concentration, it is unlikely that developing country companies are able to get a relevant market share. Market leader Det Norske Veritas (DNV) has a share of 56 per cent and the three leading companies together achieve 88 per cent (Michaelowa 2005).

Forestry and CDM in India

The CDM framework has been very active in India by timely creation of National CDM Authority (NCDMA) at Ministry of Environment and Forestry, Government of India, the Designated National Authority (DNA) for India. India is the leading CDM country in the world with 31 per cent of the total share of global CDM projects. At international level so far 268 CDM Projects have been registered out of which 83 projects are from India. The Ministry of Environment and Forests has adopted a proactive approach in facilitating CDM project processing. However to facilitate the CDM project activities in India the high risk and transaction costs associated with the CDM approval process must be taken control of for the proactive involvement of various CDM project developers.

The CDM project portfolio currently approved by National CDM Authority (NCDMA) is dominated by small scale projects mainly from the renewable energy sectors. As on August 2006, NCDMA has issued Host Country approval to 239 CDM projects in different sectors. The sector wise distribution is: Renewable Energy: 123 (Out of which 32 are related to biomass); Energy Efficiency: 72; Industrial Process: 28; Municipal Solid Waste 3 and Fuel switching: 13. (http://cdmindia.nic.in/).

Forestry Potential

India has been witnessing continuing improvement in existing policies and formulation of new policies and thus trying to keep pace with the activities and programmes that ensure the achievement of the country's national goal of 33 per cent forest/tree cover by the year 2012. To realize this goal India has to bring about 109 Mha (33 per cent of its total geographic area) of 328 Mha of land under forest/tree cover. The latest statistics of the State of Forest Report (SFR 2003) shows that the existing forest cover in India is 67.8 Mha (20.6 per cent) besides an additional 9.9 Mha (3.0 per cent) of tree cover outside forests in the country. Thus the total land under forest/tree cover in India currently is 77.8 Mha or 23.6 per cent of its geographic area (SFR, 2003). Out of it 1.56 per cent (5.1Mha), 10.32 per cent (33.9.0 Mha) and 8.8 per cent (28.8 Mha) is under very dense, moderately dense and open forests respectively. An extent of 4.0 Mha (1.2 per cent) of scrub forests is also available in addition to the forest cover of 67.8 Mha. The difference in existing forest cover and the target area suggests that an additional 31.2 Mha has to be brought under the tree cover to achieve 33 per cent tree cover as envisaged in National Forest Policy. This can be done by establishing plantations on non-forestland and through agro-ecosystems in existing agricultural lands.

Table 7.1: A List of CDM Projects Developed in India

Sl.No.	*CDM A and R Project*	*Developer*	*CO_2 Removal/ yr (kt)*	*Size (ha)*	*Remarks*
1.	Begapalli CDM Afforestation Programme by Women for Sustainable Development (WSD), Kolar District, Karnataka	Women for Sustainable Development (WSD), Bangalore	8.00	1383	Baseline Methodology not approved by CDM EB
2.	The International Small Group and Tree Planting Group (TIST), Tamil Nadu	Active TIST Small Group Tamil Nadu	51.6	2 million trees	Not submitted to CDM EB
3.	Improving Rural Livelihood through Carbon Sequestration by Adopting Environmentally Friendly Technology based Agroforestry Practices by VEDA-MACS Limited, Hyderabad	VEDA MACS Hyderabad	20.0	3500	Supported by WB Carbon Finance Yet not submitted to CDM EB
4.	Afforestation of the Crop-land through Agroforestry Practices in 3658 ha. Area in Khammam District of Andhra Pradesh, India under ITCs Farm Forestry Project	ITC Bhadrachalam	47.6	3658	not approved by CDM EB

Most of the forest in the "Open Forest" (28.8 Mha) category of 10-40 per cent canopy density is either patchy or degraded and needs restoration efforts. Thus there is a potential to bring a total of 59.6 M ha (28.8 Mha open forest + 031.2 Mha additional land area). This land can be proposed for A and R or other LULUCF activities to achieve a 33 per cent forest cover target. Assuming a conservative Carbon sequestration rate of 5 tonnes/ha/yr, these activities are expected to sequester 300 million tones of CO_2 per annum per ha. Ravindranath (2002) has estimated the mitigation potential from India under two different scenarios. Under Technical potential scenario when all the land available for forestry is revegetated is estimated to be 175 million tonnes annually. Under the demand driven scenario with significant feasible enhancement of the area afforested, the annual carbon sequestration potential is 78 million tones annually.

Barriers to the CDM A and R Projects

For the successful implementation of CDM Forestry Projects, the barriers must be reasonably identified and attempts be made to overcome them. For successful implantation of CDM A/R projects, Ravindranath (2002) has identified following significant barriers:

Technological Barriers

Lack of high yielding genetic planting material and improved Silvicultural practices, demarcation of eligible land area and use of modern tools like remote sensing. Non-availability of database Barriers.

Market-based Barriers

Very limited and unorganised market institutions for Carbon Trading, higher transaction cost, low rates of CERs and predominantly buyers dominated market.

Institutional Barriers

Need for indigenous Designated Operational Entity (DOEs) cut transaction costs, Weak institutional capabilities of Forest departments, absence of Agroforestry Co-operatives, inadequate extension, Governmental and Policy Barriers.

Physical Barriers

Existing land-use pattern, projected land-use change, land tenure systems, land-titling requirements, political and economic structures.

Miscellaneous Barriers

Long gestation period of trees, exploitation of farmers by middlemen, non-remunerative prices of timber by the user industries.

Concerned with the above barriers, questionnaires were prepared and circulated to various stakeholders like State Forest departments, leading NGOs having stake in CDM A/R Projects and leading farmers. On the basis of views obtained from various respondents, the major concerns that need to be addressed for promotion of CM A/R Projects in the country have been analyzed and following is recommended to create an enabling environment for CDM A/R projects:

1. *Delineation of Potential CDM A/R Area*: For development of a quality A/R PDD, eligibility of land as per defined by CDM M and P is a prerequisite. Identification of eligible areas to be undertaken for CDM A/R Projects is a big barrier in this direction. Through land records and other tools like remote sensing, data on forest area that deforested after 31 Dec. 1989 need to be identified for reforestation activities and areas historically not forested for past 50 years need to be identified for afforestation activities. As the country forest definition submitted by Govt. of India is a minimum area of 30 per cent crown density, agencies like FSI/NRSA should delineate the

area less that this threshold at District, State and National level. FSI in its biennial SFR can redefine its category of "Open Forest" of 10-40 per cent crown cover to 10-30 per cent. Such an attempt will drastically reduce the cost of PDD development and would be a great help to facilitate CDM sink projects at all levels.

2. *Inclusion of CDM Component in Forest Working Plans*: Some of the respondents have given the idea of inclusion of CDM component in forest Working Plans. Delineating the areas of 10-30 per cent will also facilitate State Forest Departments (SFDs) to identify potential areas to be included in CDM. This will require capacity building of Working Plan Officers in the area of CDM. Although ICFRE in its various programs is active in capacity building programs some more concerted efforts are needed with the support from Government of India. Some of such potential area could be planned for developing CDM A/R Projects in the Forest Working Plans. State Forest Departments can facilitate in this endeavor by opening a CDM Cells in their research and extension wings. For example the Uttarakhand Govt. has opened its CDM cell with its Forest Department as Nodal Office. Many other State Governments has already opened CDM Cells in their states with one or other department as nodal Office. Where Forest Department is not a nodal office, their forest department should establish synergy with their nodal offices for capacity building and promotion of CDM in Forestry sector.

3. *Incorporation of CDM Component in Joint Forest management (JFM) Programmes*: Strong support for inclusion of CDM A/R projects under ongoing JFM programs is proposed by many of the respondents. The analysis of CDM A/R Projects at National and International level reveals slow initial progress of these programs. Small Scale CDM A/R M and P explicitly states that these projects should be essentially developed by low income communities. Involvement of communities for success of such programs has been strongly supported by respondents from SFDs and NGOs. JFM structure offers opportunity in this area. CDM A/R projects so far developed in India suggest that only with the involvement of communities such projects are feasible. All the CDM A/R projects that have been developed so far in India are with the active participation with communities. Such experiment can be replicated in JFM programs as well.

4. *Need for Forestry Based CDM Projects other than CDM A/R*: As has been analyzed in this report, there are many barriers to the CM A/R Projects at international and national level owing to stringent modalities and procedures and lack of sufficient methodologies and registered projects. There is a vast scope of forestry related projects to be developed under CDM project other than CDM A/R Projects where well established methodologies exists and some already registered and functional projects are available as benchmark. Under renewable energy by biomass co-generation can give CDM benefits to the forestry sector. A good number of such projects have got host country approval and some have already been registered with

CDM EB. Wherever possible energy plantations for power generation can be a good option. Such option can be explored under JFM Programs as well. Raising Jatropha for oil and its replacement with fossil fuel is another example where land based Projects can be registered as CDM Project without involving into cumbersome modalities and Procedures of CDM A/R Projects. Here one need not prove the eligibility of land or confirming to forest definition.

5. *Development of Baseline for Different Eligible Land Use Type*: Baseline information on carbon content for different eligible land use types can be developed to determine the most potential advantageous area for undertaking CDM activity. Area with low baseline carbon content will economically more profitable for CER generation. An area with higher baseline will give less return. Proposed CDM Cells or Working Plan Officers can generate this information for the eligible areas under their control.
6. *Making the Finance Available for CDM Projects*: While CDM projects earn additional revenue due to the sales of CERs, it often have difficulty in attracting initial investment. This may be because CDM projects bring in revenue through the as-yet relatively little known trade in CERs. Paucity of funds, lack of investments under plan for such projects has been identifies as one the barrier by many respondents. In India banks and financial institutions are not aware with the globally developing carbon market. At present WB biocarbon fund, WB Community Development Funds that are raise solely to fund CDM A/R Project are over subscribed. There is need to involve the financial institutions for funding viable CDM A/R Projects. Public sector banks, rural banks and institutions like NABARD can be approached. This further needs educating financial institutions about this emerging carbon market that provide community benefit with sustainable development and other co benefits. Risk guarantee mechanisms provided by the government could be one way to increase the comfort-level of financial institutions to provide loans for CDM projects.
7. *Institutional Strengthening and Capacity Building*: Several consultants have been able to establish themselves in the CDM market. There are big multinational players functioning as consultants as well as "Designated Operational Entity" (DOE) for validation and verification of these projects. As reported earlier in this study developing countries have only 0.5 per cent market share of "DOE/OEs". Unfortunately there is not a single operational entity accredited by CDM EB for LULUCF sector. ICFRE is pursuing its application of "DOE" in forestry sector. An indigenous DOE will reduce the transaction cost of the CDM A/R Projects. There is lack of good consultants and low level of capacity building in this newly emerging area.
8. *Creating Market for CDM A/R Projects*: Market for CERs from CDM A/R projects is limited mainly because of cap on CERs from LULUCF projects and exclusion criteria of EU emission Traxding Scheme. Hopefully, at the end of 2007 EU ETS will revise its policy and shall be open for CERs from LULUCF projects.

Negotiators from CDM active countries like India need to rethink their strategy and lay emphasis on fixing a minimum quota for CERs to be purchased from A and R projects to meet the Kyoto targets of the Annex 1 countries. Preference be given to small scale A and R projects developed by low income communities that add to the sustainable development and meet the ultimate objective of CDM.

Conclusion

Contrary to the belief of some observers that that development of land use benchmarks for the definition of forestry projects is an almost insurmountable challenge, the CDM A/R methodologies are now seeing the light of the day. The specific reasons for rejection of CDM A/R projects were subsequently improved and now we have four methodologies approve for LULUCF Projects. India inspite of being number one CDM country in the world is far behind as far as the CDM A/R Projects are concerned. So far only two CDM Sink projects from have submitted to CDM EB. Present study has identified some of the barriers which are hindering the development of these projects in LULUCF sector and possible remedies to overcome these barriers have been suggested.

Presently the rehabilitation of forest and land, in the form of Afforestation and Reforestation, solely rely on governmental initiatives, with public participation. For actively involving the communities in rehabilitation of the degraded forest cover the efforts could be triggered through various efforts, like appropriate economic incentive, guarantee of community/society rights, local institutional reinforcement, and also reinforcement of the capacities of local governments, which are managing directly the activities of rehabilitation.

The development of CDM A/R projects can be a tough task for smaller groups and local communities. The high transaction costs, required information for PDD development, negotiation, monitoring and risk management requirements make it difficult for smaller organizations and community groups to access the benefits of the CDM unless there is some active intervention from governments and agencies wishing to promote people oriented forestry. This can be addressed through appropriate institutional arrangements and facilitation through NGOs, SHGs, rural banks or even by confederations of village committees/JFM institutions etc. The institutional arrangements should be put in place to ensure information flow to the stakeholders, consistency of management regime, and financial and administrative systems for cost-benefit sharing amongst many small and marginal players. In order to involve smallholders/communities in the CDM A/R projects, their land-use systems need to be recognized by the sustainable development criteria applied to the CDM nationally and also the transaction costs need to be overcome. National laws regarding property ownership may need to be reviewed and amended to address the emerging carbon market.

If India offers a well tailored CDM infrastructure for afforestation and reforestation projects then the transaction costs can be lowered down and an enabling environment for CDM A/R project will be created, which may give India a competitive edge over others. Simply working out a combination of economic, governmental and investment factors could make it happen.

Acknowledgments

This paper is as a results of studies under taken under EU ICFRE SPF project "Beyond Kyoto EU- India CDM Partnership: Promoting Stakeholders Dialogue and Analysis of Barriers to Forestry Mitigation Projects" (Contract Ref. No. 95720), the outcome of two International workshops and a stakeholders meeting conducted under the project. The authors are grateful to Shri Jagdish Kishwan, DG, ICFRE for rendering all possible support for implementing the project. The financial assistance from EU under its SPF program is also gratefully acknowledged.

References

ECON. 2005. CDM Methodology Status Report 2005 FINAL REPORT Commissioned by World Bank Carbon Finance Business. ECON-Report No. 2005-046, ECON Analysis AS Oslo, Norway.

Eco Securities Consult. 2006. Should Temporary CERs be Included in the EU ETS Linking Directive? and other questions concerning the potential demand for CDM Forestry CERs. Eco Securities Consult 21 Beaumont Street Oxford OX1 2NH, UK (available at www.climate-standards.org/images/pdf/EcoSecurities-demand-survey-March06.pdf.)

Forner C., Jotzo F. 2002. Future restrictions for sinks in the CDM How about a cap on supply? *Climate Policy* 2, 353–365.

Grubb, M., Brewer, T., Muller, B., Drexhage, J., Hamilton, K., Sugiyama, T, Aiba, T with contributions from Sharma, A., Michaelowa, A., Azar, C. and Karas, J., 2003. *A Strategic Assessment of the Kyoto Marrakech System.* The Royal Institute of International Affairs, Briefing paper no 6. London

IPCC, 2001. *Climate Change 2001: Synthesis Report.* Cambridge University Press Cambridge, CB2 2RU, England.

IPCC. 2007. *Climate Change 2007: Part I Contribution of Working Group I (Scientific Basis) to the Fourth Assessment Report (AR4) of the Intergovernmental Panel on Climate Change (Summary for Policy Makers).*

Jotzo, F. and Michaelowa, A., 2002. Estimating the CDM market under the Marrakech Accords. *Climate Policy* 2: 179–196.

Michaelowa, A., Stronzik, M., Eckermann, F. and Hunt, A., 2003. Transaction costs of the Kyoto Mechanisms. *Climate Policy*, 3: 261–278.

Michaelowa, Axel, 2005. *CDM: Current Status and Possibilities for Reform.* HWWI Research Paper 3, Hamburg, 27 pp.

NSS, 2005. *CDM Implementation in India: The National Strategy Study.* The Energy and Resource Institute (TERI), New Delhi

Pearson, T., 2006. Projects under the CDM: experience to date. Presentation at UNFCCC Technical Workshop on Joint Implementation, 9-10 March, 2006. (From: http://www.winrock.org/ecosystems/publications.asp?BU=9086.

Ravindranath, N.H., 2002. Climate change, forests and India: CO_2 emissions, impacts, mitigation and global negotiations. In: *Climate Change and India Issues, Concerns and Opportunities*. Tata McGraw-Hill Publishing, Company, New Delhi, pp. 173-195.

Michaelova, A. and Rawat, V.R.S., 2007. *CDM Afforestation and Reforestation Baseline Methodologies: An Analysis of Submission and Approval Process*. Hamburg Institute of International Economics (HWWA), Discussion Paper 4-9, Hamburg.

Schlamadinger, B., 2005. Experiences from review of AR methodologies. Presentation at 'Gaining approval for LULCF projects and project methodologies under the CDM: Lessons learned'. Winrock Side Event Presentations, 2nd December 2005, UNFCCC–COP 11/MOP 1–Montreal 2005. (from: http://www.winrock.org/ecosystems/publications.asp).

SFR, 2003. *State of Forest Report 2003*. Forest Survey of India, Dehradun. Ministry of Environment and Forests, Govt. of India,

UNFCCC, 2001. *Report of the Conference of Parties on its Seventh Session*, Addendum Part 2: Action taken by Conference of parties Vol II, (FCCC/2001/13/Add.2).

UNFCCC, 2003. *Modalities and Procedures for Afforestation and Reforestation Project Activities Under the CDM–Decision* 19/CP.9 (FCCC/CP/2003/6/Add.2).

Ecosystem Diversity and Carbon Sequestration: Climate Change Challenges and a Way Out for Ushering in a Sustainable Future *Pages 71–80*
Editors: P.L. Gautam, Vir Singh and Uma Melkania
Published by: DAYA PUBLISHING HOUSE, 2009

Chapter 8

Microbial Diversity as an Indicator of Soil Organic Carbon Status: Redevelopment of Humid Subtropical Perturbed Ecosystem

Saurindra Nr. Goswami and Soneswar Sarma

The study makes an attempt to observe the relationship between the microbial diversity and soil organic carbon status during hill agro ecosystem redevelopment after perturbation in Dimoria Block and Tribal Belt of Assam. The experiment was laid during 2003-2006 on the mid hills, which are disturbed repeatedly after every 4-years by *jhum kheti*. Soil physico-chemical and hydraulic characteristics, plant community analysis, microbial diversity and microbial biomass carbon were estimated at 0-20cm of soil depth during the study and compared with a nearby forest ecosystem considered sacred by the Karbi (a traditional community of Assam). Soils in both the abandoned jhum fallow and the nearby forest ecosystem were acidic and sandy loam. Significant relationship was observed between microbial diversity and soil bulk density. Soil temperature was higher in the jhum fallow when compared with the forest ecosystem. The wilting point and the field capacity in cm^3cm^{-3} of the soils in the *jhum* fallow and the forest ecosystem was 0.09 and 0.21 and 0.12 and 0.46, respectively. Soil moisture content was lower in the *jhum* fallow and it might be one of the reasons for recording less microbial diversity in the perturbed ecosystem. Soil organic matter content in jhum fallow increased as fallow period increased but still the values were statistically lower when compared with the forest ecosystem. Microbial diversity steadily increased as the plot was left fallow after *jhum kheti*. It was interesting to observe statistically significant linear relationship between the soil organic carbon and microbial diversity during ecosystem redevelopment. The study suggests that microbial diversity is linked to, among other properties, soil organic carbon one of the important parameter for increasing soil fertility during redevelopment of the degraded ecosystems. Therefore, microbial diversity can be used as a biological indicator to determine the recovery of soil fertility in the repeatedly perturbed ecosystem in the humid tropics.

Keywords: *Jhum, Microbial diversity, Fertility redevelopment, Perturbation, Organic matter.*

Introduction

The traditional communities residing in the biodiversity rich Dimoria Block and Tribal Belt in Assam, Northeast India are directly and indirectly attached with the forests for their sustenance since generations. Members of these traditional communities prefer slash-and-burn agriculture commonly called *jhum kheti*. Perhaps and maybe due to increase in human population, which limits the availability of land as reported elsewhere by Ramakrishnan (1992) and Singh (1996) the *jhum* farmers of Dimoria are nowadays cultivating in the same plot of land after every 4-years. Many workers criticize jhum as an inefficient form of agriculture (Stracey 1967; Borah and Goswami 1973; Valdiya and Bartarya 1991; Nakano and Miyauchi 1996). However, in contrast, many workers have also viewed *jhum* favourably (Conklin 1969; Rappaport 1971; Ramakrishnan 1984, Altieri and Merrick 1997).

It is very much important to mention here that *jhum kheti* is an integral part of the custom, tradition and practice among the traditional communities not only of Dimoria but also of other parts in the entire Northeast India. Therefore, jhum is considered as a significant land use practice, which cannot be directly suppressed by so called developmental programmes and instead efforts should be taken to make the jhum farmers aware about the soil fertility conditions. Scientific findings will definitely be useful to the communities if it is disseminated respecting the traditional knowledge.

Microbial characteristics of soil are being evaluated increasingly as sensitive indicators of soil health because of the clear relationship between microbial diversity, soil and plant quality and ecosystem sustainability. Soil nutrient maintenance and fertility buildup is dependent on microbes. Microbes are the major agent of nutrient transfer and release in soil and are especially important for supplying nutrients to crop plants. Moreover, many mineral nutrients including micronutrients held in an unavailable organic combination are also released in an available form by microbial decomposition. Microbial communities in soil underpin the vital ecosystem functions of nutrient cycling (Robinson *et al.*, 1992). Plant species also acts as important drivers of soil organic microbial biomass and soil microbial activity by influencing net primary productivity and litter input to the soil (Beare *et al.*, 1995, Wardle and Lavelle 1997). Under nutrient limiting conditions, species effects could be significant on soil microbial biomass (Hungate *et al.*, 1996).

Looking at the importance and keeping in mind the highly significant role of microbes and of course its diversity in soil organic matter buildup the present study was carried out to find out a way to use microbial diversity as an indicator to somewhat know the status of soil fertility in perturbed ecosystem in the humid tropics so that the findings can be conveyed to the members of the traditional communities and together the scientific and the traditional communities members of which are very rich in Traditional Ecological Knowledge (TEK) can work together to push the perturbed ecosystem towards its natural recovery trajectory.

Materials and Methods

Site Description

The study area is located at the mid hills just 2000 meters away from Garo Basti at Dimoria Block of Assam, which lies between 92°3′ to 92°5′ E Longitude and 26°6′ to 26°6′ N Latitude. For the experiment a jhum plot of 2.0 hectares abandoned during December 2002 and another plot of 2.0 hectares inside a natural forest, regarded as sacred by the Karbi (a traditional community of Assam) was selected.

The altitude of the *jhum* fallow and the forest is 246m and 254m above mean seal level, respectively. The slope of the jhum fallow and the forest area measured using an Abney's levels were 23° and 5°, respectively. The average rainfall was 117cm and the average maximum and minimum temperatures were 28° and 13°, respectively during the study. Heavy rainfalls were recorded during July to September. Occasional winter rains were also recorded during December to February.

Methods

A soil auger was used to collect soil samples from 0-20 cm depth from both the ecosystems from. 50 samples from each experimental site were collected after every four months. Undesired materials for our analysis like stones and other hard substances if any were then removed and the samples were kept inside the polythene bags and marked. Soil texture was measured yearly and this value was used to calculate other soil hydraulic properties *i.e.* wilting point and field capacity in cm^3cm^3 following the software developed by Saxton (1986). Soil pH, moisture content, temperature and water holding capacity were then estimated following the procedure described by Singhal (1996) whereas, soil organic matter was calculated after estimating the soil organic carbon following Nelson and Sommers (1982). Microbial biomass carbon was determined using the fumigation extraction method as described by Vance *et. al.* (1987).

Plant Community and Microbial Diversity Analysis

After the jhum cultivators left the plot 50 quadrats of $1m^2$ were laid randomly on the entire 2.0 hectares of that plot. Similarly 10 quadrates of $1m^2$ were laid on the scared forest to estimate the coverage of the prominent exotic weed species. After carrying out reconnaissance survey inside the sacred forest we selected only 10 quadrats on the periphery of the sacred forest where disturbances seem to occur to compare our results. We selected to record the coverage of the prominent exotic weeds only as then it becomes easier for the traditional communities to understand and accept our findings. Similar procedure was applied to collect data for microbial studies.

Statistical Analysis

Data were subjected to analysis of variance (ANOVA) to study the soil redevelopment in shortened jhum fallows with time. A possible mean difference of the cumulative data was tested with statistical package Statistica, StatSoft (1995) and was correlated to the different soil properties.

Experimental Results

Occurrence of Prominent Plant Community in 4-yr Old Abandoned Jhum Plot and the Sacred Forest

Table 8.1 shows the coverage of the prominent weeds in $m^2/100\ m^2$ that comes just after a jhum plot is abandoned in the Dimoria Block, data was recorded upto the 4th year. For comparison prominent highly invasive exotic weeds on the periphery of a sacred forest was also recorded. Interestingly during the fourth year the abandoned jhum plot was covered with *M. micrantha* and the values were statistically significant at $p<0.05$.

Table 8.1: Coverage m²/100m² of Prominent Plant Community on the 4-yr Old Abandoned Jhum Plot and the Sacred Forest During the Study

Sl.No.	*Prominent Plant Species*	*2003*		*2004*		*2005*		*2006*	
		Jhum Plot	*Sacred Forest*	*Jhum Plot*	*Sacred Forest*	*Jhum Plot*	*Sacred Forest*	*Jhum Plot*	*Sacred Forest*
1.	*Eupatorium odoratum*	32	10	68*	12	4	12	1	2
2.	*Lantana camera*	18	4	54*	7	2	10	0.2	13
3.	*Mikani micrantha*	0	32*	10	36*	60*	40*	95*	46*

Note: * Marked values are statistically significant at p<0.05 when compared with the sacred forest.

Microbial Diversity in 4-yr Old Abandoned Jhum Plot and the Sacred Forest

Table 8.2 shows the different types of microbial diversity that comes just after a jhum plot is abandoned in the Dimoria Block for 4-years. The results shows that diversity of bacteria is more than fungus and actinomycetes, respectively during the natural recovery of the perturbed ecosystems. However, all the values were much lower when compared statistically with the scared forest.

Table 8.2: Microbial Diversity on the 4-yr Old Jhum Plot and the Sacred Forest During the Study

Sl.No.	*Different Types of Soil Microbes*	*2003*		*2004*		*2005*		*2006*	
		Jhum Plot	*Sacred Forest*	*Jhum Plot*	*Sacred Forest*	*Jhum Plot*	*Sacred Forest*	*Jhum Plot*	*Sacred Forest*
1.	*Bacteria*	9	32	15	34	20	34	24	38
2.	*Fungus*	4	27	8	27	11	29	14	31
3.	*Actinomycetes*	1	14	3	16	6	16	9	18

Note: All the values for jhum plot are statistically significant at p<0.05 when compared with the sacred forest.

Soil Physical, Hydraulic, Chemical and Microbial Biomass of the 4-yr Old Abandoned Jhum Plot and the Sacred Forest

Table 8.3 shows the average soil physical, hydraulic, chemical and microbial biomass during 4-years of jhum redevelopment. Bulk density of the jhum soils after 4-years was 1.53 gcm^{-3} whereas it was 1.44 gcm^{-3} for the forest.

Table 8.3: Soil Physical and Hydraulic Properties of the 4-yr Old Jhum Plot and the Sacred Forest

Estimated Soil Properties	*Abandoned Jhum Plot*				*Sacred Forest*			
	2003	*2004*	*2005*	*2006*	*2003*	*2004*	*2005*	*2006*
Physical								
Bulk density (gcm^{-3})	1.53	1.53	1.52	1.51	1.44	1.44	1.44	1.43
Hydraulic								
Wilting point (cm^3cm^{-3})	0.09	0.09	0.1	0.1	0.12	0.14	0.14	0.14
Field capacity (cm^3cm^{-3})	0.21	0.21	0.22	0.22	0.25	0.25	0.25	0.26
Chemical								
pH	4.8	5	5.1	5.4	5.69	5.71	5.84	5.88
Soil temperature (°C)	28.7	27.3	26.1	25.7	18.9	19.2	19.4	19.2
Soil moisture content (%)	16.1	18.2	20.6	23.1	29.4	29.5	29.8	29.5
Water holding capacity (%)	36.14	42.1	54.74	59.27	64.58	64.57	64.63	64.5
Organic matter (ppm)	3.2	3.71	4.05	4.69	5.3	5.34	5.56	5.78
Microbial biomass estimation								
Microbial biomass carbon (μgg^{-1} of soil)	93	102	114	128	202	207	209	214

Note: All the values for jhum plot are statistically significant at $p<0.05$ when compared with the sacred forest.

Hydraulic properties that were estimated which included the wilting point, field capacity, in cm^3cm^{-3} of the jhum land were 0.09 and 0.21, respectively, whereas the values of the sacred forest were 0.12 and 0.24 in cm^3cm^{-3}, respectively. These values were statistically significant at $p<0.05$

Soil chemical properties like soil pH was 4.8, 5.0, 5.1 and 5.4 during 1st, 2nd, 3rd and 4th year of jhum redevelopment, respectively. However the values were statistically much lower than that of the nearby sacred forest. Average pH of the sacred forest was 5.7. Yearly mean soil temperature in °C during 1st, 2nd, 3rd and 4th year of jhum redevelopment were 28.7, 27.3, 26.1 and 25.7, respectively. The average soil temperature recorded in the forest soil was 29.4 °C. The values were significant at $p<0.05$. Yearly mean water holding capacity in percent during 4-years of jhum redevelopment was 36.14, 42.11, 54.74 and 59.27. However, the value recorded was 64.63 for forest.

Microbial biomass carbon in ugg^{-1} of soil was 93, 102, 114 and 128, respectively 1st, 2nd, 3rd and 4th year of jhum redevelopment, respectively whereas it was 202, 207,

209 and 214, respectively in 2003, 2004, 2005 and 2006 in the sacred forest. Comparison of the two ecosystem showed that the values were highly significant from each other.

Relationship between Soil Organic Matter and Microbial Diversity during Redevelopment of Perturbed Ecosystem

Data pertaining to soil organic matter (ppm) and microbial diversity during jhum land redevelopment and sacred forest ecosystem is provided in Figure 8.1. From the figure it is evident that soil organic matter steadily increases with increase in microbial diversity. The values when compared with the sacred forest ecosystem were statistically significant. Moreover, significant linear relationship (at $p<0.05$) existed between soil organic matter ($R = 0.97$), the co-efficient was very much close to 1.0 indicating that every microbe individual is contributing towards soil organic matter buildup during soil redevelopment following perturbation in the mid hills.

Discussion

The object of the experiment was to study the relationship between microbial diversity and soil organic matter buildup so that the traditional communities

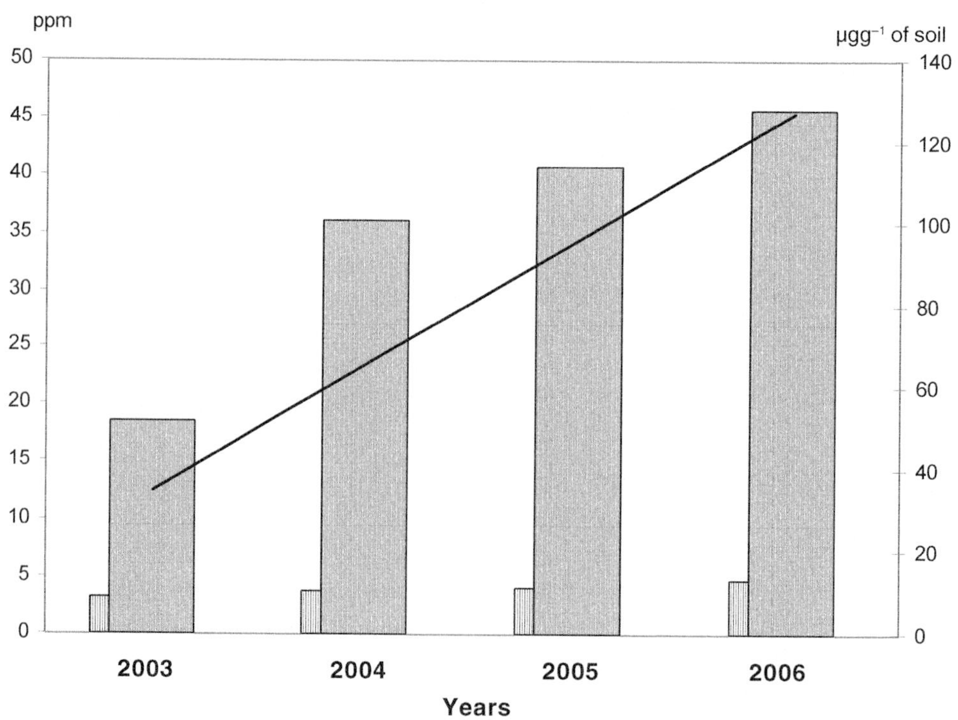

Figure 8.1: Relationship Between Microbial Diversity with Organic Matter Content (ppm) and Microbial Biomass Carbon (μgg^{-1} of soil) During Abandonment of Jhum Plot for 4-years after Perturbation for its Natural Recovery at Dimoria During the Study

practicing jhum can use this measure as a biological indicator for knowing the fertility status of an already disturbed ecosystem that was left fallow for its natural recovery. Ramakrishnan (1985) emphasized the importance of maintenance of soil fertility in the humid tropics. Syers and Springett (1984) have showed the major role of biological factors in litter decomposition and maintenance of soil fertility in the surface layers of soil.

During jhum perturbations that occur in the physico-chemical properties of the soil are all confined to surface soil layers. Soil physical properties like bulk density affected during *jhum kheti* as revealed from Table 8.2 might be the reason that affects the hydraulic properties of the soil, thereby decreasing biological organism activity may be due to prominence dominance of the highly invasive exotic weeds.

On the other hand as jhum abandonment period reaches the third year only *Mikani micranth*, which is a weed actually remains and most probably it removes the more undesired *E. odoratum* and *L. camera* as *M. micrantha* cuts of the sunlight required by the above mentioned two menacing weeds thereby making the perturbed ecosystem somewhat suitable as then only cattle starts visiting these mid-hills. Rapid increase of *M. micrantha* cover as observed on the 4th year reveals that soil became less acidic. Perhaps *M. micrantha* litter makes the soil pH more favourable for microbial activities. Lack of plant cover and only *E. odoratum* and *L. camera* during initial stages of abandonment is another factor, which leads to increase in soil temperature in the jhum fallow and decrease in microbial diversity. As the soil temperature decreases with increase in the fallow period microbial diversity increases. Soil moisture content is another important factor that regulates soil fertility pool. Similarly water holding capacity an important factor that determines soil health also showed positive relationship with microbial diversity. Lavella (1996) reported that the numbers of earthworm species are generally lowered after replacement of the original ecosystem by human disturbances in the tropics and Ramakrishnan (1992) too highlighted the importance of earthworms in jhum agro ecosystems and the beneficial role earthworm plays in this highly leached humid soil by rapid addition of detritus into the mineral soil, which ultimately offers potential for biological maintenance of soil fertility. *Mikania micrantha* are abundant in the sites where the above-mentioned workers worked therefore, it might be one of the prime contributor towards increase in the earthworm density too.

Strong relationship between microbial diversity and soil organic matter during the study reveals the fact that there exists a strong relationship between soil fertility status of an ecosystem and the *M. micrantha* and microbial diversity. In the present study recording of significant correlation coefficient reveals that each and every individual of microbes contributes to the soil nutrients.

Results of this experiment suggests that perhaps microbial diversity can be used as a biological indicator for assessing the soil organic matter status of jhum agro-ecosystems left fallow for natural recovery of soil fertility where use of other methods becomes difficult. Determination of microbial diversity is cheaper and bit easier and therefore, observing the microbial diversity can perhaps make an indication about the soil organic matter status which is the important and essential factor for soil fertility status of the naturally recovering and repeatedly exploited ecosystems. By utilizing this biological indicator traditional communities depended on forests can be made aware about the status of similar perturbed ecosystems, which are common in Northeast India and elsewhere in the tropics. Because of the existence of the major roles of the traditional societies in ecosystem functions in the tropics, acceptance of such biological indicators like microbial diversity in this case is an effective tool and is very much necessary during ecosystem restoration.

Acknowledgements

Financial assistance received from UGC is duly acknowledged. We would like to thank the authority of Dimoria College, Khetri for providing all the necessary facilities.

References

Altier, M.A. and Merrick, L.C., 1997. *In-situ* conservation of crop genetic resources through maintenance of traditional farming systems. *Economic Botany*, 41: 86–91.

Bea, M.H., Coleman, D.C., Crossley, Jr., D.A. Hedrix, P.F. and Odum, E.P., 1995. A hierarchical approach to evaluating the significance of soil biodiversity to biogeochemical cycling. *Plant and Soils,* 170: 5–22.

Borah, D. and Goswami, N.R., 1973. A comparative study of crop production under shifting and terrace cultivation: A case study in the Garo hills, Meghalaya. Ad hoc Study 35, Agro-economic Research Centre for North East India, Jorhat.

Conklin, H., 1969. An ethno-ecological approach to shifting agriculture. In: *Environment and Cultural Behaviour*, (Ed.) A.P. Vayda. Academic Press, New York. pp. 221–233.

Ecocrop, 2003. *Ecocrop Info*. Food and Agricultural Organisation. http://ecocrop.fao.org.

Edwards, C.A., 1983. Earthworm ecology in cultivated soils. In: *Earthworm Ecology from Darwin to Vermiculture,* (Ed.) J.E. Satchell. Chapman and Hall, London, pp. 123–136.

Hungate, B.A., Canadell, J. and Chapin III, F.S., 1996. Plant species mediate changes in soil microbial N in response to elevated CO_2. *Ecology*, 77: 2505–2515.

ISSG, 2005. *Global Invasive Species Database*. Compiled by Invasive Species Specialist Group. http://www.issg.org/database/species/ecology.asp.

Keeney, D.R. and Nelson, D.W., 1982. Nitrogen inorganic forms. In: *Methods of Soil Analysis, Part II 2nd edn.*, (Eds.) Page, A.L., Miller, R.H., and D.R. Keeney. American Society of Agronomy and Soil Science Society of America, Madison, Wisconsin, USA, pp. 643–698.

Kale, R.D. and Krishnamoorthy, R.V., 1981. What affects the abundance and diversity of earthworms in soils? *Proc. Indian Academic Science*, 90(1): 117–121.

Lavella, P., 1996. Diversity of soil fauna and ecosystem. *Biol. Int.*, 33: 3–16.

Lavella, P., Dangerfield, M., Fragosa, C., Eschenbrenner, V., Lopez-Harnandez, D., Pashanasi, B. and Brussard, L., 1994. The relationship between soil macro fauna and tropical soil fertility. In: *The Biological Management of Tropical Soil Fertility*, (Eds.) P.L. Wommer and M.J. Wiley Chichester, UK, p. 137–169

Nelson, D.W. and Sommers, L.E., 1982. Total carbon, organic carbon, and organic matter. In: *Methods of Soil Analysis, Part II 2nd edn.*, (Eds.) Page, A.L., Miller, R.H., and D.R. Keeney. American Society of Agronomy and Soil Science Society of America, Madison, Wisconsin, USA, pp. 539–579.

Ramakrishnan, P.S., 1984. The Science Behind Rotational Bush Fallow Agricultural System (Jhum). *Proc. Indian Academy of Science. Plant Science*, 93: 397–400.

Ramakrishnan, P.S., 1985. Humid Tropical Forests. In: *Research and Humid Tropical Forests*. Regional Meeting National MAB Committee of Central and South Asian Countries. MAB-India. MoEF, New Delhi, pp. 39.

Ramakrishnan, P.S., 1992. Shifting agriculture and sustainable development: A case study from North Eastern India. In: *Man and Biosphere*, U.K. UNESCO, Paris and Parthenon Publishing, Caernforth, Lancs. (Republished by New Delhi; Oxford University Press), pp. 424.

Rappaport, R.A., 1971. The flow of energy in an agricultural society. *Science American*, 225: 117–132.

Robinson, C.H., Ineson, P., Piearce, T.G. and Rowland, A.P., 1992. Nitrogen mobilization by earthworms in limited soils under *Picea sitchensis*. *Journal of Applied Ecology*, 29: 226–237.

Sharpley, A.N. and Seyers, J.K., 1976. Potential role of earthworm casts for the phosphorus enrichment of run-off water. *Soil Biology and Biochemistry*, 8: 341–346.

Singh, D., 1996. *The Last Frontier: People and Forests in Mizoram*. Tata Energy Research Institute, New Delhi.

Singhal, R.M., 1996. *Soil and Vegetation Studies in Forests*. ICFRE Publication, New Forest, Dehradun, pp. 91–107.

StatSoft, Inc., 1995. *STATISTICA for Windows* (Computer program manual). Tulsa, OK: StatSoft, Inc., 2325 East 13th Street, Tulsa, OK, 74104, (918) 583–4149, fax: (918) 583–4376.

Stracey, P.D., 1967. A note on Nagaland. *Journal of the Bombay Natural History Society*, 64: 440–446.

Syers, J.K. and Springett, J.A., 1984. Earthworms and soil fertility. *Plant and Soil*, 76: 93–104.

USDA, 2006. *National Genetic Resources Program*. Germplasm Resources Information Network (GRIN). Online searchable database. USDA. Res. Serv.

Valdiya, K.S. and Bartarya, S.K., 1991. Hydrological studies of springs in the catchments of Gaula river, Kumaon Lesser Himalaya, India. *Mountain Research and Development*, 11: 239–258.

Vance, E.D., Brookes, P.C. and Jetkinson, D.S., 1987. An extraction method for measuring soil microbial biomass C. *Soil Biology and Biochemistry*, 19: 703–707.

Wardle, D.A. and Lavelle, P., 1997. Linkages between soil biota, plant litter quality and decomposition. In: *Driven by Nature: Plant Litter Quality and Decomposition*, (Eds.) G. Cadish and K.E. Giller. CAB International, Wallingford, UK, pp. 107–124.

Ecosystem Diversity and Carbon Sequestration: Climate Change Challenges and a Way Out for Ushering in a Sustainable Future *Pages 81–88*
Editors: **P.L. Gautam, Vir Singh and Uma Melkania**
Published by: **DAYA PUBLISHING HOUSE, 2009**

Chapter 9

Ecosystem Diversity and Sustainability: Towards Middle Path

B. Mohan Kumar

The inherent resilience in nature and the inevitably ubiquitous constancy of subtle or drastic alterations tend to subsume, even counter-balance the various human interventions in the form of resource-utilization in the course of civilizational ascension. Ecological biodiversity implies coexistence of multiple, often divergent forces and organisms. The crux of the problem lies in the prevailing development premise that most of the natural resources are for the exploitation and utilization of human beings. There are diametrically opposing perspectives to perceive the issue. One view often implies irrational and unsustainable exploitation of even depleting bio-physical resources for short-term gains causing long-term ecological damage. This seemingly realistic view is essentially pessimistic critically blaming human efforts of growth for ecological problems amounting to self-flagellation. The other more optimistic view counters the first one with the argument that such an idea amounts to amentia or non-recognition of human intelligence, creativity and ingenuity generating a wide range of astounding techno-scientific and aesthetic milestones, multidimensional achievements and search for sustainable alternatives. In reality, human beings are still continuing their struggle for survival in the face of adversities and uphill challenges. It is, however, wiser to follow a middle path as some human efforts bring in certain desirable (or culturally approved) and intended (or institutionally mandated) consequences while others undesirable and unintended ones. The former is more functional to ecological sustainability while the latter may be mal(dys)functional in their effects. Variations in terms of temporal (periodic), spatial and situational dimensions have to be accorded due cognition that itself nullifies any premise having universal applicability.

Keywords: *Development paradigm, Evolution, Pessimistic/optimistic views, Middle range.*

Introduction

The incremental progression of human development spurred by the accumulated and accreted scientific knowledge and technological know-how may seem to have outpaced but essentially subsumed in the subtle evolutionary process inherent in nature. However, it may be reiterated that human beings being just one of the organisms in the biodiverse and multi-variate nature play a miniscule albeit significant role in the ongoing universal triadic scheme of creation, sustenance and destruction. There prevails divergent perception related to possible role and impacts of human intervention on the composite ecology. In fact, there is a broad spectrum of perceptions. There is a pessimistic view that presupposes and blames human efforts for ecological degradation at the one end; while the optimistic view glorifies material development and control over nature at the other end. There are contingent views in the middle range either tilting towards one of the extreme perspectives or postulating a balanced stance highlighting the positive points and abnegating the negative ones of both. The present analysis is an assertion of relative advantage in the middle range perspective.

It is, however, unjustified to highlight only the negative points of any perspective. The views tilted to pessimistic extreme has a brighter side as they highlight the risks to natural equilibrium due to excessive exploitation of resources that can hardly be replenished. These bring into focus the serious issue of rapid depletion without full swing committed search for alternative as in the case of fossil fuel that may not last more than a couple of decades. The search, for instance, in lasting energy sources are still experimental and sporadic. Such a view essentially forewarns the intellectual, economic and political leaders sensitizing them to the looming calamity of multiple pollutions and long-term adverse impacts. These are inevitably being brought about by the natural or manmade processes of erosion, contamination and degradation of biotic-abiotic diversity making them unsustainable. Such realization may hopefully spur formation of policies and matching endeavours attuned to sustainability of the planet earth securing it for the coming generations. It does not be little the onward march of human specie as this can not be wished away, but simply implies that moderation is required in utilization of depletable resources while focusing on rational exploitation of non-depletable ones. However, the outcry over a broad range of adverse impacts on air, water, soil, biodiversity, human health and climate often corroborated empirically (albeit sporadic and inconsistent) betrays an alarmist view of doomsayer. One has to recall the scientific notion of constancy of evolutionary process that continues in a subtle manner and perceive the issue as a persistence of human efforts to survive, strive and thrive in the face of varied adversities.

The whole issue, however, is too complex and varied to be analyzed and comprehended simplistically. It is related to the techno-scientific advancements creating more comforts and amenities on the one hand, and concomitant ecological degradation and depleting bio-physical resources on the other. The middle path is a saner approach that focuses on the continuation of positives and avoidance of negatives in terms of sustainability and for the sake of leaving a habitable ecology for the coming generations.

There are multiple substantive issues and premises involved in the problem. Some of these having middle range thrust may be critically perused:

In-built Unsustainability in the Prevailing Development Premise

There seems to be certain inherent presumptions in the overarching development paradigm being followed the world over that tantamount to overexploitation of ecological resources for enhancing productivity and improving human comforts assisted by a broad range of techno-scientific innovations without taking into account the long-term adverse impacts. The increasing needs of food, energy, water, fuel, shelter, civic amenities and a host of other related consumer necessities accelerate the ongoing process of exploiting natural resources. The fallout is observed in depleting reserves of water, fossil fuel, forest on the one hand; and degradation/contamination of air, water, soil, biodiversity and the likes on the other. These invariably lead to green house effect, depletion of protective ozone layer, unpredictable climatic conditions and increasing susceptibility of human being to various psychosomatic diseases.

There seems to be a vicious circle of increasing population, their unlimited needs and consequent overexploitation of resources in which the humanity has got mired. Unless the development paradigm spurring unrestrained consumerism is altered to imbibe the sustainable agenda, there may not be an easy way out. Such refocusing requires sensitization of the people in ecosystem diversity envisioned with sustainability.

Continuation of Subtle Evolutionary Process

It would be unscientific to presume that evolution has ceased to continue. Certain physiological human traits (*i.e.*, opposing thumbs, bipedal movement, advanced brain, genetic configuration etc.) and acquired manipulating skills and ingenuity have enabled the humanity to ascend higher on the imagined civilization ladder and excel in many realms. The mind-boggling advancements in science and technology have accelerated the control of nature and wise utilization of resources. There has been stupendous progression in cultural and aesthetic realms. Yet, the efforts to enhance comfort level and create facilities/infrastructure ceaselessly continue. That some of these have generated unintended and undesirable impacts may be indicative of emerging challenges that humanity is striving to face.

The increasing concern is reflected in serious deliberations for evolving measures to counter or minimize the adverse effects of environmental pollution, preservation of biodiversity and search for appropriate alternatives in order to ensure sustainable course of development. It would be interesting to hypothesize that even subtle changes may be occurring in the human physiology and psychology attuned to the new challenges and opportunities. These initially seem to be contrary to existing notions of rational existence, but we must not doubt the inherent/acquired capacity, resilience and ingenuity of humanity to emerge unscathed in this struggle for survival.

Inevitability of Epochal Changes and Concomitant Alterations

It is a truism, yet needs to be reiterated that the earth system with its ecological diversity has witnessed multiple changes within it without human intervention (*i.e.* shift in continental plates, destruction/formation of terrains and water bodies in different eras, climatic changes, and other geological upheavals) and emergence of variegated life forms, survival of some through adaptation and extinction of many. Somewhere along the line, human beings evolved about three million years ago and emerged the most resilient and wisest to survive and strive ahead in the struggle for existence. They have proved to be more manipulative and intelligent to carve niches and evolve mechanisms to make life more comfortable. The quest and striving continues with leaps and bounds in many sectors. These can not be wished away or belittled. We may be witnessing a miniscule part of the epochal changes in our times. The cynicism of doomsayers will get relegated to the backwaters as it largely tends to underrate the human ingenuity and capacity to adapt. No doubt, some of these endeavours have generated adverse conditions, but humanity is striving to find way out of any imbroglio and is undertaking massive global projects towards sustainability despite initial hiccups.

Search for Sustainable Alternatives

There is a set of multi-pronged agenda with a vision to minimize ecological imbalance and ensure secure future within the limits of human efforts and intelligence. Some of these may be cited to substantiate optimistic confidence in human ingenuity. For instance, search for viable sources of energy besides the existing hydel, thermal and nuclear ones in the forms of exploiting solar, wind, sea-waves, bio-resources etc. are indicative of sustainable efforts. Striving to focus on eco-friendly bio-control, biofertilization, organic farming, genetically engineered crops and the likes may prove to be positive moves towards sustainability. Moreover, extensive reforestation, conservation and preservation of biodiversity may reinstate the equilibrium suited to new challenges and ethos of sustainable future.

Ecological Zone-specific Projects of development

The diversity in the extant ecological systems has largely been identified and working database formulated to facilitate region/sub-region-specific projects in tune with the needs and priorities of a particular region and inhabitants. The earlier thrust on uniform policies has given way to contextualized ones that are attuned to geographical conditions, human resources and prevalent styles of livelihood. These tend to elicit positive response and consequent involvement of local populace. For instance, the projects for resource-scarce zones of hills, deserts, arid and semi-arid areas inhabited by indigenous and under privileged people engaged in unproductive works have been focused for mainstreaming through efforts of public and voluntary agencies of development. However, there has been perceptional divergence in priority-setting primarily due to near-absence of *ex ante* assessments of existing livelihood sources, infrastructure and world-view of concerned people. The sustainability issue has remained sidelined to a greater extent, thus minimizing the expected and desirable impacts.

Ensuring Access to Sustainable Livelihood Opportunities

The more urgent and pressing concerns of the humanity are to ensure rational provision of reasonable access of the masses to drinking water, nutritional food, shelter, health care and civic amenities besides adequate enablement and empowerment to sustain survival and civilized existence. The major efforts are indeed directed towards these projects. The problem may be marginalization of issues like judicious distribution and equitable access to enhancing avenues. Somewhere the priorities have got bogged down in efforts for procuring mundane luxuries for the privileged few causing negligence of basic necessities for the underprivileged majority. The increasing dissent and violent unrest being witnessed today throughout the world needs to be accorded topmost priority, otherwise the humanity may get mired in resolving small or large scale conflicts rather than focusing capabilities on issues of long-term implications such as global warming, conserving and preserving ecosystem diversity and so on.

Elitist Bias and Lip Service to the Cause of Sustainable Livelihood

There is a widespread feeling that such concerns betray elitist bias in thinking amounting to idealized and formalized policy-formation and negligible implementation at the ground level. This may be curtailed by addressing and prioritizing the unfulfilled needs of more than half of humanity in the widely scattered tribal, rural and slum zones that remain largely sidelined in the course of global development. In fact, the scientists, technologists, other intellectuals and policy makers may be blamed if they fail to undertake such onerous responsibility with urgency for the sake of justifying massive public/private expense on them, if not for loftier cause of humanity. Increasing practice of enforcing accountability through evaluation and monitoring of scientific and development works is one of the mitigating measures for streamlining. Indeed, it must not be presumed to be a case of misplaced optimism. They have been able to find ways to bring incremental progression in nearly all the possible realms, and they can definitely find ways to the specific concerns of marginalized lot. Sporadic attempts have proved to be impacting inadequately and marginally in the face of large scale and magnitude of unfinished tasks of sustainability.

Ecological Concerns as Subterfuge/Distraction from Real Issues and Perpetrators

One is tempted to opine that modern and privileged sections and regions have got too engrossed in sensate (*i.e.* pleasure-seeking pursuit) culture to get concerned with the lurking catastrophic hazards to humanity and the earth that their consumption and life choices tend to generate and aggravate. But the enchanting lure of luxuries and comforts overshadow any sensible and sustained efforts towards mitigation of ecological or judicious development issues. Can these be wished away or has mass sensitization been adequately undertaken to wean away? The answer is an emphatic No. There seems to be little light at the end of the tunnel of self-destruction the humanity perforce has tumbled into. Reliable data vindicate the point that the disproportionately greater consumption of energy and search for maximizing profits through production of luxury items has intensely magnified the ecological disasters in the forms of global warming, depletion of protective ozone layer, climate change, melting of polar caps and river-sustaining glacial ice, deforestation etc. These inevitably threaten the sustainable existence of life forms and accelerate the process of indispensable destruction.

But these harsh realities need not turn us into hardcore pessimists and doomsayers as the current world-view seem to be incorporating corrective measures for mitigation.

Appendix-I: Global Environmental Change

Large-scale global environmental changes include

- ✰ Climate change
- ✰ Stratospheric ozone depletion
- ✰ Loss of biodiversity
- ✰ Changes in hydrological systems
- ✰ Changes in the supplies of freshwater
- ✰ Land degradation and
- ✰ Stresses on food-producing systems

Appreciation of this scale and type of influence on environment especially human health requires a new perspective which focuses on ecosystems and on the recognition that the foundations of long-term good health in populations rely in great part on the continued stability and functioning of the biosphere's life-supporting systems. It also brings an appreciation of the complexity of the systems upon which we depend.

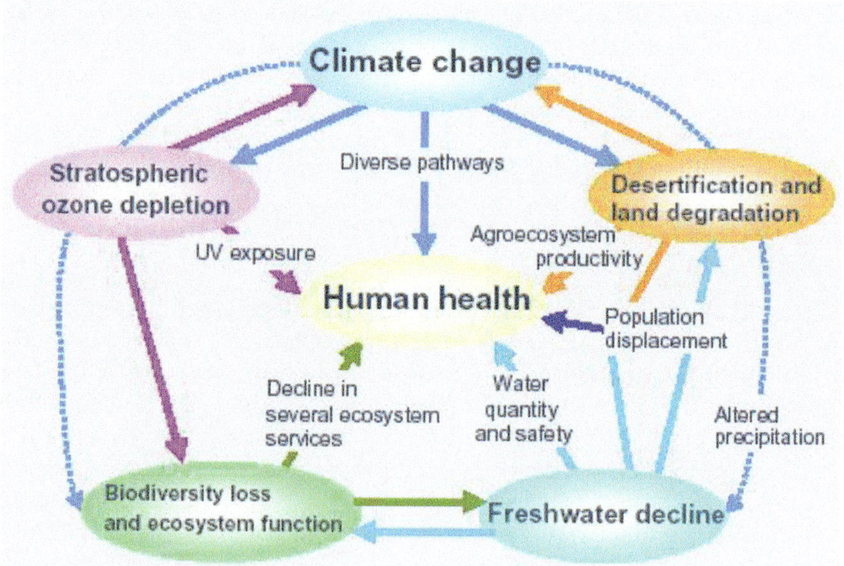

Appendix-II: Rising Temperature and Greenhouse Effect

Global mean surface temperatures have increased 0.5-1.0°F since the late 19th century. The 20th century's 10 warmest years all occurred in the last 15 years of the century. Of these, 1998 was the warmest year on record. The snow cover in the Northern

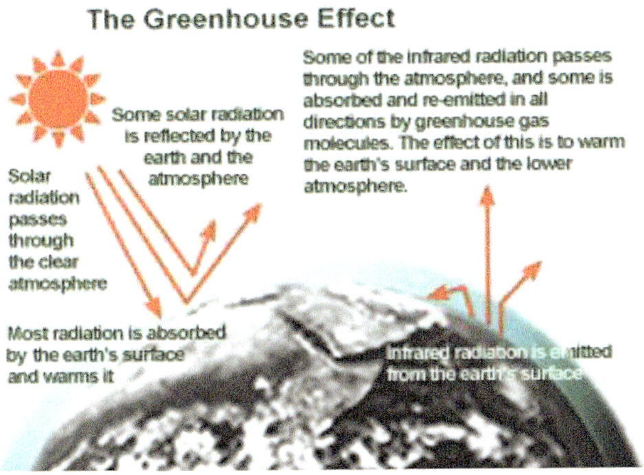

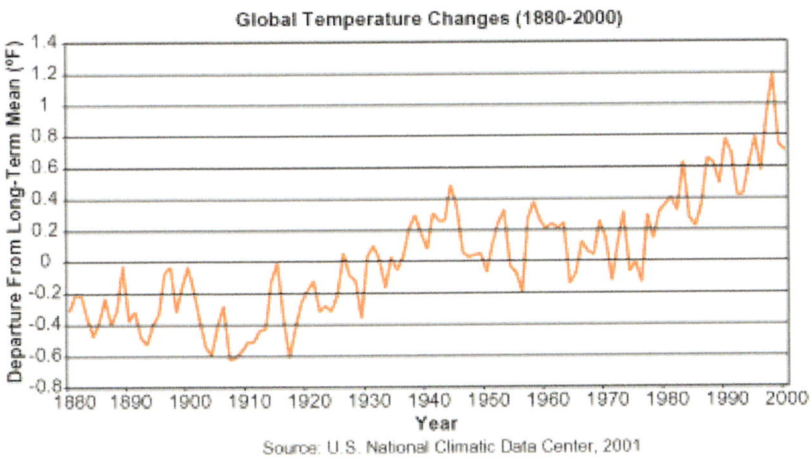

Hemisphere and floating ice in the Arctic Ocean have decreased. Globally, sea level has risen 4-8 inches over the past century. Worldwide precipitation over land has increased by about one percent.

Increasing concentrations of greenhouse gases are likely to accelerate the rate of climate change. Scientists expect that the average global surface temperature could rise 1-4.5°F (0.6-2.5°C) in the next fifty years, and 2.2-10°F (1.4-5.8°C) in the next century, with significant regional variation. Evaporation will increase as the climate warms, which will increase average global precipitation. Soil moisture is likely to decline in many regions, and intense rainstorms are likely to become more frequent. Calculations of climate change for specific areas are much less reliable than global ones, and it is unclear whether regional climate will become more variable.

Appendix-III: Health Impacts of Air Pollution

Pollutant	*Origin of Pollutant*	*Effects*
Sulphur dioxide	Industries, especially where coal or oil are used as fuel	Irritation of eyes, and respiratory system, increased mucus production, cough and shortness of breath
Carbon monoxide	Automobile exhaust and industries	Reduction in oxygen-carrying capacity of blood
Oxides of nitrogen	Automobile exhaust	Irritation of pulmonary tract affecting functioning of lungs
Hydrocarbons Chlorine	Automobile exhaust chloralkali industry	Lung cancer and irritation of mucous membrane
Ammonia	Fertilizer industry, agriculture and in poultry farming	Irritation of mucous membrane
Hydrogen sulphide	Manufacture of coke, viscose rayon, Distillation of tar and petroleum	Inhalation leads to death
Acids and aldehydes	Chemical industries	Eyes, nose and throat irritation
Suspended particulate matter	Industries, automobile exhaust	Respiratory diseases
Dust	Industries and automobile exhaust	Silicosis
Asbestos	Roofing, brake linings	Asbestosis
Lead	Automobile exhaust	Cumulative poison, impairment of central nervous system
Beryllium	Aerospace industry, hold appliances	Fatal to heart and manufacture of house-lungs
Manganese	Mining operations	Damages nerves and reproductive systems
Benzene	Automobile exhaust and manufacture of chemicals	Leukemia, chromosomal damage
Pesticides	Manufacture and application of pesticides	Depression; leads to death if inhaled in excess
Arsenic	Thermal power plants	Toxic

Courtesy by: C. P. R. Environmental Education Centre, Chennai.

Ecosystem Diversity and Carbon Sequestration: Climate Change Challenges and a Way Out for Ushering in a Sustainable Future *Pages 89–101*
Editors: P.L. Gautam, Vir Singh and Uma Melkania
Published by: DAYA PUBLISHING HOUSE, 2009

Chapter 10

Sacred Groves in India: Celebrating Sanctity of Life through Biodiversity Conservation

Anubhav, Kundan Singh, Akanksha Rastogi and Vir Singh

Introduction

Biodiversity conservation in diverse ecosystems holds the key to mitigate vicious environmental crises gradually leading to environmental catastrophes. Increasing concentration of carbon dioxide in the atmosphere followed by global warming that culminates into climate change is perhaps the worse catastrophe in the history of mankind and appears to take heavy toll of life on the living planet in due course of time. Some of the traditional ways evolved by devout people still stand as a lesson for us to learn. The way communities have been managing sacred groves and conserving biodiversity in situ to this date is a glaring example of containing and averting ecological crises.

Sacred groves are not only the sacred ecosystems functioning as a rich repository of nature's unique biodiversity, but also a product of the socio-ecological philosophy our forefathers have been cherishing since olden days. Sacred groves do not just help conserve valuable biodiversity, soil and water, but are also critical in regulating weather and climate cycles so vital for life to blossom and flourish on the planet. Our forefathers held reverential attitude towards nature and its all life forms. Sacred groves are the potent symbol of the same. People cherish sanctity of life through conservation of biodiversity. This ecological philosophy is a potent tool to avert any crises in the society.

Sacred groves have well developed forest ecosystems and high degrees of species richness and a rich biodiversity in general. This, however, depends on the extent of preservation of the grove. Many, rather most, of these pristine ecosystems have either vanished or are disturbed to a great extent. But those still existing are living instances of carbon pools and nature's preserved uniqueness.

The Sacred Groves

Sacred groves have been defined variously by various scholars. According to Saikia of Department of Geography, Gauhati University, India, "Sacred groves are small groves that vary in size from a few hectares to a few kilometers protected by local communities as being the sacred residences of local deities and sites for religio-cultural rituals, have served as valuable storehouses of biodiversity". M.S. Swaminathan Research Foundation, Chennai defines sacred groves as a biological heritage and a system that has helped to preserve the representative genetic resources existing in the surrounding regions for generations.

The historical links of sacred groves have been traced to the primitive state of human society. The first instances of Indian traditional conservation of our biological wealth in various pockets of natural forest are often referred to "sacred groves". In the past, sacred groves were indicators of the phenomenon of ethno-environmental management. Sacred groves are forests that have been protected for ages by traditional societies. In the past these were present in numerous parts of the world, nearly every continent, and were entities held sacred by communities with different religions and different forms of economic and social organizations. The sacred groves also exist in Ghana, Syria, Nigeria, Turkey and Japan. Most of the world's sacred groves, unfortunately, have disappeared and only few are reported to occur to date. And yet, some groves have passed down the generations and are flourishing in their full form. The hills of the north eastern region (NER) of India are the living testimony of the same. These groves are relatively undisturbed (http://wgbis.ces.iisc.ernet.in/biodiversity/newsletter/issue8/; Sherring1974; Rao and Nayudu 1979; Khoshoo 1986; Ramesh and Pascal 1997; Pushpangadan 1998; Ramakrishnan 1998, 2001; Sinha and Maikhuri 1998).

Concepts of Biodiversity

Biodiversity, in essence, is the totality of all life on earth. Both in wild and domesticated forms, biodiversity is the source of food, medicine, clothing and housing, most of the cultural diversity, and most of the intellectual and spiritual inspirations. It is, without doubt, the very basis of man's being. Its conservation therefore is of utmost significance. Convention of Biological Diversity (CBD) 1992 offers a major opportunity to safeguard and conserve the biological capital at global and regional level.

According to CBD, *"Biological diversity" means the variability among living organisms from all sources including, inter alia, terrestrial, marine and other aquatic ecosystems and the ecological complexes of which they are part; this includes diversity within species, between species and of ecosystems.*

It is estimated that there exist 5-30 million species of living forms on our earth. Of these, only 1.5 million have been identified. The tropical forests are regarded as the richest in biodiversity. Scientists are of opinion that whatever be the absolute numbers, more than half of the species on the earth live in moist tropical forests which is only 7 per cent of total land surface (McNeely *et al.*, 1990).

Sizable percentage of endemic flora and fauna, on account of immense variety of climatic and altitudinal conditions coupled with varied ecological habitats, makes India quite rich in terms of natural biodiversity. Biodiversity varies from the humid tropical rain forests to alpine vegetation, hot semi-arid desert of Rajasthan to the cold desert of Ladakh and icy mountain of Himalayas to the warm coasts of peninsular India. India is classified as one of the world's 12 centres of origin of cultivated plants identified by Vavilov. There are 167 cultivated species and 273 wild relatives of crop plants in India (Ghosh and Tiwari 1984; Pearce and Moran 1994; Agarwal, Tiwari, and Dubey 1996; Abe *et al.*, 1997; Ananthakrishnan 1999).

Among the 25 hot spots of the world identified by Meyers (1988), two lie in India, *viz.*, Eastern Himalayas and Western Ghats. In addition, India has 26 recognized endemic centres that are home to one-third of flowering plants of the world identified to date (Khiewtam and Ramakrishnan 1989; McNeely, Miller, Reid and Werner 1990).

Loss of Biodiversity

The loss of biological diversity is a global crisis. There is hardly any region on Earth that is not facing ecological catastrophes. Of the 1.5 million species known to inhibit the Earth, one fourth to one third is likely to go extinct within the next few decades. Biological extinction has been a natural phenomenon in geological history. But the rate of extinction was perhaps one species every 1000 years. But man's intervention has speeded up the extinction rates to an alarming scale. Between 1600 and 1950, the rate of extinction went up to one species every 10 years. Currently it is perhaps one species every year.

India's biodiversity is one of most significant in the world. An assessment of wildlife habitat loss in tropical Asia in 1986 showed that the country had lost about 20 per cent of its original wildlife habitat. In last few decades India has lost at least half of its forests, polluted over 70 per cent of its water bodies, built on or cultivated much of its grasslands and degraded most of its coasts. Under such circumstances, none can say how many species have been done to extinction. According to IUCN 2000 Red list data, India contributes nearly 3 per cent of world's total number of threatened species. This includes 86 species of mammals, 70 species of birds, 25 species of reptiles and 3 species of amphibians. Among plants, 19 species are extinct, 44 are critically endangered, 113 are endangered and 87 are vulnerable (Khoshoo1986, 1994; Khurana 1998; Kotwal and Banerjee 2000).

Causes for the Loss of Biodiversity

There are a number of the causal agents that lead to the destruction of biodiversity. Some are listed in Table 10.1.

There are a number of causes that harm biodiversity or even lead to total annihilation of many a species of crucial value. Often a complex of factors operates on an area or a system and health of the whole biodiversity or of rare or vulnerable species depends on the intensity of the operating factors.

Table 10.1: Various Causes for the Loss of Biodiversity and their Explanation

Cause	*How Does it Lead to Biodiversity Loss?*
Habitat destruction	
Natural habitats are destroyed for settlement, grazing grounds, agriculture, mining, industries, highway construction, drainage, dam building etc.	Species must adapt to the changes, move elsewhere or may succumb to predation, starvation or diseases and eventually die.
Deforestation	
Forests are felled and cleared to give way to agriculture, population settlement, shifting cultivation, development projects, demand for fuel wood, demand of wood for industries and other commercial purposes.	Numerous species the natural forests would harbour would be lost for ever.
Over exploitation	
Commercial exploitation of biodiversity has invariably meant its overuse and eventual destruction.	Biological curiosities like insectivorous and primitive species and other taxa needed for teaching or laboratory work (like *Nepenthes*, *Gnetum*, *Psilotum*, etc.) are lost.
Collection for zoos and research	
Animals and plants are collected throughout the world for zoos and biological reserves for studies and research in science and medicines.	Species eventually squeeze in limited area and come on the brink of extinction.
Hunting	
Wild animals are hunted for their products such as hides and skin, tusk, antlers, fur, meat, pharmaceuticals, perfumes, cosmetics, decoration purposes and sports.	Rare animal species get exterminated over long period of time.
Introduction of exotic species	
Plantation of non-native species want only for commercial ends.	Native species are subjected to competition for food and space due to introduction of exotic species.
Pollution	
Mixing of undesirable elements and deleterious chemical compounds in environmental components	Pollution alters the natural habitats. Pollution of water bodies is especially injurious to the biotic components of estuary and coastal ecosystem. Insecticides, pesticides, sulphur and nitrogen oxides, and acid rain, ozone depletion and global warming too, affect adversely the plant and animal species.
Control of pests and predator	
Use of synthetic pesticides for plant protection in agriculture.	Pest control measures generally kill predators that are a component of balanced ecosystem and may also indiscriminately poison non-target species.
Other factors	
Other ecological factors like distribution range, specialization, reproductive rate etc., and socio-economic and political factors overlooking conservation of resources.	Natural forests or virgin habitats are destroyed. Rare, endangered and threatened species succumb to biotic pressures.

Increasing human population is not the only factor that cripples our biodiversity conservation programmes. However, any plan attempting at conservation must take socio-economic development into consideration. India's efforts of biodiversity conservation deserve applause because of the fact that it has world's largest network of protected areas.

The conservation plan for biodiversity has an holistic approach and encompasses whole spectrum of biota and activities ranging from ecosystems at macro level (*in situ conservation*) to DNA libraries at the molecular level (*ex-situ conservation*). *In-situ* conservation refers to protection and management of biodiversity in their natural habitat while conservation of components of biological diversity outside their natural habitat, like gene bank, seed bank, zoos etc. comes under *ex-situ* conservation.

Conservation-Religion Relationship

Conservation of biodiversity has been at the heart of Indian civilization from the very beginning. The entire premise of right conservation of biodiversity on global scale is based on the cultural values. India harbours one of the oldest and most diverse cultural traditions associated with conservation. Conservation of some animals and plants was considered as sacred by Indus valley civilization. The Hindus still follow ancestral worshiping and animism in the form of deity worship. Animism means the belief that a supernatural force animates and organizes the universe, or the belief that things in nature, *e.g.* trees, mountains, and the sky, have souls or consciousness. Affection towards nature as a zoolatry (worshipping of animals) and totem (considering plants and animals sacred) proves religious beliefs, traditions and customs of the Hindus. Sanatana Dharma has linked some animals with the specific Gods and Goddesses. Among the animals considered sacred are Lion, Cow, Bullock, Cobra, Rat *etc.* Trees of *Cedrus deodara* are believed to be the tree of God and are planted around temples in the mountains.

In Kerala, it is common practice among *Hindus* to assign a part of their land near ancestral home as an abode of Goddess Durga or serpent God Naga. On a rough estimate there are hundreds of Sarpakavu (*Sarpa* means snake and *Kavu* means jungle) in Kerala. These Sarpakavu preserve more than 200 species of angiosperms (20 per cent of whole state) out of which 150 plants are medicinal and 40 per cent plants are rare or endangered.

Sacred seeds, sacred trees, sacred forests, sacred cow, sacred river, sacred mountain etc. are the crucial notions prevailing among Hindus. A tree near a temple is never trimmed and is used as a breeder tree. Peepal (the Banyan tree) is worshipped by devout *Hindus*. River Ganga is held with reverential attitude and the *Ganga-jal* (water of the Ganga) is regarded as sacred. Tribals especially hold reverential attitudes towards different components of nature. It is this region-conservation relationship that in spite of depletion of forest in many parts of country, virgin vegetation units like Sacred Groves, still remain intact and serve as rich pool of biological diversity (Malhotra *et al.*, 1998; Pushpangadan *et al.*, 1998; Ramakrishnan 1998, 2001; Sinha and Maikhuri 1998).

Sacred Groves and Biodiversity Conservation

Man is an integral part of nature. Any impact on nature, therefore, has influence on man. Nature flourishes, man flourishes too. Nature gets destroyed; man is bound to be destroyed. Sacred groves represent the major effort to recognize and conserve ethnic biodiversity in traditional ways. Our fundamental social-cultural, ethical, aesthetic and economic values are directly and indirectly linked with nature and its biological resources. Throughout the history of agriculture Indian farmers have been increasing the level of biodiversity in agro-ecosystems. Diversity in genes, species and ecosystem has contributed immensely to the productivity of agriculture.

The existence of sacred groves in India most likely dates back to an ancient pre-agrarian hunter-gathering era, and their presence has been documented since the early 1800s. Believing trees to be the abode of gods and ancestral spirits, many communities set aside sanctified areas of forest and established rules and customs to ensure their protection. These rules varied from grove to grove but often prohibited the felling of trees, the collection of any material from the forest floor, and the killing of animals. Presiding deities administered punishment, often death, to individuals who violated the rules, and sometimes to the entire community in the form of disease or crop failure. As a result of these protective restrictions, preserved over countless years, sacred groves are now important reservoirs of biodiversity.

Sacred groves are often the last refuge for endemic and endangered plant and animal species. They are storehouses of medicinal plants valuable to village communities as well as modern pharmacopoeia, and they contain wild relatives of crop species that can help to improve cultivated varieties. Sacred groves also provide for the water needs of nearby communities. Many sacred groves contain water resources such as ponds and streams, and the vegetative mass that covers the floor of a grove can absorb water during rainy season and release it during times of drought. Trees also improve soil stability, prevent topsoil erosion and provide irrigation for agriculture in drier climates. There has been no comprehensive survey of sacred groves in India, so their exact number and area are unknown. At least 13,720 sacred groves have been reported in various regions of the country, but experts estimate that the actual number is likely 100,000 to 150,000 (Pushpangadan *et al.*, 1998; Ramakrishnan 1998, 2001; Sinha and Maikhuri 1998).

Sacred groves perform several other functions besides biodiversity conservation. For instance, they maintain micro-climate, recharge aquifers, enhance nutrient cycling, prevent soil erosion and they can be used as a unique source of *in-situ* conservation.

Plants Related to Religion

Several studies bring to the fore some interesting accounts of the environmental problems of the ancient Greeks and Romans that give us a rare portrait of the sacred groves of the Mediterranean region, glimpses of which are given in the following few paragraphs (Sherring 1974; Sen 1989; Ramesh and Pascal 1997; Hughes and Chandran 1998; Malhotra *et al.*, 1998; Pushpangadan *et al.*, 1998; Ramakrishnan 1998, 2001; Sinha and Maikhuri 1998):

- ☆ The Greek and Roman landscapes were dotted with hundreds of sacred places. Sacred enclosures formed one of the major categories of land use. These usually contained groves of trees and springs of water; within them the environment was preserved, as a rule, in its natural state. As Seneca (1st century ad) remarked, "If you come upon a grove of old trees that have lifted up their crowns above the common height and shut out the light of the sky by the darkness of their interlacing boughs, you feel that there is a spirit in the place, so lofty is the wood, so lone the spot, so wondrous the thick unbroken shade."
- ☆ Ovid said, "Here stands a silent grove black with the shade of oaks; at the sight of it anyone could say, 'there is a spirit here!' "
- ☆ In Virgil (1st century BC) is found a reference to the rural folk wondering on seeing an old tree-covered Capitoline hill: "Some god has this grove for He also noted that the dwelling!" Gods favour wild trees "unsown by mortal hand".
- ☆ Pliny the Elder (ad 23-79) indicated that trees were the first temples of the gods, and "even now simple country people dedicate a tree of exceptional height to a god...."
- ☆ Aquelus spoke of travellers praying under the trees on "a little sacred hill fenced all around". But the grove of Daphne was ten miles in circumference. A grove near Lerna stretched all the way down a mountainside to the sea. Alexander the Great found an entire island dedicated to a goddess identified as Artemis.
- ☆ The Romans often personified the spirit of plant life. The 'forest king' was the human personification of the spirit of the sacred tree, the oak, his living double.
- ☆ The ancient Greeks represented the spirit of conservation in the shape of a formidable protector of animals and plants, the goddess Artemis. She was the protector not only of wildlife but of the wilderness itself. In ancient religious thought, Artemis endowed the wilderness with sacredness.

There could be many reasons why the groves vanished from Europe. A kind of multiple uses was allowed in groves. Although they were strictly protected in Greece and Rome, religious use of their resources was allowed. As much wood might be taken as was necessary for sacrifices.

Animals in the grove, such as goats and deer, might be captured and offered to the deity. Trees in the grove could be used in building a temple inside it or even away from it. Wood from the sacred trees was believed to keep its magical powers when fashioned into other objects and was used for making a variety of objects like statues of gods, staffs, scepters, etc. Wood was even supplied to private persons at a fixed price for sacrifices (Hughes and Chandran 1998).

It seems the groves also suffered from the pressures of urbanization, as baths, stadiums, gymnasiums, schools, etc., were established in them. At times they also had to cater to the timber needs of the military. What caused the final downfall of the groves in Europe? In the words of Hughes and Chandran (1998):

> The groves lasted as places of economic and religious importance down through the Christianization of the Roman Empire. As centres of pagan worship, they became the objects of Christian zeal, again both religious and economic. The emperor Theodosius II (5th century ad) issued an edict directing that the groves be cut down unless they had already been appropriated for some purpose compatible with Christianity. A few of them became monastery gardens and churchyards, and here and there in the Mediterranean even today one will find a gigantic old tree that is supposed, however unlikely it might seem, to have sheltered Plato or Hippocrates, and has been spared for that reason.

Celebrating Nature and its Biodiversity

Hindu religion is replete with the events of celebration involving several trees and herbaceous plants. A glimpse of such religious festivals is shown in Table 10.2.

Table 10.2: Religious Festivals Associated with Sacred Trees

Festivals	*Month of Festival*	*Species Associated*
Sheela Asthami	March	*Azadirachta indica*
Neem Saptami	April	*Azadirachta indica*
Vat Savitri	May	*Ficus bengalensis*
Bilva Mangal	May–June	*Aegle marmelos*
Sawan ke Somvaar	Mid July–Mid August	*Bael*
Kadii Vrat	September	*Musa paradisica*
Somvari Amavasya	15 of all months	*Ficus religiosa*

Every festival in India is virtually associated with a plant species and foods derived from different species. This is one of the very special ways of celebrating biodiversity of nature. Use of Amaranth and buckwheat on the occasion of observing a fast is very common. This reveals that Indians knew for ages that amaranth and buckwheat are not cereals. Today's biological sciences also prove that these two crops are pseudo cereals rather than cereals which are prohibited during fast.

Sacred Groves of India

Green patches of woodlands in India are as diverse as Indian cultures. They, however, share an important commonality: they are all held sacred. India has a long and unique history of nature worship, and this practice continues to date, especially through the veneration of forest groves. The sacred groves have been preserved over generations. Dedicated to local deities or ancestral spirits, these sacred groves are protected by local communities through social traditions and taboos that incorporate spiritual and ecological values. The sacred groves represent native vegetation in a natural or near-natural state. These are often exceptionally rich in biodiversity and also often embrace rare species of plants and animals. Thousands of sacred groves remain and many villages continue to observe traditional practices. Moreover, in the face of degradation, conservationists and local communities are recognizing that traditional knowledge and sacred practice are important elements in the conservation and management of these ecological treasures.

The sacred groves are located in a variety of habitats ranging from resource rich forested landscape, such as Western Ghats and North-East to the extremely resource poor desert conditions in western and central India. More than 30 sacred groves exist in the East Khasi Hills district of Meghalaya. Nearly two thirds of these are in excess of 100 hectares in size. In the neighboring Karbi Anglong in Assam, in the eastern part of the district more than 100 sacred groves proliferate. These two districts, both inhabited by indigenous tribal populations portray different contexts of conservation of sacred groves. In East Khasi Hills sacred groves are fairly well documented and strictly governed; this is less so in eastern Karbi Anglong (Khiewtan and Ramakrishna 1989; Ramakrishnan 1998, 2001; Sinha and Maikhuri 1998; Tripathi *et al.*, 1995; Jamir and Pandey 2002).

Indian sacred groves demonstrate unique diversity. Some contain only a few trees, while others are spread over hundreds of acres. Sometimes groves overlap with larger forested areas, while others exist as islands in open plains or desert. Even their names vary from region to region. Most sacred groves in India are associated with the almost 40,000 endogamous groups within the *Hindu* caste system and other major religions such as Buddhism and Islam, along with other religious communities and traditional tribal groups.

In many groves, villagers perform annual rituals and ceremonies to appease the presiding deity and ensure the well-being of the community. It is also common for people to make individual offerings, often in the form of terracotta figures, in exchange for wishes such as good health or harvest or the birth of a child. Sacred groves help to define the cultural identity of the communities that revere and protect them. They are also closely linked to the politics and economies of their communities, and their legal status and management vary among regions and individual villages.

Some groves are associated with and managed by separate caste groups within a community, some by a village as a whole, and some by neighboring districts within a larger geographical area. There are also large "pan-Indian" groves that involve people from many parts of the country. In some groves, all forms of resource extraction are strictly prohibited, while in others people may collect material such as fallen branches and leaves from the forest floor or fruit from the trees.

The Western Ghats–a mountain range that runs along India's west coast, through the states of Maharashtra, Goa, Karnataka, Tamil Nadu and Kerala–is one of the world's biodiversity "hotspots." Its ecosystems include tropical wet forest, mountain evergreens, moist deciduous forest and scrub grassland. An estimated 2,000 plant species and 300 vertebrate animal species live in this region and nowhere else. Karnataka harbors nearly 1,500 documented sacred groves, which are most commonly called *devarakadus* or *devarkans.* In smaller groves, tree cutting or other resource extraction is strictly prohibited, while larger groves often function as "resource forests," in which people are allowed to gather products for sustenance. In Kerala, sacred groves are called *kavus* and at least 2,000 have been documented. Hundreds are dedicated to snakes, which protect agricultural crops by controlling insect and rodent populations. The Bishnoi tribes manage sacred groves called *orans* in the arid and desert regions of Rajasthan, in northwest India. Despite sparse vegetation and limited water resources, the area reportedly supports a higher density of human and animal populations than any other desert region in the world because of the conservation practices of its people. The basic philosophy of the Bishnoi faith is that all living things have a right to live and share resources, and the group has a set of abiding laws including a ban on killing animals and on felling trees, especially their most sacred *khejadi* tree, which has numerous life-sustaining properties. Its dropped leaves can be used for fodder, its dropped branches for fuel and its fruit for food. *Khejadi* trees stabilize sand dunes and they are said to increase yields of crops that grow close by. The *orans* also provide a protective habitat for the Indian gazelle and blackbuck. There are only a handful of documented sacred groves in Rajasthan, but their areas often exceed thousands of acres (Khiewtan and Ramakrishnan, 1989; Tripathi *et al.*, 1995; Malhotra, 1999; Jamir and Pandey, 2002).

Table 3: Sacred groves in India

State	*Number of Documented Sacred Groves*
Andhra Pradesh	750
Arunachal Pradesh	58
Assam	40
Chhattisgarh	600
Gujarat	29
Haryana	248
Himachal Pradesh	5,000
Jharkhand	21
Karnataka	1,424
Kerala	2000
Maharashtra	1,600
Manipur	365
Meghalaya	79
Orissa	322
Rajasthan	9
Sikkim	56
Tamil Nadu	448
Uttarakhand	1
West Bengal	670
Total	13,720

Source: Khiewtan and Ramakrishnan (1989); Tripathi *et al.* (1995); Malhotra (1999); Jamir and Pandey (2002).

India is believed to have nearly 14,000 sacred groves spread among different states. Most of the groves are located in Himachal Pradesh, Kerala, Karnataka, Andhra Pradesh, Maharashtra, West Bengal and Chhatisgarh (Table 3).

Threats to Sacred Groves

India's sacred groves appear to be on the verge of depletion. Modernisation followed by liberalization, privatization and globalization is leading to erosion of the values that were vital for the conservation of sacred groves in India. A mad rush for development which is often demonstrated through mega river valley projects, large-scale mining works, road construction, tourism, urbanization, industrialization, etc. is experienced almost everywhere. Government and local governing bodies are insensitive towards local people's rights over and access to natural resources. Many groves are suffering what is called "Sanskritization," the transformation of primitive nature worship into formal *Hindu* practice. In the earlier system people gave more value to nature and natural objects rather than to idol worship.

In the era of liberalization, privatization and globalization, communities are losing their traditional bases of sustainable livelihoods, such as forests and rangelands and are being forced to enter into market system that generally regards nature merely as a commodity. This compels communities deviate from the customs which were guiding them to protect and maintain sacred groves. The market-oriented social systems do not hold reverential attitude towards nature. Sanctity of life has no place in the market-linked livelihood systems.

Conclusion

Sacred groves have been flourishing in India from time immemorial. This is the wonderful response of people to the loss of biodiversity. The sacred groves have been one of the living examples of rich and vibrant biodiversity. A number of plant and animal species described as rare, endangered, vulnerable or threatened are existing today thanks to the these sacred groves, which have also been pivotal in protecting and conserving endemic species. Protection of nature's biodiversity and especially the rare types of life forms through sacred groves is regarded as an obligation on the part of the society. Sacred groves have been an embodiment of an act of faith, fear or social obligation as well as eco-philosophical outlook of the people living amidst the bounty of nature. Of late, decline in traditional beliefs, economic development and other socio-economic factors have undermined the status of the sacred groves across our country.

Sacred groves also serve as a meaningful fusion of religion and ecology. People's protection, preservation and conservation of species through sacred groves is people's way of celebrating the sanctity of life.

References

Abe, T., Levin, S.A. and Masahiko, H., 1997. *Biodiversity: An Ecological Perspective.* Springer-Verlag Inc., New York, pp. 277–289.

Agarwal, S.K., Tiwari, S. and Dubey, P.S., 1996. *Biodiversity and Environment.* A.P.H. Publishing Corporation, New Delhi, pp. 125–135.

Ananthakrishnan, T.N., 1999. Multidimensional links in biodiversity research: An integrated exercise. *Current Science*, 77(3): 356–358.

Ghosh, A.K. and Tiwari, K.K., 1984. Biodiversity conservation in North-East India: Problems and realities. In: *Resources Potentials of North-East India Vol. II: Living Resources*, (Ed.) R.S. Tripathi. Meghalaya Science Society, Shillong, pp. 105–109.

http://wgbis.ces.iisc.ernet.in/biodiversity/newsletter/issue 8/.

Hughes, J.D. and Chandran, M.D.S., 1998. Sacred groves around the earth: An overview. In: *Conserving the Sacred for Biodiversity Management*, (Eds.) Ramakrishnan, P.S., Saxena, K.G. and U.M. Chandrashekhara. Oxford and IBH Publishing Co. Pvt. Ltd., New Delhi, pp. 69–86.

Jamir, S.A. and Pandey, H.N., 2002. Status of biodiversity in the sacred groves of Jaintia Hills, Meghalaya. *The Indian Forester*, 59(5): 738–744.

Khiewtam, R.S. and Ramakrishnan, P.S., 1989. Socio-cultural studies at the sacred groves at Cherrapunji and adjoining areas in north-eastern India. *Man in India*, 69(1): 64–71.

Khoshoo, T.N., 1994. Census of India's biodiversity: Tasks ahead. *Current Science*, 67(1): 577–582.

Khoshoo, T.N., 1986. *Environmental Priorities in India and Sustainable Development*. Presidential address, 73rd session, India Science Congress Association, Jan. 3–7, New Delhi.

Khurana, I., 1998. Unnatural decline. *Down to Earth*, 6(23): 34–38.

Kotwal, P.C. and Banerjee, S., 2000. *Biodiversity Conservation in Managed Forests and Protected Areas*. Agribios (India) Publications, Jodhpur, pp. 217–224.

Malhotra, K.C., Gokhale, Y., Chatterjee, S. and Srivastava, S., 1999. *Sacred Groves in India: An Overview*. National Museum of Mankind. Bhopal.

McNeely, J.A., Miller, K.R., Reid, W.R. and Werner, T.B., 1990. *Conserving the World's Biological Diversity*. IUCN, Gland, Switzerland.

Pearce, D.W. and Moran, D., 1994. *The Economic Value of Biodiversity*. Earthscan, London, p. 24.

Pushpangadan, P., Rajendra Prasad, M. and Krishnan, P.N., 1998. Sacred groves of Kerala synthesis on the state of the art of knowledge. In: *Conserving the Sacred for Biodiversity Management*, (Eds.) Ramakrishnan, P.S., Saxena, K.G. and U.M. Chandrashekhara. Oxford and IBH Publishing Co. Pvt. Ltd., New Delhi, pp. 193–209.

Ramakrishnan, P.S., 2001. Increasing population and declining biological resources in the context of global change and globalization. *Journal of Bioscience*, 26(4):465–479.

Ramakrishnan, P.S., 1998. Conserving the sacred for biodiversity: the conceptual framework. In: *Conserving the Sacred for Biodiversity Management*, (Eds.) Ramakrishnan, P.S., Saxena, K.G. and U.M. Chandrashekhara. Oxford and IBH Publishing Co. Pvt. Ltd., New Delhi, pp. 3–15.

Ramesh, B.R. and Pascal, J.P., 1997. *Atlas of Endemics of the Western Ghats (India)*, Institute Francais, Pondicherry.

Rao, K.N. and Nayudu, M.V., 1979. *Plant Wealth of Seven Hills*. T.T.D. Publications, Tirupati.

Sen, S., 1989. Urbanisation and erosion in traditional values and practices in the tribal setting of North-east India. In: *Environment Conservation and Wasteland Development in Meghalaya*, (Eds.) Gupta, A. and D.C. Dhar. MSS, Shillong, pp. 179–188.

Sherring, M.A., 1974. *Hindu Tribes and Castes, Vol. III*. Cosmo Publications, New Delhi, India, p. 140.

Sinha, B. and Maikhuri, R.K., 1998. *Conservation Through Socio-Cultural-Religious Practices in Garhwal Himalaya: A Case Study of Haryali Sacred Site*. APH Publication Corporation, New Delhi, pp. 289–299.

Tripathi, R.S., Tiwari, B.K. and Barik, S.K., 1995. *Sacred Groves of Meghalaya: Status and Strategy for their Conservation*. NAEB, NEHU, Shillong, pp. 112–125.

LIFE ON EDGE
Climate Change, Critical Environmental Problems, Alarming Trends, Species Extinction and the Likes

Ecosystem Diversity and Carbon Sequestration: Climate Change Challenges and a Way Out for Ushering in a Sustainable Future *Pages 105–112*
Editors: P.L. Gautam, Vir Singh and Uma Melkania
Published by: DAYA PUBLISHING HOUSE, 2009

Chapter 11

Climate Change and its Effects on Global Biodiversity: Evidences of Alarming Trends and Species Extinction in Different Eco-Regions of the World

Ragupathy Kannan

There is very little doubt today among the scientific community that anthropogenic, or human-induced, greenhouse gases pollution has contributed significantly to the global warming (climate change) experienced since the beginning of the industrial era (Houghton *et al.*, 2001; Santer *et al.*, 2003). Over the past 100 years, the earth's surface has warmed by approximately 0.6°C (Walther *et al.*, 2002). Human activities such as the large-scale burning of fossil fuels to operate power plants and automobiles are releasing greenhouse gases like CO_2 into the atmosphere at an unprecedented rate. We are currently releasing 70 million tonnes of CO_2 per day into the atmosphere. Greenhouse gases are named so because they trap heat and impede its radiation back into the atmosphere, much like a greenhouse does.

Scientists warn that if the current rate of greenhouse gases emissions continues, global air temperatures could rise between 1.5 to 4.5°C by the year 2100 (Houghton *et al.*, 2001). The direct and more tangible consequences for human civilizations of such a temperature increase range from melting of polar ice caps and the subsequent rising of sea levels and flooding of coastal areas, to increase in transmission of tropical diseases, to large-scale disruptions in global climatic patterns resulting in unusual droughts and flooding world-wide. Carbon dioxide levels (which have been clearly tied to temperature increases) were under 300 ppm (parts per million by volume) for the past 160,000 years[4,27]. The levels touched 300 ppm in 1900s and have been steadily increasing since then. They are currently at an all-time high of 360 ppm and are expected to reach at least 550 ppm by the year 2100 if current rates of emissions continue (Revkin, 1992; An Inconvenient Truth, DVD, 2006). The worst case scenario predicts year 2100 levels as unimaginably high as 1,000 ppm. Clearly, we are headed for a CO_2-rich world with all its deleterious consequences, and are faced with a morally obligatory task of mitigating these levels by controlling greenhouse gas emissions.

This lecture focuses on the impact of such climate change on global biodiversity and it draws from the wide body of scientific evidence that has accumulated over the past few years (current through 2007). It is intended for students and the lay public to convince them that global warming is a real and growing problem that threatens the existence of human and other life forms. My hope is that the lecture would increase awareness of this critical issue, will convince skeptics of the reality of the problem, and will stimulate people to start thinking of ways they can reduce their own carbon footprint on the planet.

My strategy in this lecture is to take the audience on a tour of the planet and highlight what global warming has done to a variety of life forms. I will focus in-depth on some examples and briefly mention many others. I will first highlight some case studies at the individual and population levels. I will then present some studies that have done some meta-analyses of data to discern a globally coherent fingerprint of global warming across a suite of taxa.

Amphibians Raise the Alarm: Global Warming and Evidence of Extinctions from Costa Rica

In the mountain forests of Central America, strikingly colored Harlequin frogs (*Atelopus* sp.) were the first charismatic conspicuous vertebrates to succumb to global warming. Sixty-seven per cent of 110 endemic species have become extinct in just two decades (Blaustein and Dobson, 2006; Pounds *et al.*, 2006). A pathogenic Chytrid fungus which produces aquatic flagellated zoospores was first implicated in this large-scale demise of the frogs. The fungus infects the skin and causes widespread diseases among the amphibians. Chytrids were first discovered as pathogens in distant Africa in the 1930s, among African-clawed frogs (*Xenopus*). Unfortunately, global trade of *Xenopus* peaked in the 1950s because of the use of *Xenopus* tissue in pregnancy tests, and that is how the fungal pathogen is thought to have colonized the New World (Blaustein and Dobson, 2006).

But why did it take nearly three decades for the Chytrids to seriously affect Central American frog populations? Prior to the 1980s, the Chytrids did not make serious inroads into frog populations. J. A. Pounds and his colleagues (Pounds *et al.*, 2006) made the connection between global warming over the past 20 years and the fungal pathogen in decimating the amphibians. They demonstrated that human-caused climate change induced by global warming has altered transmission rates of the pathogen and thus has lead to widespread population declines and local extinctions.

Using long-term meteorological and climate data, Pounds *et al.*, showed that global warming has increased daytime cloudiness in the mid elevations and thus has resulted in decreased daytime temperatures. At the same time, the increased cloud cover reduced heat loss at nights and thus resulted in night-time warming. This reduction in range of temperatures, they showed, has favored Chytrids because the range fell right within the optimum range for the growth, reproduction and transmission of the fungal pathogens.

Owing to the fact that global warming made it conducive for Chytrids in mid elevations to thrive, mid elevations witnessed the maximum number of species extinctions. Unfortunately, the greatest diversity of frogs existed in mid elevations and this resulted in a double jeopardy for the frogs: the region with the highest species richness suffered the most, and hence the plethora of extinctions. This contradicted previously held assumptions that high elevations would suffer the most under projected global warming scenarios.

Pounds *et al.*, showed that the optimal range for the pathogen is between 17–25°C; at around 30°C, the Chytrids die. Harlequin frogs in the lowlands (0-1,000m) did not suffer as much extinctions because daytime (maximum) temperatures were too hot (>25°C) for Chytrids; similarly, most higher elevation (>2400m) frog species escaped extermination because minimum temperatures were below the optimal range (<17°C) for the fungi.

The case of the Central American frogs reinforced long-held fears that global warming can increase the transmission rates of pathogen driven diseases. Other problems forecasted with increased global warming are the pole-ward or altitudinal shift in the ranges of many species, and the advancement of spring phenological events in life histories of various organisms. Such evidence poured in from various other corners of the world, some of which are mentioned below.

Evidence from the United Kingdom

With earlier onset of warm temperatures in spring, many organisms advanced spring life-history activities significantly earlier than before. Eighteen butterfly species advanced their spring arrival dates by 2.8-3.2 days per decade over the past 23 years[3,7]. Amphibians did the same, but not all species responded the same. Predatory newts entered ponds earlier than before, but some traditionally early breeding frogs did not, and hence they suffered increased predation to newts[3,8]. Such differential responses to global warming are expected to alter community and trophic dynamics in various other food webs.

Several species of birds arrived earlier than they did in pre-warming times. Spring first-time arrivals advanced an average of 1.3 to 4.4 days per decade; and subsequent breeding activities too hastened by an average of 1.9 to 4.8 days per decade over a time frame of 30-60 years (Walther *et al.*, 2002). Over a span of 25 years, 20 species of birds were recorded to lay eggs significantly earlier than they did before (Crick and Sparks, 1999; Crick *et al.*, 1997). In the "business as usual" scenario of global warming gas emissions (*i.e.*, if current rates continue unmitigated) it is predicted that the average egg laying dates will be even earlier for 75 per cent of the species by the year 2080

(Crick and Sparks, 1999). The consequences of such early breeding can be varied. If not synchronized with prey abundance, such early breeding could result in decreased lifetime reproductive success and recruitment rates (see evidence from Europe, below).

Northward shifts in bird species ranges occurred as predicted. Twelve species of UK birds have shifted their ranges by an average of 18.9 km over the past two decades (Walther *et al.*, 2002; Thomas and Lennon, 1999).

Evidence from Europe

Late arrival in the spring of migratory birds has resulted in increased competition for optimal nest sites with species that arrived earlier (Both and Visser, 2001). In Great Tits (a bird; *Parus major*) it was shown that global warming has jeopardized the synchrony between prey availability and a critical phase of breeding phenology. Maturity dates of caterpillars (a chief prey for tit hatchlings) occurred much earlier than the hatching of tit chicks, and this has lead to food deprivation for the chicks and has affected reproductive success and population recruitment (Walther *et al.*, 2002; Visser and Holleman, 2001; Visser *et al.*, 1998).

Several butterfly species have shown marked northward shifts in ranges as a direct consequence of warming. For thirty-eight species (in Europe and North America), the shift in range has been up to 200 km over a span of 27 years (Walther *et al.*, 2002; Parmesan *et al.*, 1999).

The evidence has mounted from the plant kingdom as well. Flowering and leafing among several European plant species has occurred earlier in the year by an average of 1.4-3.1 days per decade, in a study covering a time period of 30-48 years (Walther *et al.*, 2002; Menzel and Estrella, 2001). Treelines on mountain slopes have moved up significantly (Kullman, 2001; Meshinev *et al.*, 2000). Alpine plants too have displayed such a trend. Some species have shifted upwards in elevational distribution by an average of 1-4 m per decade (Walther *et al.*, 2002; Grabherr *et al.*, 1994). Even lichens (in Netherlands) have shown long term responses to global warming (Parmesan and Yohe, 2003; van Hark *et al.*, 2002).

Global warming can affect community composition by encouraging invasion by exotic species. A remarkable study from Switzerland has documented how the drastic reduction in the number of frost days has lead to a consequent increase in the number of exotic evergreen broad-leaved plants in a landscape that was originally predominantly deciduous (Walther *et al.*, 2002). In the marine environment, warm water species of plankton have increased in areas that previously had few of them (Nehring, 1996).

Evidence from North and Central America

In Canada, global warming trends have resulted in a northward expansion of the range of the red fox, and a subsequent retreat of the range of the Arctic fox (Walther *et al.*, 2002; Hoffman and Parsons, 1997).

The Edith's Checkerspot Butterfly of western United States has also shown such an expansion. The species has been reported to have experienced a 124 m upward elevational and a 92 km northward shift since the beginning of the 20th century (Parmesan *et al.*, 1999; Parmesan, 1996).

Such range expansions may not be possible for species (like the Florida Panther and several sedentary species in that state's Everglades National Park) whose habitats are hemmed in by development (Revkin, 1992).

One of the more pronounced evidences in the North American continent and elsewhere is the increased abundances of warm water species like zooplankton, jellyfish and other cnidaria, and some reef fishes in areas where they were relatively rare [for *e.g.*, the Monterrey coast off California (Parmesan and Yohe, 2003; Sagarin *et al.*, 1999)]. Jellyfish in particular have become a nuisance with their burgeoning numbers[27], as they tend to get sucked up power plant cooling systems and have proven to be a menace for beach bathers in many parts of the world.

Shifts in ranges have been reported from the neotropics too. Lowland birds in Costa Rica have extended their altitudinal distributions from mountain slopes to higher elevations, largely attributable to anomalies in dry season mist frequencies induced by global warming (Pounds *et al.*, 1999).

Evidence from Antarctica

With the melting of large areas of the Antarctic ice shelf, more liquid and warmer water is available for colonization, and therefore several Antarctic plants and invertebrates have shown striking distributional changes (Smith, 2001). Macroscopic plants like mosses have especially been influenced by the change in climate.

But perhaps the Antarctic phenomenon that had not only regional but also global effects is the drastic reduction in populations of Krill (*Euphausia superba*) in the seas near the continent[26]. Reduction in sea ice area near the Antarctic peninsula has affected krill recruitment rates. With antarctic krill being a prime staple diet for numerous oceanic species (including whales) this reduction in krill numbers has had profound effects on the regional food web and subsequently human economics.

More Evidence from the Oceans

One of global warming's most destructive effects is seen today in the coral reefs of the world's oceans. Coral polyps face retarded growth and reproductive rates in an increasingly warm world. But in addition to that, elevated CO_2 levels increase acidity of the oceans by dumping more carbonic acid into the waters [*An Inconvenient Truth* (DVD), 2006]. Of the 70 million tonnes of CO_2 released per day into the atmosphere, 25 million tonnes are soaked up by the oceans and converted into carbonic acid. This increasing acidification has made it difficult for coral polyps to capture calcium from the waters. These polyps and their dinoflagellate symbionts need the calcium to make calcium carbonate, a chief ingredient in shells of marine invertebrate organisms. Coral reefs therefore face bleaching (whitening and death as a result of death of the algal symbionts and the polyps) at large-scale levels.

A recent report revealed the gravity of this problem. Areas with optimal conditions for calcium uptake has shrunk precipitously in the world's oceans in the past two decades, and it is estimated that, with all the acidification, such favorable areas can virtually vanish by the year 2050, leading to mass extinctions of not only coral polyp species but the entire coral reef ecosystem with it phenomenally high biodiversity.

Coral reefs are especially sensitive to even miniscule amounts of warming. An increase in long-term summer averages in the tropics by 1.0ÚC for several weeks in enough to trigger mass bleaching (Walther *et al.*, 2002). There have been six such mass bleaching events since 1979, and the frequency and intensity of mass bleaching has increased as a direct result of global warming. The worst was in 1998 when an estimated 16 per cent of the world's reef building corals died[27], mostly attributed to the loss of dinoflagellate symbionts.

Global warming could directly affect human fishery industry. Variations in atmospheric air circulation over the Bering Sea impede the movement of juveniles away from adults of a commercially and ecologically vital species of fish, the Walleyed Pollock (*Theragra chalcogramma*) (Cushing, 1995). This makes the juveniles increasingly vulnerable to cannibalism. Considering the fact that Pollock is an important forage fish for a wide variety of mammals (including humans) and birds, negative consequences to the entire food web may be inevitable.

It's not only Pollock that face the scourge the global warming. Similar climate-induced changes in prey abundance has affected the migration and spatial distribution of other pelagic fish like Bluefin Tuna (*Thunnus thynnus*) (Polovina, 1996), which is one of the world's most sought-after table fish species for humans.

Global warming increases the frequency and intensity of oceanic upwelling, a phenomenon in which nutrients from the bottom are periodically brought up to surface waters (Walther *et al.*, 2002; McGowen *et al.*, 1998). Such natural upwelling is excellent for many wildlife species in the marine ecosystems (like the coasts of Peru, which is known for its abundant marine life). But the increased upwelling is feared to precipitate a global fish production decline "due to reduction in the concentration and retention processes".

Global warming along with human overexploitation can exacerbate the situation of depleted fish stocks. Another species of commercially and ecologically important fish, the North Sea Cod (*Gadus mohrua*), faces such a "double jeopardy" situation. The cod's long adult life span usually buffers it against times of poor recruitment, but over harvesting of adult fish by humans has increased the species' vulnerability to population crashes in face of climate change (Walther *et al.*, 2002; Plancue and Fredou, 1999).

Acknowledgements

I am grateful to Prof. Douglas A. James of the University of Arkansas for help in procuring key reference papers and to Amrit Kannan for his invaluable help in the reference collating process. Dr. P. L. Gautam, the Honourable Vice Chancellor of G.B. Pant University invited me to make this lecture and thoughtfully suggested this timely topic. Dr. Uma Melkania, Chair of Environmental Science Department invited me for the 'Fulbright Visiting Lecturership' and provided all the other facilities to make this happen. I am thankful to the United States government and my fellow Americans for awarding me the prestigious Fulbright Scholarship that enabled me teach in India.

References

An Inconvenient Truth (DVD), 2006. Paramount pictures, U.S.A.

Beebe, T.J.C., 1995. Amphibian breeding and climate change. *Nature*, 374: 219–220.

Blaustein, A. and Dobson, A., 2006. A message from the frogs. *Nature*, 439: 143–144.

Both, C. and Visser, M.E., 2001. Adjustment to climate change is constrained by arrival date in a long distance migrant bird. *Nature*, 411: 296–298.

Crick, H.Q.P. and Sparks, T.H., 1999. Climate change related to egg laying trends. *Nature*, 399: 423–424.

Crick, H.Q.P. *et al.*, 1997. UK birds are laying eggs earlier. *Nature*, 388: 526.

Cushing, D.H., 1995. *Population Production and Regulation in the Sea: A Fisheries Perspective.* Cambridge Univ. Press, Cambridge.

Grabherr, G. *et al.*, 1994. Climate effects on mountain plants. *Nature*, 369: 448.

Hoffman, A.A. and Parsons, P.A., 1997. *Extreme Environmental Change and Evolution.* Cambridge Univ. Press, Cambridge.

Houghton, J.T. *et al.* (eds.), 2001. *Climate Change 2001. The Scientific Basis.* Third Assessment Report of the Intergovernmental Panel on Climate Change (Cambridge Univ. Press, Cambridge.

Kullman, L., 2001. Twentieth century climate warming and tree limit rise in the southern Scandes of Sweden. *Ambio*, 30: 72–80.

Loeb, V. *et al.*, 1997. Effects of sea ice extent and krill or salp dominance on the Antarctic food web. *Nature*, 387: 897–900.

McGowen, J.A. *et al.*, 1998. Climate ocean variability and ecosystem response in the northeast pacific. *Science*, 281: 210–217.

Menzel, A. and Estrella, N., 2001. In: *Fingerprints of Climate Change: Adapted Behaviour and Shifting Species Ranges*, (Eds.) G.R. Walther, C.A. Burga and P.J. Edwards. Kluwer Academic, New York, pp. 123–137.

Meshinev, T. *et al.*, 2000. Influence of warming on timber line rising: A case study on *Pinus peuce* in Bulgaria. *Phytocoenologia*, 30: 431–438.

Nehring, S., 1996. Establishment of thermophilic phytoplankton species in the North Sea: Biological indicators of climate changes? *ICES J. Mar. Sci.*, 55: 818–823.

Parmesan, C. and Yohe, G., 2003. A globally coherent fingerprint of climate change impacts across natural systems. *Nature*, 421: 37–42.

Parmesan, C. *et al.*, 1999. Poleward shifts in geographical ranges of butterfly species associated with regional warming. *Nature*, 399: 579–583.

Parmesan, C. *et al.*, 1999. Poleward shifts in geographical ranges of butterfly species associated with regional warming. *Nature*, 399: 579–583.

Parmesan, C. In: *Evolution and Ecology Taking Flight: Butterflies as Model Systems*, (Ed.) Boggs, C.L., W.B. Watt and P.R. Ehrlich. University of Chicago Press, Chicago (In Press).

Parmesan, C., 1996. Climate and species' range. *Nature*, 382: 765–766.

Plancue, B. and Fredou, T., 1999. Temperature and the recruitment of Atlantic Cod (*Gadus mohrua*). *Can. J. Fish. Aquat. Sci.*, 56: 2069–2077.

Polovina, J.J., 1996. Decadal variation in the trans-Pacific migration of northern Bluefin Tuna (*Thunnus thynnus*) with climate induced change in prey abundance. *Fish Oceanogr.*, 5: 114–119.

Pounds, J.A. *et al.*, 2006. Widespread amphibian extinctions from epidemic disease driven by global warming. *Nature*, 439: 161–167.

Pounds, J.A., *et al.*, 1999. Biological response to climate change on tropical mountain. *Nature*, 398: 611–615.

Revkin, A., 1992. *Global Warming: Understanding the Forecast*. Abbeville Publ., New York.

Roy, D.B. and Sparks, T.H., 2000. Phenology and British butterflies and climate change. *Global Change Biology*, 6: 407–416.

Sagarin, R. *et al.*, 1999. Climate-related change in an intertidal community over short and long time scales. *Ecol. Monogr.*, 69: 465–490.

Santer, B.D. *et al.*, 2003. Contributions of anthropogenic and natural forcing to recent tropopause height changes. *Science*, 301: 479–483.

Smith, R.J.L., 2001. Plant colonisation response to climate change in the Antarctic. Masarykianae Brunensis. *Geographia*, 25: 19–33.

Thomas, C.D. and Lennon, J.J., 1999. Birds extend their ranges northwards. *Nature*, 399: 213.

van Hark, C.M. *et al.*, 2002. Long-term monitoring in the Netherlands suggests that lichens respond to global warming. *Lichenologist*, 34: 141–145.

Visser, M.E. and Holleman, L.J.M., 2001. Warmer springs disrupt the synchrony of oak and winter moth phenology. *Proc. R. Soc. Lond. B.*, 268: 289–294.

Visser, M.E. *et al.*, 1998. Warmer springs lead to mistimed reproduction in Great Tits (*Parus major*). *Proc. R. Soc. Lond. B.*, 265: 1867–1870.

Walther, G., Post, E., Convey, P., Menzel, A., Parmesan, C., Beebe, T.J.C., Fromentin, J., Hoegh-Guldberg, O. and Bairlen, F., 2002. Ecological responses to recent climate change. *Nature*, 416: 389–395.

Walther, G.R., 2000. Climate forcing on the dispersal of exotic species. *Phytocoenologia*, 30: 409–430.

Ecosystem Diversity and Carbon Sequestration: Climate Change Challenges and a Way Out for Ushering in a Sustainable Future *Pages 113–119*
Editors: P.L. Gautam, Vir Singh and Uma Melkania
Published by: DAYA PUBLISHING HOUSE, 2009

Chapter 12

Climate Change and its Effects on Global Biodiversity: Triggering Effects and Frightening Prospects

B.S. Mahapatra, A.P. Singh, A.K. Chaubey and D.K. Shukla

In large part of Asia, agricultural production is mainly dependent on the vagaries of monsoon. Studies indicate that there are large-scale climatic variations at micro-regional level in different parts of Asia. The causes of these regional climate changes vary from global to regional level. It is evident that there was, there is and there will be climate variability at global, regional and local levels. Over the last 200 years, human activities have resulted in emissions of greenhouse gases (primarily CO_2), which have altered the composition of the atmosphere and caused an 'enhanced' greenhouse effect. As a result, earth's temperature is rising which, in turn, is changing the climate patterns. Studies indicate that if no corrective measures are taken, the atmospheric temperatures may increase by 1.4 to 5.8°C by the year 2100 (IPCC, 2001). This will have serious impacts on day-to-day life. Likely impacts of climate change on agricultural productivity in India is causing a great concern to the scientists and planners as it can hinder their attempts for achieving household food security. Despite the limitations, sequestration of carbon in soil organic matter has a role to play as part of a raft of measures to mitigate climate change. In addition to carbon mitigation, the measures leading to soil carbon accumulation generally have a positive influence on soil quality and have wider environmental benefit.

Anthropogenic interventions in global climatic systems through increased concentration of greenhouse gases in the atmosphere has led to serious concerns to the mankind because of their induced impacts which include climate change and climate variability. The gradual build up of the greenhouse gases in the atmosphere during the last century, resulting from increased human activity had already brought about many visible environmental impacts across the world. These activities not only enhanced the processes of climate change, increasing the mean global temperatures by 0.6°C during the past 100 years but have also induced increased climatic variability in many parts of the world. Studies indicate that the last decade (1990-2000) across the globe have recorded the warmest years during the past century, the three years, *viz.*, 1997, 1998 and 1999 recording more warmer conditions, increasing in sequence.

In large part of Asia, agricultural production is mainly dependent on the vagaries of monsoon. Studies indicate that there are large-scale climatic variations at micro-regional level in different parts of Asia. The causes of these regional climate changes vary from global to region level. It is evident that there was, there is and there will be climate variability at global, regional and local levels. Since climate is closely related to human activities and economic development including agricultural production systems, there is a serious concern about its stability (Sinha *et al.*, 2000).

The influence of climatic variability is much more in the fragile ecosystems like the rainfed agro-ecosystem where the balance between the availability and demand exists at a very low level. Any deviation persisting over a longer duration is bound to trigger the process of degradation in soil health and productivity. While there are uncertainties on the nature and extent of the impending changes, it is certain that climate induced variability and change will affect the agricultural production and food security.

Global Climate Change Scenario

Global climate change is one of the greatest threats to the mankind. The natural greenhouse effect keeps the planet 33°C warmer than it would otherwise be, allowing the earth to sustain life. Over the last 200 years, human activities have resulted in emissions of greenhouse gases (primarily CO_2), which have altered the composition of the atmosphere and caused an 'enhanced' greenhouse effect. As a result, earth's temperature is rising which, in turn, is changing the climate patterns. The composition and changes in concentration of greenhouse gases in the atmosphere is given in Table 12.1.

Table 12.1: Composition and Changes in Concentration of Greenhouse Gases in the Atmosphere

Gas	*Concentration 1985*		*Annual Increase Since 1986 to Present (%)*	*Contribution to Global Warming (%)*
CO_2	345 ppm	(1)	0.5	50
CH_4	90 ppb	(32)	0.8	19
N_2O	1.65 ppm	(150)	1.0	5
CFC	0.24 ppb		3.0	15
Others	–		–	11

Source: Bouwman (1990); USEPA (1990) Values in the parenthesis indicate relative impact.

Studies indicate that if no corrective measures are taken, the atmospheric temperatures may increase by 1.4 to 5.8° C by the year 2100 (IPCC, 2001). This will have serious impacts on day-to-day life. Some of the projections made are as follows:

- Inundation of coastal areas and seashore recession due to rise in sea level
- Disruption of ecosystem complexes
- Shifts in rainfall pattern and length of growing season
- Shifts in cropping pattern, geographical crop preferences
- Higher ET rates

Climate Change and Agriculture

Several predictions have been made on the possible changes in climate due to increased concentrations of GHG and their impacts on agriculture. The general circulation models (GCM) indicate that doubling of CO_2 concentration by 2030 will increase average global temperature by 1-3°C. Govinda Rao *et al.* (1994) indicated a decrease in cereal production and increase in tropical cyclonic activity, posing serious flood damages along their tracks. Similarly, increase in rainfall by 5 to 15 per cent in South Asian region was predicted by Houghton *et al.* (1990) due to enhancement in GHG concentration.

The effect of temperature rise will lead to an increase in biological activity as well as the physical and chemical processes. Increase in CO_2 concentration can lower pH, thereby, directly affecting both nutrient availability and microbial activity. Climate change is also expected to bring about the changes in the amount, intensity and distribution pattern of rainfall.

The changes in the crop yield depend not only on the change in rainfall but also on the changes in CO_2 concentration. While positive changes in the above two parameters can lead to increase in crop yield, the impact of temperature is generally negative in the tropics. Thus it is necessary to ascertain whether the impact of increased rainfall and CO_2 will overwhelm the impact of increased temperature and hence, evapotranspiration. However, it is not clear that over the Indian region whether temperature and rainfall will simultaneously increase (Gadgil 1996). Rosenzweig and Parry (1994) estimated that the net effect of climate change may lead to reduce global cereal production up to 5 per cent. They further indicated that production in the developed world would increase while it may decline in developing countries. Doubling of CO_2 concentration may increase the photosynthetic rates by as much as 30-100 per cent. C_3 plants such as wheat, rice, soybean etc. will become more water efficient as they quickly grow. However, the response in C_4 plants such as maize, sorghum, sugarcane, millets, etc. may not be spectacular (IUCC 1992). The effect of increased CO_2 and temperature on Indian agriculture was carried by Aggarwal (1993) and predicted a shift in wheat production belt.

The average temperatures are expected to increase more near the poles than at equator, the shift in climatic zones will be more pronounced in the higher latitudes. In mid-latitudes, the shift is expected to be 200-300 km for every increase of 1°C (IUCC, 1992). These projections indicate that increased temperature resulting from global warming is likely to reduce the profit from wheat cultivation and will compel farmers of lower latitudes to opt for maize and sorghum which are better adapted to higher temperature. Morey and Sadaphal (1981) reported a decrease of wheat yield by 400 kg ha^{-1} for a unit increase of 1°C maximum temperature and 0.5 hr sunshine. The effect of temperature on the phenology and yield of important crops grown in Punjab has been worked out (Hundal and Kaur 1996) using crop weather simulation models. In similar studies, Ramakrishnan *et al.* (1996) have estimated a fall in productivity of pearl millet in Rajasthan by 10-15 per cent due to increase in temperature by 2°C.

Studies of Lal *et al.* (1998) using wheat and rice models showed higher yields of both crops under elevated CO_2 levels (28 and 15 per cent, respectively, for a doubling of CO_2). A 3°C (2°C) rise in air temperature nearly cancels out the positive effect of elevated CO_2 on the wheat (rice) yields. They concluded that acute water shortage conditions combined with the thermal stress would adversely affect both the wheat and more severely the rice productivity in NW India even under the positive effects of elevated CO_2 levels in the future.

Likely impacts of climate change on agricultural productivity in India is causing a great concern to the scientists and planners as it can hinder their attempts for achieving household food security. Food grain requirements in the country (both human and cattle) would reach about 300 m tonnes in 2020. The question that is of great concern is that, with the alarming increase in GHG concentration and its expected impact on climate, will it be possible to achieve the targeted production? Sinha *et al.* (1998a) observed that during the past 25 years, significant changes in climate are observed over different regions of the country. For example, many parts of northern India show increase in minimum temperature by about 1°C in Rabi cropping season. However, mean temperatures are misleading as some of the individual regions could exhibit a larger variation with a larger impact on Rabi crop production. For example, Sinha and Swaminathan (1991) presented a case study of actual change in temperature in North India. They brought out that while the mean air temperatures over the wheat growing regions were high by 1.7°C over a period of 15 days (January 16 to February 1), the actual temperature rise was 2.3 to 4.5°C in the major wheat-producing region of Punjab and Haryana (Figure 12.1). Through these studies they projected the serious effects of regional temperature on productivity of major crops. Sinha *et al.* (1998b) also observed a 10 per cent decline in solar radiation in NW India. Since solar radiation influences the photosynthesis and productivity, such a decrease will adversely affect the productivity and to an extent unnoticeably, nullifies the positive effects due to improved management practices. They further added that in view of the proportionate production changes in major food crops, *viz.*, rice and wheat, over the years, the dependence on rice and wheat has increased considerably. Therefore, any factor that would influence the productivity through climatic change would affect the food security of the Nation, as both these crops are sensitive to temperature variations.

Conclusions

There is a need to analyse the available evidences of climate change and variability more critically to identify the specific regions, where such changes are causing

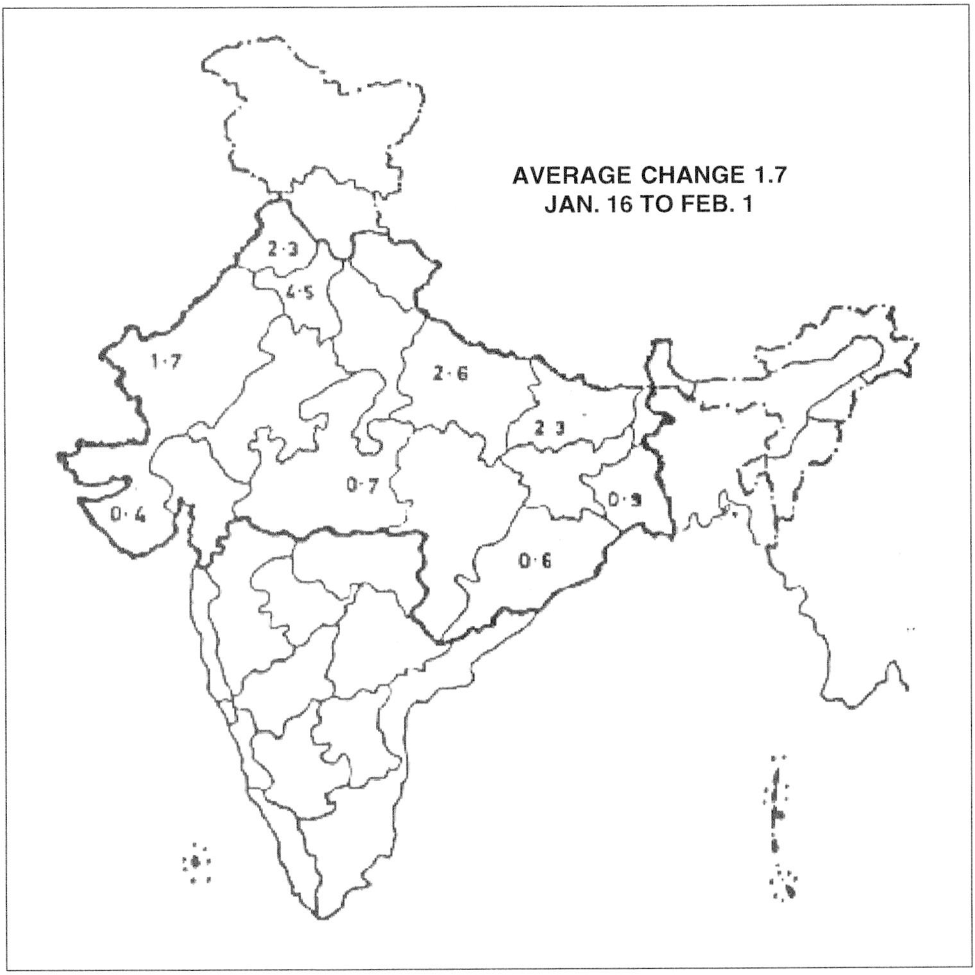

Figure 12.1: Change in Temperature in Regions of India for an Average Increase of 1.7°C from January 16 to February 1 in Regions Delineated by Thick Lines (*Source*: Sinha and Swaminathan, 1996)

hindrance to enhance agricultural productivity and posing a threat to food security. Results from these can help in planning ameliorative measures that can be adopted to attain sustainability in agricultural production.

Sequestering carbon in soils or vegetation is a less favourable option than cutting emissions of CO_2 from fossil fuels–that is, it is better to prevent the problem than to apply a cure that is rather inefficient due to its finite size and possible reversibility. Substituting a part of current and future fossil fuel use with biofuels is a better option as the benefit continues indefinitely. Converting some land currently used for agriculture to perennial biofuel crops gives a double benefit: in addition to the ongoing replacement of fossil carbon emission, additional carbon will be sequestered in the soil.

Despite the limitations, sequestration of carbon in soil organic matter has a role to play as part of a raft of measures to mitigate climate change. In addition to carbon mitigation, the measures leading to soil carbon accumulation generally have a positive influence on soil quality and have wider environmental benefits.

References

Aggarwal, P.K., 1993. Agro-ecological zoning using crop growth simulation models: Characterization of wheat environments of India. In: *Systems Approaches for Agricultural Development*, (Eds.) F.P. De Vries, Paul Teng, and Klaas Metselaar. Kluwer Academic Publishers, London, pp. 97–109.

Gadgil, Sulochana, 1996. Climatic change and agriculture: An Indian Perspective. In: *Climate Variability and Agriculture*, (Eds.) Y.P. Abrol, Sulochana Gadgil and G.B. Pant. Narosa Publishing House, New Delhi.

Govind Rao, P., Kelly, M., Hulme, M and Srinivas, G., 1994. *Centre for Social and Economic Research on Global Environment*, University of East Anglia, UK., Working Paper, GEC, pp. 94–22.

Houghton, J.T., Jenkins, G.J. and Ephraums, J.J., 1990. *Climate Change*. The IPCC scientific assessment WMO/UNEP. Cambridge Press, Cambridge, U.K., pp. 365.

Hundal, S.S. and Kaur, Prabhjyot, 1996. Climate change and its impact on crop productive in Punjab. India. In: *Climate Variability and Agriculture*, (Eds.) Y.P. Abrol, Sulochana Gadgil and G.B. Pant. Narosa Publishing House, New Delhi, pp. 377–393.

IUCC, 1992. Information unit on climate change, Fact Sheet No. 101. Impacts of Climate Change. UNEP, Polaisdes Nations. CH 1211, Geneva.

IPCC, 2001. Climate Change 2001: Impacts, adaptation and vulnerability. Inter–Governmental Panel on Climate Change. *Report of the Working Group II*. Cambridge, UK, p. 967.

Morey, D.K. and Sadaphal, M.N., 1981. Effect of weather elements on yield of wheat at Delhi. *Punjabrao Krishi Vidyapeeth Res. Journal* 1:81–83.

Ramakrishnan, Y.S., Victor, U.S. and Ramana Rao, 1996. Moisture regime variability and its impact on agriculture in the Indian Arid Zone. In: *Climate Variability and Agriculture*, (Eds.) Y.P. Abrol, Sulochana Gadgil and G.B. Pant. Narosa Publishing House, New Delhi.

Rosennzweig, C. and Parry, K.L., 1994. Potential impact of climate change on world food supply. *Nature*, 367: 133–138.

Sinha, S.K., Kulshreshtha, S.M., Purohit, A.N. and Singh, A.K., 1998a. *Climate Change and Perspective for Agriculture: Base Paper*. National Academy of Agricultural Sciences, 20 pp.

Sinha, S.K., Singh, G.B. and Rai, M., 1998b. *Decline in Crop Productivity in Haryana and Punjab: Myth or Reality?* Indian Council of Agricultural Research, New Delhi, India, 89pp.

Sinha, S.K. and Swaminathan, M.S., 1991. Deforestation, climate change and sustainable nutrition security: A case study of India. *Climate Change*, 19: 201–209.

Sinha, S.K., Kulshrestha, S.M. and Ramakrishnan, Y.S., 2000. Climatic variability and climate-change impact on agriculture. Invited Plenary Lecture at the Intl. Conference on Managing Natural Resources for Sustainable Agricultural Production in the 21st Century, New Delhi, February 14–18. *Extended Summaries Vol. 1: Invited Papers*, 13–15.

Ecosystem Diversity and Carbon Sequestration: Climate Change Challenges and a Way Out for Ushering in a Sustainable Future *Pages 120–127*
Editors: P.L. Gautam, Vir Singh and Uma Melkania
Published by: DAYA PUBLISHING HOUSE, 2009

Chapter 13

Impact of Climate Change on Crop Productivity: Need of Adjustments in Agriculture

S.K. Saini, Yogendra Pal and Amit Bhatnagar

Climate change will affect crop productivity directly because of increased atmospheric CO_2 concentration and greenhouse gases and indirectly through stratospheric ozone layer depletion. Increased CO_2 level could increase photosynthesis and water use efficiency. However, high temperature and greenhouse gases will modify rainfall, evaporation runoff and soil moisture storage and will adversely affect crop growth and productivity. The increased amount of ultra violet (UV) radiation due to depletion of stratospheric ozone layer will exert its deleterious effect on crop growth and productivity by destruction of chlorophyll and reducing photosynthetic rate. To cope up with change in regional climate due to climatic changes, agriculture may have to adopt some changes. These adjustments will depend on future development in technology, demand, prices and national policy.

Introduction

Since the end of the 19th century earth's average surface temperature has increased by 0.3–0.6°C. Over the last 40 years, the rise has been 0.2–0.3°C. It is predicted that over the next 100 years, the earth's average temperature could further rise by 1.4–5.8°C. Climate change will affect crop productivity directly because of alterations in temperature and rainfall and indirectly through changes in soil quality, pests and diseases. Extreme weather conditions such as high temperature, heavy rainfall, floods, draughts etc. will also affect crop production.

The effect of climate change on crop production may be classified into two broad groups.

A. Direct Effect

1. Increased atmospheric CO_2 concentration can have direct effect on growth rate of crop plants and weeds.
2. Green house gas induced changes of climate may alter level of temperature and moisture limit to agriculture.

B. Indirect Effect

Stratospheric ozone layer depletion and increase in UV-B radiation.

Effect of CO_2 Enrichment

It is estimated that even if man-made emissions of CO_2 could be kept at present rate, atmospheric CO_2 would increase (from present level to approximately 382 ppm) to about 450 ppm by the year 2060, and about 520 ppm by the year 2100 (IPCC 1991). Because CO_2 is the substance for photosynthesis, it is self-evident that this will have major impacts on plants.

Effect on Plant Growth

(*a*) Photosynthesis: Plant species with C_3 photosynthetic pathway (*e.g.* wheat, rice, soybean, etc.) tend to respond more positively to increase CO_2 because it suppresses the rate of photorespiration. C_4 plants (*e.g.* maize, sorghum, sugarcane and millets) are less responsive to increased CO_2 level since they are largely CO_2 saturated at current ambient levels and also have negligible photorespiration.

Clifford *et al.* (1993) observed that in groundnut and sorghum yield per plant increased due to CO_2 enrichment but percentage yield increase at elevated CO_2 was higher in groundnut (C_3 plant) as compared to sorghum (C_4 plant) (Table 13.1).

(*b*) Increased leaf area: This allows greater interception of incident radiation and greater whole plant photosynthesis.

(*c*) Increased number of tillers and branches.

(*d*) Reduced transpiration due to decline in stomata opening.

(*e*) Increased N fixation: The CO_2 concentration is an important controller of the symbiotic relationships between plants and nitrogen fixing organisms.

(*f*) More/longer plant root.

Table 13.1: Yield of Groundnut and Sorghum Grown in Controlled Environment (Glass Houses) at Ambient and Enriched Carbon Dioxide Levels

Crop	*Pod/Seed Yield (g/plant)*		*% Yield Increase at Elevated CO_2*
	350μ mol mol^{-1}	*700μ mol mol^{-1}*	
Groundnut (C_3)	9.06	13.29	30.9
Sorghum (C_4)	25.78	27.37	6.2

Source: Clifford *et al.* (1993)

Effect on Water Use Efficiency (WUE)

Increased CO_2 has an effect on the closure of stomata and thus reduces water loss through transpiration and thereby give in improved WUE.

Effect on Salt Tolerance

Salt tolerance increases as CO_{-2} concentration goes up. There are two possible explanations: (*i*) extra supply of photosynthates may help to offset increased respiration demands, and (*ii*) less water throughout in the transpiration stream could lessen the quantity of salt taken up.

Protection from Air Pollutants

The narrowing of stomata by increased levels of CO_2 immediately infers the possibility of protection from air pollutants that enter leaves by this route.

Effects on Crop-Weed Competition

C_3 crops in temperate and subtropical regions could also benefit from reduced weed infestation. Fourteen of world's seventeen most troublesome terrestrial weed species are C_4 plants in C_3 crops. The difference in response to increased CO_2 may make such weed less competitive. In contrast, C_3 weeds in C_4 crops, particularly in tropical regions could become more of a problem, although the final outcome will depend on the relative response of crops and weeds to climate change as well.

Effect of Crop-Insect Interaction

Under increased CO_2 levels, nitrogen content of plants is likely to decrease while C content increase implying reduced protein level. This may reduce nutritional value of plants for insect-pests.

Effect on Crop-Pathogen Interaction

Elevated CO_2 generally promotes plants growth and, as plants structure is modified, the affected plants may sustain a higher level of infection without reduction in yield.

Plants which are more vigorous are able to resist infection from weak pathogens, such as facultative parasites, resulting in lower disease incidence and severity. However, larger plants provide more surface area for infection, and diseases (particularly those caused by obligate parasites) may increase in incidence and severity.

Elevated CO_2 may benefit plant health and productivity by altering the morphology and physiology of plants to the detriment of pathogenic microbes.

Effect of Changes of Climate due to Greenhouse Gases

Effect of High Temperature

In middle and high latitudes global warming will extend the length of the potential growing season, allowing earlier planting of crops in the spring, earlier maturation and harvesting and the possibility of completing two or more cropping cycle during the same season.

Crop producing areas may expand pole ward in countries such as Canada and Russia, although yield in higher latitude will likely be lower due to the less fertile soils that lie there.

Effect on Moisture Availability to Crops

Climate change will modify rainfall, evaporation, run off and soil moisture storage. In general, at mid latitudes, evaporation increase by about 5 per cent for each degree centigrade mean annual temperature rise. Increased evaporation from the soil and accelerated transpiration in the plant themselves will cause moisture stress.

Effect of Drought, Heat Stress and Other Extremes

1. Hurricane intensities will increase with climatic warming.
2. Small increase in mean annual temperature in sub-tropical regions could sufficiently increase heat stress on temperate crops such as wheat so that these are no longer suited to such areas.
3. In temperate mid latitudes, there will be reduction of winter chilling (varnalization). Many temperate crops require a period of low temperature in winter to either initiate or accelerate the flowering process. Low vernalization results in low flower bud initiation and ultimately in reduced yield.
4. As a result of high rate of evapo-transpiration, there would be higher frequency of droughts in the tropics and sub-tropics.

Effect of Distribution of Agriculture Pests and Diseases

Temperature increase may extend the geographic range of some insect pests presently limited by low temperature. Longer growing season will enable insects to complete more number of reproductive cycles during the season.

Effect on Crop Growth and Productivity

In most instances, particularly at low and mid latitudes, the increase in temperature results in reducing the total duration of the crop, by inducing early flowering and shortening the grain-filling period and thereby decreases productivity.

Mavi (1994) simulated the variability in the rice yields in Punjab under different climate scenario (Table 13.2).

Effect on Soil Fertility and Erosion

Higher temperature is likely to speed the natural decomposition of organic matter and to increase the rates of other soil processes that affect fertility. The continual cycling of plant nutrients-carbon, nitrogen, phosphorus, potassium and sulphur in the soil-plant-atmosphere system is also likely to accelerate in warmer conditions, enhancing CO_2 and N_2O greenhouse gases emissions.

Table 13.2: Rice Crop Response (%) to Change in Temperature

Temperature Change	*Crop Duration (Days)*	*Grain Yield (Kg/ha)*	*Maximum LAI*
+2.0°C	-3.3	-8.4	-3.9
+1.5°C	-2.6	-8.2	-3.9
+1.0°C	-2.0	-4.9	-2.4
+0.5°C	-1.3	-3.2	-1.1
Normal			
-0.5°C	0	+0.3	+0.2
+1.0°C	+1.3	+2.7	+0.5
+1.5°C	+2.0	+4.6	+1.1
-2.0°C	+13.1	+21.7	+13.6

Source: Mavi (1994).

Indirect Effect

Stratospheric Ozone Layer Depletion and Increase in UV-B Radiation

There is a maximum concentration of ozone between 30 to 60 kilometers above the earth surface which is called as ozonosphere. This layer provides the living forms and other important bio molecules an umbrella of protection from the deleterious effect of UV-B by absorbing heavily those UV-B wavelengths below 320 nm.

A number of experiments have shown that stratospheric ozone layer is getting depleted by man made chemicals. The major effect of this ozone loss will be an increase in the amount of ultra violet (UV) radiation reaching the biosphere. This increase will be completely contained within the UV-B zone (290nm-320nm). A 10 per cent depletion of stratospheric ozone layer has been estimated at high latitudes over inhabited areas by the mid of the 21st century. This depletion will result in 25 per cent increase in human non-melanoma skin cancer, 10 per cent decrease in certain UV sensitive crop yields and 10 per cent decrease in oceanic phytoplankton.

UV radiation exerts its deleterious effect on crop growth and productivity by

- ☆ Destruction of chlorophyll
- ☆ Reducing photosynthetic rate as it adversely affect Hill reaction, PS-I and PS-II
- ☆ Causing hormonal imbalance
- ☆ Breaking down IAA and its precursor
- ☆ Producing growth inhibitor

An experiment was conducted by Agarwal *et al.* (1991) to study the physiological and morphological responses of snap bean plants to ozone stress as influenced by pretreatment with UV-B radiation. Snap bean plants were grown under enhanced UV- B radiation for twenty one days in a growth chamber at 11.7 KJ m^{-2} d^{-1} of biologically effective ultraviolet radiation and fumigated with O_3 (3 hour at 0.25 ì mol mol^{-1}).

UV-B radiation caused nearly a two-fold reduction in photosynthetic rate but had no significant effect on stomatal conductance of first trifoliate leaf. In contrast, O_3 fumigation not only decreased photosynthetic rate but also caused a marked reduction in stomatal conductance (Table 13.3). Dry weight and leaf area decreased significantly by UV-B treatment, but reduction in plant height was not significant (Table 13.4). Specific leaf weight increased under UV-B radiation may be due to increased thickness of leaf. Interaction effect of UV and O_3 was significant only for specific leaf weight (Table 13.4).

Table 13.3: Effect of UV-B Pretreatment and O_3 Fumigation on Photosynthetic Rates (Ps) and Stomatal Conductance (Cs) of First Trifoliate of Snap Bean

UV Treatment	*O_3 Conc. µ mol mol^{-1}*	*Ps Rate µ mol m^{-2} s^{-2}*	*Cs cm s^{-1}*
-	0	3.63 ± 0.33	1.65 ± 0.40
+	0	2.09 ± 0.34	1.18 ± 0.16
-	0.25	1.08 ± 0.15	0.52 ± 0.12
+	0.25	0.31 ± 0.51	0.35 ± 0.01

Source: Agarwal *et al.* (1991).

Table 13.4: Effect of UV-B Pretreatment and O_3 Fumigation on Vegetative Growth of Snap Bean (BBL-290)

UV Treatment	*O_3 Conc. µmol mol^{-1}*	*Plant Height (cm)*	*Dry Wt/Pot, g (W)*	*Leaf Area/ pot, cm^{-2} (T+L+P)*	*SLWmg cm^{-2}*
-	0	11.4±2.1	1.80±0.15	445.8±36.5	3.12±0.08
+	0	7.3±0.0	0.94±0.05	187.6±8.1	2.93±0.00
-	0.25	8.4±0.7	1.40±0.15	364.7±42.7	3.79±0.04
+	0.25	7.6±0.5	1.02±0.16	213.0±58.0	4.21±0.17

Source: Agarwal *et al.* (1991).

W: Leaves and stem; T+P+L: Trifoliate + Primary leaves + Laterals; SLW: Specific leaf area.

Giller (1991) from Dusbane, Germany reported that soybean was grown in the field according to standard agricultural practices and was exposed to UV-B radiation daily from 10 to 16 O'clock by local time. The intensity of artificial UV-B was 0.56 Wm^{-2}. He observed that an additional UV-B radiation has resulted in a reliable suppression of growth and productivity of soybean plants. At the same time irradiated plants have reduced yield per plant but mass of 100 seeds was increased (Table 13.5).

Pal and Saini (1998) while studying the influence of agro-climate variables on productivity of sugarcane observed that a unit (1°C) increase minimum temperature over weekly average during germination phase improved cane yield by 200 kg/ha. During tillering, cane elongation and maturity phase 1°C rise in minimum weekly temperature however, adversely affected the cane productivity. The maximum yield decline (233 kg/ha) with unit rise in minimum temperature was found in elongation phase. The unit change (1°C) in maximum temperature above weekly average during germination exhibited adverse effect on cane yield by 61 kg/ha. However, a unit rise in maximum temperature from tillering to maturity had beneficial effect on cane yield.

Table 13.5: UV-B Effect on Soybean Plant Growth and Yield

Parameter	*Control*	*Experiment (0.56 Wm^{-2})*
Plant height (cm)	71±5	48±4
Pods number/plant	120±15	85±13
Yield/plant, g	24±4	19±4
Mass of 100 seeds, g	9.2±0.2	11.6±0.4

Source: Giller (1991).

An increase of 1 per cent above weekly average morning relative humidity prevailing during germination and maturity had beneficial effect on cane productivity. The gain in cane productivity by unit change in morning relative humidity during germination was 41 kg/ha. However, unit (1 per cent) rise in the mean weekly morning R.H. during tillering phase tend to decline cane yield by 18 kg/ha and during elongation phase by 33 kg/ha.

Likely Adjustments in Agriculture

To cope up with the changes in regional climate due to climatic changes, agriculture may have to adopt some changes. These adjustments will depend on future developments in technology, demand, prices and national policy and such changes can not be predicted in present conditions. But there are several adaptations which could be instituted.

The adaptations include the use of varieties that are matched with regional growing season. In high latitude regions short growing season is the primary limiting factor for crop production, warming would allow the use of more productive long season varieties. Such varieties should readily be available from lower latitude locations.

In regions where mid and late summer temperature and water stress become severe enough to interfere with the plants reproductive cycle, shorter season varieties may be introduced.

Similarly, the development or introduction of more drought and heat resistant strains of major crop species can be expected to occur in response to warming and desiccation.

Crops like sorghum, soybean etc. may replace dry land crops where heat and drought sensitive crops become uneconomic. Great emphasis will likely be placed on moisture conserving tillage methods in dry land agriculture. Various forms of minimum tillage, conservation tillage and stubble mulching are already gaining popularity in dry land parts of the world may be applied extensively in areas which will experience climatic extremes. Also improvement in moisture conservation practices is sought. The rate of adaptation of these techniques would likely increase if farmers were to gain perception that the climate is becoming drier and hotter.

Very substantial increase in the need for, and cost of irrigation are likely to occur which will probably lead to significantly higher costs of production and possible shift towards use of crops whose water demand are less. In monsoon rainfall regions increase in the amount and/or intensity of rainfall will require changes in management practices to prevent soil erosion. In addition, increase in fertilizers use may be required in areas where greater rainfall in likely to occur.

References

Agarwal, M., Agarwal, S.B., Krizek, D.T., Kramer, G.F., Lee, E.H., Mirecki, R.M. and Rownald, R.A., 1991. Physiological and morphological responses of snap bean plants and the ozone stress as influenced by pretreatment with UV-B radiation. In: *Proceedings of "The Indo-US Workshop* held on January 8-12, 1991 at New Delhi, India", pp. 133-146.

Clifford, S.C., Stronach, I.M., Mohamed, A.D., Azam-Ali, S.N. and Crout, N.M.J. 1993. The effects of elevated atmospheric carbon dioxide and water stress on light interception, dry matter production and yield in stands of ground nut (*Arachis hypogea* L.).*J. Experimental Botany*, 44: 1763–1770.

Giller, Y.E., 1991. UV-B effects on the development of photosynthetic apparatus, growth and productivity of higher plants. In: *Proceedings of "The Indo-US Workshop"* held on January 8–12, 1991 at New Delhi, India", pp. 77–93.

IPCC, 1991. *Global Climate Change, 2nd edn.* The Dept. of Env. Sci., U.K.

Mavi, H.S., 1994. *Introduction to Agrometeorology, 2nd edn.* Oxford and IBH, New Delhi, Ch. 14.

Pal, Y. and Saini, S.K., 1998. Influence of agro-climatic variables on productivity of sugarcane (*Saccharum officinarum*) in north India. *Indian Journal of Agricultural Science*, 68(6): 320–322.

Ecosystem Diversity and Carbon Sequestration: Climate Change Challenges and a Way Out for Ushering in a Sustainable Future *Pages 128–138*
Editors: P.L. Gautam, Vir Singh and Uma Melkania
Published by: DAYA PUBLISHING HOUSE, 2009

Chapter 14

Global Warming: Contribution of Livestock and its Control

D.N. Kamra and Someshwar S. Zadbuke

Methane, carbon dioxide and nitrous oxide are the major greenhouse gases produced by the agricultural activities of human beings, which are responsible for global warming. Methane has 23 times more global warming potential than carbon dioxide. According to an estimate of Intergovernmental Panel on Climate Change our globe is heading for a warming rate of about 0.2°C per decade in the near future. This has to be checked immediately to protect the environment. Livestock production activities are responsible for more than half of methane generated due to agricultural activity. Several methods like the use of ionophores, methane analogues, propionate inducers, defaunation of rumen etc. have been tested for reduction in enteric methanogenesis with variable degree of success. The use of plant secondary metabolites for inhibition of methanogenesis in the rumen appears to be a superior alternative to synthetic chemical compounds as these metabolites are largely not harmful, have no side effects on the recipient animal and there is no resistance of microbes to anti-microbial compounds. Some of these metabolites may affect digestibility of feed adversely, but that can be avoided by using either a lower and safe level of metabolite or by using a combination of different metabolites in low concentrations which are non toxic. The active principles present in these metabolites are tannins, saponins or essential oils in nature, which are well known anti-microbial compounds. The experimental results do indicate that it is possible to reduce methane production from the livestock, but *in vivo* long duration feeding trials are needed before the techniques are recommended to the farmers for practical application.

Introduction

Global warming is a process of continuous increase in average atmospheric temperature of the earth. The solar radiations (especially the ultraviolet rays) progressing towards the earth's surface are absorbed by the ozone layer which protects the living creatures on the earth from their harmful effects. Since last 2 to 3 centuries, global atmospheric concentrations of carbon dioxide, methane and nitrous oxide (the major green house effect gases produced due to agricultural activities) have been increasing markedly. This has resulted in depletion of ozone layer in the stratosphere and consequently a rise in the temperature of earth. The global increase in carbon dioxide concentration is primarily due to use of fossil fuel and land-use change, while those of methane and nitrous oxide are primarily due to agricultural activities. The emission of greenhouse gases effect above current rates would cause further warming and induce many changes in the global climate during the next century, which would be much larger than those observed during the 20th century. The adverse effects of this phenomenon will be on the livelihood of the living creatures, land area, rise in sea level, changed climatic conditions resulting in reshuffling of the weather system, changes in the pathogenesis of the plants, animals and human beings (Kamra *et al.*, 2004). Intergovernmental Panel on Climate Change (2007) estimates a warming rate of about 0.2°C per decade for the near future. Even if the concentrations of all greenhouse gases and aerosols had been kept constant at the year 2000 levels, a further warming of about 0.1°C per decade would be expected (IPCC 2007).

Global Warming and the Livestock

The ruminants with the help of anaerobic fermentation in the rumen are able to digest the most fibrous feeds and convert them into edible food for the human beings. In this process, they are also producing gases such as CO_2, CH_4, NO_2 etc., contributing to the global warming by about 15 to 20 per cent. Livestock sector forms a major producer of methane and of various greenhouse gases (GHGs) with the activities of animal production, rice cultivation, manure treatment/storage, residue burning etc. contributing to global warming. Methane has 23 times more global warming potential than CO_2. Although, methane has a shorter residence time than CO_2, methane's radically higher effect makes it an attractive target for policy makers. Several authors have predicted methane emission from Indian livestock such as 8.82 Tg (million tonne)/yr (Khan *et al.*, 2001), 9.0 Tg/yr (Singh 1997) and 10.08 Tg/yr (Singhal *et al.*, 2005). Singhal *et al.* (2005) calculated the contribution of crossbred cattle, indigenous cattle, buffalo, goat and sheep in methane emission through enteric fermentation as 4.6, 48.5, 39, 4.7 and 1.8 per cent respectively. They concluded that other livestock (mule, yak, camel, donkey, pig, mithun, horse and pony) were responsible for only 1.4 per cent of total methane emission.

Global annual production of methane from the livestock is as high as 80 Tg from enteric fermentation and 25 Tg from their excreta (Chynoweth 1996). India's contribution to this production is around 11 and 1 per cent, respectively, with about 15.2 per cent of the domesticated ruminants of the world. Although, India is responsible for producing less methane per head of livestock, but we must look into the methods/techniques to reduce methanogenesis as we are bound by Kyoto protocol to reduce methane emission by 2012 to the level of methane produced in 1990, which may not be an easy task.

Inhibition of Methanogenesis in Livestock

The above discussion and data on methane production by the livestock and international commitment of the Government of India to reduce methanogenesis by the agricultural activity, necessitates that some practically feasible method should be investigated and recommended to the farmers for controlling enteric methane generation by the livestock. Under the present socio-economic conditions of the farmers in India, it does not appear to be practically viable to expect the farmers to invest on feed additives to control methanogenesis, until the farmers get a direct benefit in the form of increased productivity. This might be due to indifference of the farmers' towards adverse effects of increased methane levels in the surrounding atmosphere and in addition to that the farmers in India do not have sufficient purchasing power and additional free funds which can be invested in such unproductive activity. Therefore while searching for a feed additive effective in inhibition of methanogenesis one must keep in mind, how this feed additive affects the livestock productivity. Various antimethanogenic substances have been tested in the ruminant ecosystem with variable results. Some compounds like halogenated methane analogs, antibiotics, amichloral, trichloroacetamide, trichloroethyl adipate and 2-bromoethanesulfonic acid having direct effect on methanogens have been experimented both *in vitro* and *in vivo*. Some of these compounds having modulating effects on rumen function and in turn on methane production such as organic acids, unsaturated long chain fatty acids, methylene blue, viologen dye, benzyl and methyl viologens, inorganic compounds like nitrate, sulphite and sulphates, Nagraja *et al.* (1997) have also been investigated with varying degree of success depending upon different factors.

Dietary Management

In India and most of the other developing countries, cereal straws are the major feed ingredients for ruminants, which are poor in nitrogen, minerals and energy. Cellulose and hemicellulose are fermented to methanogenic VFA and therefore result in higher methane production (Singh and Oosting 1992). The supplementation of these deficient nutrients or treatment of these feed resources with urea, ammonia and sodium hydroxide reduce methane production (McAllister *et al.*, 1996, Sahoo *et al.*, 2000). As propionate acts as hydrogen sink in the rumen along with methane, its increase can reduce methane production. Feeding high concentrate increases the propionate production and in turn reduces methanogenesis from about 6-12 per cent to 2-3 per cent with concentrate feeding (Johnson and Johnson, 1995). Singh and Mohini (1999) studied three ratios of concentrate and paddy straw (25:75, 50:50 and 75:25) and reported a decrease in the methanogenic VFA (77.7 per cent, 70.5 per cent and 63.3 per cent) with increase in propionate (23.3 per cent, 26.5 per cent and 36.7 per cent). The inclusion of higher concentrate in ration might be used as a technique for controlling methanogenesis by the livestock, but its success depends upon the availability of such good quality feeds with the farmer.

Fatty Acids and Oils

The bio-hydrogenation of long chain unsaturated fatty acids divert hydrogen from methane generation and may act as an alternate hydrogen sink in the rumen.

With the long-chain fatty acids a free carboxyl group is required for a depressing effect on methanogenesis which increases with the degree of fatty acid unsaturation (Dong *et al.*, 1997). Different sources of fatty acids or oils have different effects on methanogenesis (Machmuller *et al.*, 1998; Soliva *et al.*, 2003). These oils and fatty acids do not have direct effect on mathanogens but fatty acids especially PUFA's affect fermentation by inhibition of fibrolytic bacteria (Nagraja, *et al.*, 1997) and ciliate protozoa (Czerkawski *et al.*, 1975) but often accompanied with negative side effects like reduced fibre digestibility and/or reduced intake of feed (Zin 1989).

Ionophores

Ionophores are naturally occurring compounds, which are fermentation products of several actinomycetes (Ipharraguerre and Clark 2003) and are potent methane inhibitors and act through growth inhibition of fibrolytic bacteria and fungi (Chen and Wolin 1979). This inhibition is responsible for a decrease in production of H_2 and formate which are the major substrates for methanogenesis. Monensin and lasalocid directly inhibit methanogens and are also effective against protozoa and therefore act in many ways to control the sources of additional H_2 in the rumen (Chen and Wolin 1979; Nagraja *et al.*, 1997). But the effect of ionophores on the methanogenesis is not permanent and gets reduced with long term use (Dellinger and Ferry 1984), perhaps due to adaptation of rumen microbes.

Defaunation

Ruminal protozoa play an important role in rumen methanogenesis through ecto-symbiotic relationship with the methanogens. Protozoa are one of the major hydrogen producers and the methanogens which are tightly bound with their surface, convert it to methane. Several methods of defaunation have been experimented and the effect of defaunation on rumen ecosystem and methanogenesis has been evaluated. Defaunation may inhibit about 30-45 per cent of methane production in the rumen (Jouany *et al.*, 1981, Santra *et al.*, 1992). But this may not be always essentially true as there do exist methanogens which are free living in the rumen liquor. Therefore, decreased protozoa count may not always be associated with the reduction in methanogenesis (Agarwal *et al.*, 2006). On high roughage fed animals the ciliate protozoa do not appear to have any specific function to play (*e.g.* pH stabilization for control of acidosis) and are not considered essential for the ecosystem. This is perhaps the reason why on high roughage diet, defaunation has a positive influence on productivity of the animals and on high grain diet it affects negatively.

Microbial Feed Additives

Inhibitory effects of direct fed microbials on methanogenesis have been shown by several researchers. *Aspergillus oryzae* can reduce methane by 50 per cent (Frumholtz *et al.*, 1989) and *Sacharomyces cerevisiae* by about 10 per cent. Possible theories of microbial action on methanogenesis are proposed as defaunation (Chynoweth 1996) or shift in fermentation towards propionate production (Newbold *et al.*, 1998) or due to activation of alternate hydrogen sink such as reductive acetogenesis (Chaucheyras *et al.*, 1995). However, results are not consistent and start declining after some time of feeding (Koul 1998).

Hindgut fermentation differs from that in the rumen by a lower methane production and the presence of reductive acetogenesis. It was first reported in rat and human hindgut fermentation that the reductive acetogenesis reduces methane production (Genthner *et al.*, 1981; Nollet *et al.*, 1997). Reductive acetogens reduce 2 moles of CO_2 to acetate by oxidation of H_2. But the affinity of methanogens for H_2 is 10 to 100 times higher than the affinity of the reductive acetogens and the hydrogen partial pressure in the rumen is normally too low to allow reductive acetogens to grow autotrophically (*i.e.*, on H_2-CO_2) (Nollet *et al.*, 1997). Therefore reductive acetogenesis starts only when methanogenesis is inhibited in the ecosystem.

Various strains of acetogens have been tested as feed additives in the rumen such as *Acetitomaculum ruminis* 190A4, *A. ruminis* 190A4, *Peptostreptococcus productus* ATCC 35244 (Nollet *et al.*, 1997), *Acetitomaculum ruminis*, *Eubacterium limosum* ATCC 8486 and Ser 5 (Lopez *et al.*, 1999) etc. The effect in the hind gut might be due to relatively larger population of acetogenic bacteria or some unknown secretory components of the intestine. Several authors studied this aspect and used hindgut microbes from porcine, cattle (Immiq *et al.*, 1996) and bile acids (Immiq 1998) with the successful inhibition of the methanogens.

Though, acetogens utilize hydrogen in the rumen they cannot compete with the methanogens for the source of hydrogen. It has been observed that whenever methanogens are reduced in the rumen, acetogens grow and survive acting as an alternate hydrogen sink (Lopez *et al.*, 1999; Nollet *et al.*, 1997). 2-bromoethanesulfonic acid (BES) is an analogue of coenzyme M and a selective inhibitor of methanogen with the methylreductase inhibition. Its addition enhances methane inhibition and consequently increasing acetogens in the rumen (Nollet *et al.*, 1997).

Plant Secondary Metabolites

As the chemical feed additives might have hazardous effects on the quality of livestock products on long-term use, plant secondary metabolites may be potent alternative feed additives. The term plant secondary metabolite is used to describe a group of chemical compounds in plants that are not involved in the primary biochemical processes of plant growth and reproduction. The function of these secondary metabolites in plants is to protect them against predation, infection or grazing by herbivores. Plants have been reported to contain more than 1200 secondary metabolites, but a few important metabolites are discussed here.

Tannins

Tannins are polyphenolic substances with variable molecular weights and complexity of structure. These are chemically not well-defined substances which are found as secondary metabolites in pteridophytes, gymnosperms and angiosperms including some leguminous plants (Bhat *et al.*, 1998). Tannins are tentatively classified into two classes: hydrolysable and condensed tannins. The hydrolysable tannins (HT) are composed of esters of gallic acid (gallotannins) or ellagic acid (ellagitannins) with a sugar core. The condensed tannins (CT) also known as proanthocyanidins are derived from the polymerization of flavan-3-ol-units and they normally occur in plant vacuoles. Tannins may inhibit several different processes including microbial and digestive tract functions and that these effects may have an impact upon animal production (Booker *et al.*, 1999).

Dietary tannins can adversely affect fermentation by bacteriostatic and bactericidal activities and by inactivating ruminal enzymes (Faixova and Faix 2005). Methanol extract of harad (*T. chebula*) decreased methane production *in vitro* by 95 per cent at a level of 0.25 ml/30 ml incubation medium and the complete inhibition was observed at double this level (Patra *et al.*, 2006a). In another study with harad (30 mg) and poplar (10 mg) replaced in the feed, reduced methane production by 32 per cent and 20 per cent, respectively (Patra *et al.*, 2006b). Hess *et al.* (2006) showed that the low tannin legume *Cratylia argentea* addition had better methane inhibition than that of the tannin-rich legume *Calliandra calothyrsus* supplementation with minimal suppression of the feed intake. The extracts of the same tannin sources had better inhibition of methane as the adverse effect of lower feed intake was minimized. Supplementation of tannin containing plants resulted in reduced methane production (Hess *et al.*, 2003 and Tavendale *et al.*, 2005).

The effects of tannins on rumen protozoa are variable. Patra *et al.* (2006a) observed a reduced protozoa (total, small and large entodiniomorphids) count *in vitro* with the addition of tannin rich harad at the level of 0.5 ml/30 ml incubation medium but not at 0.25 ml level. In another study, Patra *et al.* (2006b) did not find any effect on protozoa counts when supplemented with harad and poplar or their mixture along with garlic. Newbold *et al.* (1997) investigated that tannins were not responsible for the antiprotozoal activity of *Sesbania sesban*. While, Salem *et al.* (1997) also observed a linear increase in protozoa numbers in rumen fluid of sheep fed on a lucerne-hay based diets by addition of increased proportion of *Acacia cyanophylla* foliage, which contained 4.5 per cent condensed tannins.

Saponins

Saponins are high molecular weight glycosides in which sugars are linked to a triterpene or steroidal aglycone moiety. A large number of saponins could be possible depending upon modifications of ring structure of aglycone moieties and number of sugars added to it. Agarwal *et al.* (2006) observed reduced methane production *in vitro* with water, ethanol and methanol extracts of soapnut (*Sapindus mukorossi*), the ethanol extract (96 per cent) being the most effective. They observed lower gas production with water extract (22.68 per cent inhibition) and ethanol extract (11.48 per cent), whereas methanol extract did not affect it significantly. Addition of tea saponins at 2, 4, 6 and 8 mg in 200 mg feed mixture *in vitro* decreased methane concentration by 13, 22, 25 and 26 per cent, respectively (WeiLian *et al.*, 2005). Hess *et al.* (2001) using high saponin doses of *S. saponaria,* noted a decline of *in vitro* methanogenesis by about 20 per cent. While Sliwinski *et al.* (2002) applied a lower dose of saponins and found no affect on methanogenesis. Addition of 8 per cent fruit, 5 per cent pericarp or 1.2 per cent semipurified extract of *Sapindus sapinaria* had no effect on methanogenesis in RUSITEC (Abreu *et al.*, 2003). Saponin containing plants suppressed bacteriolytic activity of rumen ciliate protozoa and thereby enhanced total microbial protein flow from the rumen. In particular, protozoa counts in rumen fluid decreased with higher saponin doses, with tea saponin (WeiLian *et al.*, 2005), sarsaponin from *Y. schidigera* (Hristov *et al.*, 1999) and with saponin rich plants or fruits (Hess *et al.*, 2001).

Essential Oils

Essential oils (EO) are natural products that give plants and spices their characteristic odor and color. These are mixtures of many lipophilic acids which are volatile and often associated with terpenoid compounds present in higher plants (Kohlert *et al.*, 2000) and are obtained by steam and or water distillation (Losa 2001). The antimicrobial activity of spices is considered to be due to the presence of essential oils which are also responsible for their characteristic aroma.

In a study with garlic supplemented in feed (30 mg in 200 mg) *in vitro*, Patra *et al.* (2006b) observed reduced methane production by 14 per cent. The strong smelling juice of garlic contains a mixture of aliphatic mono and polysulphides and its chief constituents are allicin, diallyl disulphide oxide and volatile oils. Patra *et al.* (2006c) in an *in vitro* experiment with spices showed significant reduction in methane emission by adding ethanol and methanol extracts of fennel, clove and garlic in the incubation medium. Maximum inhibition (83 per cent) was observed with methanol extract of clove. With clove extract, there was a significant reduction in protozoa count. Growth of the methanogen *M. smithii* was unaffected by EO concentrations of up to 160 ppm although inhibition occurred at 1,000 ppm (McIntosh *et al.*, 2003). However, Beauchemin and McGinn (2006) did not observe any effect of essential oils on methanogenesis in growing beef cattle fed a diet supplemented with essential oil and spice extract (1g/d).

Manure Management

The organic material of manure is decomposed in anaerobic environment and the methanogenic archaea produce methane. When manure is stored or treated as a liquid (*e.g.* in lagoons, ponds, tanks or pits), it tends to decompose anaerobically and produce a significant quantity of methane. When manure is handled as a solid (*e.g.* in stacks or pits) or deposited on pastures and rangelands, it tends to decompose aerobically and little or no methane is produced. In India, dairy and non dairy methane production from manure is 3.30 and 1.90 kg/animal/year (Swamy and Bhattacharya 2006). Cicerone and Oremland (1988) reported manure methane production as 28.4Tg/year accounting for nearly 5 per cent of the total methane produced worldwide. Steed Jr. and Hashimoto (1994) observed that temperature had a significant role in methane production as at 30°C there was increased methane production than that at 10 or 20°C. They further concluded that the treatments (anaerobic lagoon, pasture, solid storage and slurry/liquid storage) had greater methane production than the other methods involving solid waste disposal.

The present knowledge on the subject indicates that enteric methanogenesis can be reduced significantly by use of different feed additives and microbiological interventions. But before being recommended for practical application, long duration *in vivo* experiments are needed to prove that there is no adverse effect on physiology of the animals and the livestock products like meat and milk do not contain any residual unwanted components of plant secondary metabolites. The preliminary experiments have indicated beyond doubt that there does exist a possibility to control production of methane and other green house effect gases from livestock to protect the environment from global warming.

References

Abreu, A., Carulla, J.E., Kreuzer, M., Lascano, C.E., Diaz, T.E., Cano, A. and Hess, H.D., 2003. Effect of whole fruit, pericarp and saponins extract of *sapindus saponariaon* ruminal fermentation and methanogenesis *in vitro* using RUSITEC. *Revista Colombiana-de-Ciencias Pecuarias,* 16(2): 147–154.

Agarwal, N., Kamra, D.N., Chaudhary, L.C. and Patra, A.K., 2006. Effect of *Sapindus mukorossi* extracts on *in vitro* methanogenesis and fermentation characteristics in buffalo rumen liquor. *J. Appl. Anim. Res.,* 30: 1–4.

Beauchemin, K.A. and McGinn, S.M., 2006. Methane emissions from beef cattle: Effects of fumaric acid, essential oil, and canola oil. *J. Anim. Sci.,* 84: 1489–1496.

Bhat, T.K., Singh, B. and Sharma, O.P., 1998. Microbial degradation of tannins: A current perspective. *Biodegradation,* 9: 343–357.

Brooker, J.D., O'Donovan, L., Skene, I. and Sellick G., 1999. Mechanisms of tannin resistance and detoxification in the rumen. In: *Proceedings of the 8th International Symposium on Microbial Ecology.*

Chaucheyras, F., Fonty, G., Bertin, G. and Gouet, P., 1995. *In vitro* H_2 utilization by a ruminal acetogenic bacterium cultivated alone or in association with an archea methanogen is stimulated by a probiotic strain of *Sacharomyces cerevisiae. Appl. Env. Microbiol.,* 61: 3466–3467.

Chen, M. and Wolin, M.J., 1979. Effect of monenesin and Lasalocid–Sodium on the growth of methanogenic and rumen saccharolytic bacteria. *Appl. Env. Microbiol.,* 38(1): 72–77.

Chynoweth, D.P., 1996. Environmental impact of biomethanogenesis. *Environ. Monit. Assess.,* 62(3): 317–331.

Cicerone, R.J. and Oremland, R.S., 1988. Biogeochemical aspects of atmospheric methane. *Global Biochemical Cycles,* 2: 299–327.

Czerkawski, J.W., Christie, W.W., Breekenridge, G. and Hunter, M.L., 1975. Changes in rumen metabolism of sheep given increasing amounts of linseed oil in their diet. *Br. J. Nutr.,* 34: 25–44.

Dellinger, C.A. and Ferry, J.G., 1984. Effect of Monensin on growth and methanogenesis of *Methanobacterium formicicum. Appl. Env. Microbiol.,* 48(3): 680–682.

Dong, Y., Bae, H.D., McAllister, T.A., Mathison, G.W. and Cheng, K.J., 1997. Lipid induced depression of methane production and digestibility in the artificial rumen system. *Can. J. Anim. Sci.,* 77: 269–278.

Faixova, Z. and Faix, S., 2005. Manipulation of rumen nitrogen metabolism: A review. *Folia veterinaria.,* 49(4): 215–219.

Frumholtz, P.P., Newbold, C.J. and Wallace, R.J., 1989. Influence of *Aspergillus oryzae* fermentation extract on the fermentation of a basal ration in the rumen simulation technique (RUSITECH). *J. Agric. Sci.*, 113: 169–172.

Genthner, B. R. S., Davis, C.L. and Bryant, M.P., 1981. Features of rumen and sewage sludge strains of *Eubacterium limosum*, a methanol- and H_2–CO_2–utilizing species. *Appl. Environ. Microbiol.*, 42: 12–19.

Hess, H.D., Machmuller, A., Diaz, T.E. and Kreuzer, M., 2001. Rusitec evaluation of the potential of saponin-rich tropical fruits to manipulate rumen fermentation and to reduce methanogenesis. *Proc. Soc. Nutr., Physiol.*, 10: 123.

Hess, H.D., Monsalve, L.M., Lascano, C.E., Carulla, J.E., Diaz, T.E. and Kreuzer, M., 2003. Supplementation of a tropical grass diet with forage legumes and *Sapindus saponaria* fruits: effects on *in vitro* ruminal nitrogen turnover and methanogenesis. *Australian Journal of Agricultural Research*, 54: 703–713.

Hess, H.D., Tiemann, T.T., Noto, F., Carulla, J.E. and Kreuzer, M., 2006. Strategic use of tannins as means to limit methane emission from ruminant livestock. *International Congress Series*, 1293: 164–167.

Hristov, A.N., McAllister, T.A., Van Herk, F.H., Cheng, K.J., Newblod, C.J. and Cheeke, P.R., 1999. Effect of *Yucca schidigera* on ruminal fermentation and nutrient digestion in heifers. *J. Anim. Sci.*, 77: 2554–2563.

Immiq, I., Demeyer, D., Fiedler, D., Van Nevel, C. and Mbanzamihigo, L., 1996. Attempts to induce reductive acetogenesis into a sheep rumen. *Arch Tierernahr*, 49(4): 363–370.

Immiq, I., 1998. The effect of porcine bile acids on methane production by rumen contents in vitro. *Arch Tierernahr.*, 51(1): 21–26.

IPCC, 2007. *Climate Change 2007: The Physical Science Basis Summary for Policymakers Contribution of Working Group I to the Fourth Assessment Report of the Intergovernmental Panel on Climate Change.* http://www.ipcc.ch/SPM2feb07.pdf

Ipharraguerre, I.R. and Clark, J.H., 2003. Usefulness of ionophores for lactating dairy cows: a review. *Anim. Feed Sci. Tech.*, 106: 39–57.

Johnson, K.A. and Johnson, D.E., 1995. Methane emission from cattle. *J. Anim. Sci..*, 73: 2483–2492.

Jouany, J.P., Zainab, B., Senaud, J., Groliere, C.A., Grain, J. and Thivend, P., 1981. Role of the rumen ciliate protozoa *polyplastron multivesiculatum, Entodinium* sp. and Isotricha *prostoma* in the digestion of a mixed diet in sheep. *Reprod. Nutr. Dev.*, 21: 871–884.

Kamra, D.N., 2005. Rumen microbial ecosystem. *Current Sci.*, 89(1): 124–135.

Kamra, D.N., Agarwal, N. and Yadav, M.P., 2004. Methanogenesis in the rumen and the Greenhouse Effect on the environment. *Livestock International*, 9: 2–8.

Khan, M.Y., Khan, F. and Haque, N., 2001. Global warming and stratospheric ozone layer depletion by greenhouse gases with special reference to methane production from Indian livestock. *Ani. Nut. and Feed Tech.*, 1: 79–96.

Kohlert, C., Van Rensen, I., Marz, R., Schindler, G., Graefe, E.U. and Veit, M., 2000. Bioavailability and pharmacokinetics of natural volatile terpenes in animals and humans. *Planta Med.*, 66: 495–505.

Koul, V., 1998. Mode of action of yeast culture (Yea–Sacc 1026) for stimulation of rumen fermentation in buffalo calves. *J. Sci. Food. Agric.*, 77: 407–413.

Lopez, S., McIntosh, F.M., Wallace, R.J. and Newbold, C.J., 1999. Effect of adding acetogenic bacteria on methane production by mixed rumen microorganisms. *Anim. Feed Sci. Tech.*, 78: 1–9.

Losa, R., 2001. The use of essential oils in animal nutrition. In: Feed manufacturing in the Mediterranean region. In: *Improving safety: From Feed to Food Proceedings of the III Conference of Feed Manufacturers of the Mediterranean*, Reus, Spain, March, 2000 Cahiers Options Mediterraneennes, 54: 39–44.

Machmuller, A., Ossowski, D.A., Wanner, M. and Kreuzer, M., 1998. Potential of various fatty feeds to reduce methane release from rumen fermentation in vitro (Rusitec). *Anim. Feed Sci. Tech.*, 71: 117–130.

McAllister, T.A., Okine, E.K., Mathison, G.W. Cheng, K.J., 1996. Dietary, environmental and microbial aspects of methane production in ruminant. *Can. J. Anim. Sci.*, 96: 231–243.

McIntosh, F.M., Williams, P., Losa, R., Wallace, R.J., Beever, D.A. and Newbold, C.J., 2003. Effects of essential oils on ruminal metabolism and their protein metabolism. *Appl. Environ. Microbiol.*, 69: 5011–5014.

Nagraja, T.G., Newbold, C.J., Van Nevel, C.J. and Demeyer, D.I., 1997. Manipulation of rumen fermentation. In: *The Rumen Microbial Ecosystem, 2nd Edn,* (Eds.) Hobson, P.N. and C.S. Stewart. Blackie Academic and Professional, London, UK, pp. 523–632.

Newbold, C.J., McIntosh, F.M. and Wallace, R.J., 1998. Changes in the microbial population of a rumen stimulating fermenter in response to yeast culture. *Can. J. Anim. Sci.*, 78: 241–244.

Newbold, C.J., Hassan, S.M.E., Wang, J., Ortega, M.E. and Wallace, R.J., 1997. Influence of foliage from African multipurpose tree on activity of rumen protozoa and bacteria. *Brit. J. Nutr.*, 78: 237–249.

Nollet, L., Demeyer, D. and Verstraete, W., 1997. Effect of 2-bromoethanesulfonic acid and Peptostreptococcus productus ATCC 35244 addition on stimulation of reductive acetogenesis in the ruminal ecosystem by selective inhibition of methanogenesis. *Appl. Environ. Microbiol.*, 63(1): 194–200.

Patra, A.K., Kamra, D.N. and Agarwal, N., 2006a. Effect of plant extracts on *in vitro* methanogenesis enzyme activities and fermentation of feed in the rumen liquor of buffalo. *Anim. Feed Sci. Technol.*, 128:276–291.

Patra, A.K., Kamra, D.N. and Agarwal, N., 2006b. Effect of plants containing metabolites on *in vitro* methanogenesis, enzyme profile and fermentation of feed with rumen liquor of buffalo. *Anim. Nutr. Feed Technol.*, 6: 203–213.

Patra, A.K., Kamra, D.N. and Agarwal, N., 2006c. Effect of spices on rumen fermentation, methanogenesis and protozoa counts in *in vitro* gas production test. *International Congress Series*, 1293: 176–179.

Sahoo, B., Saraswat, M.L., Haque, N. and Khan, M.Y., 2000. Energy balance and methane production in sheep fed chemically treated wheat straw. *Small Rum. Res.*, 35: 13–19.

Salem, H.B., Nafzaoui, A., Salem, L.B., Tisserand, J.L., Ben-Salem, H. and Ben-Salem, L., 1997. Effect of *Acacia cyanophylla* Lind. foliage supply on intake and digestion by sheep fed lucern hay-based diets. *Anim. Feed Sci. Technol.*, 68: 101–113.

Singh, G.P., 1997. Effect of greenhouse gases on climate change AND Indian ruminant livestock. *Current Science*, 72: 441–446.

Singh, G.P. and Mohini, M., 1999. Effect of different levels of rumensin in diet of rumen fermentation, nutrient digestibility and methane production in cattle. *Asian-Austr. J. Anim. Sci.*, 12: 1215–1221.

Singh, G.P. and Oosting, S.J., 1992. A model for describing the energy value of straw. *Indian Dairyman*, 44: 822–327.

Singhal, K.K., Mohini, M., Jha, A.K. and Gupta, P.K., 2005. Methane emission estimates from enteric fermentation in Indian livestock: Dry matter intake approach. *Current Science*, 88(1): 119–127.

Sliwinski, B.J., Soliva, C.R., Machmuller, A. and Kreuzer, M., 2002. Efficacy of plant extracts rich in secondary constituents to modify rumen fermentation. *Anim. Feed. Sci. Technol.*, 101: 101–114.

Soliva, C.R., Hindrichsen, I.K., Meile, L., Kreuzer, M. and Machmuller, A., 2003. Effects of mixtures of lauric and myristic acid on rumen methanogens and methanogenesis *in vitro*. *Lett. Appl. Microbiol.*, 37: 35–39.

Steed, Jr. J. and Hashimoto, A.G., 1994. Methane emissions from typical manure management systems. *Bioresource Technol.*, 50: 123–130.

Swamy, M. and Bhattacharya, S., 2006. Budgeting anthropogenic greenhouse gas emission from Indian livestock using country-specific emission coefficients. *Current Science*, 91(10): 1340–1353.

Tavendale, M.H., Meagher, I.P., Pacheco, D., Walker, N., Atwood, G.T. and Subathira, S., 2005. Methane production from in vitro rumen incubations with *Lutus pedunculatus* and *Medicago sativa*, and effects of extractable condensed tannin fractions on methanogenesis. *Ani. Feed Sci. Technol.*, 123/124: 403–419.

WeiLian, H., JianXin, L., JunAn, Ye, YueMin, Wu and YanQiu, G., 2005. Effect of tea saponins on rumen fermentation *in vitro*. *Anim. Feed Sci. Technol.*, 120(3/4): 333–339.

Zin, R.A., 1989. Influence of level and source of dietary fat on its comparative feeding value in finishing diets of feedlot steers. *J. Anim. Sci.*, 67: 1038–1049.

ENVIRONMENTAL MANAGEMENT
A Search for Viable Options

Ecosystem Diversity and Carbon Sequestration: Climate Change Challenges and a Way Out for Ushering in a Sustainable Future *Pages 141–147*
Editors: P.L. Gautam, Vir Singh and Uma Melkania
Published by: DAYA PUBLISHING HOUSE, 2009

Chapter 15

Role of Biofertilizers to Mitigate Environmental Problems: Soil Fertility Management in Hill Agro-ecosystems

Susheela Negi, G.K. Dwivedi and R.V. Singh

Introduction

Advances in modern agriculture to meet the increasing food grain demands have adverse effects on environment. When human population was small and technology was in its infancy, environment could easily absorbed human and industrial wastes. With increased in human population and great advancement in technology in the recent past, the waste material have multiplied in amount as well as in kind, and this has resulted in contamination of environment. During the past green revolution period, the Indian Agriculture has became a kind of agriculture in which numerous chemicals like pesticides and commercial fertilizers are being used in producing crops. This has resulted in deterioration in soil health in terms of microbial biomass and whole rhizosphere is getting polluted. Nutrients imbalance, fast depletion in soil fertility and continuous deterioration of physical properties of soil are some of the added disadvantage of chemical agriculture. Fertilizers play a key role in increasing agricultural production. Continuous, increased and imbalanced use of chemical fertilizers, has caused alarm regarding possible side effects in relation to environmental pollution in different states to our country. Repeated use of chemical fertilizers without adding adequate amount of organic manure is causing the soil to become more and more hard and impervious to water.

The fertilizer consumption of India has increased many folds in recent years (Table 15.1).

Table 15.1: All India Consumption of N_2, P_2O_5 and K_2O in million tones

Year	*N*	*P_2O_5*	*K_2O*	*Total*
1990-91	7.99	3.22	1.42	12.64
1994-95	9.51	2.94	1.06	13.51
1997-98	10.91	3.92	1.37	16.2
1998-99	11.35	4.11	1.33	16.79
1999-00	11.59	4.80	1.68	18.07
2000-01	10.92	4.21	1.57	16.70
2001-02	11.42	4.42	1.71	17.55

Source: Fertilizers News, November, 2003.

Considering the ill effects of the chemical fertilizers and in order to increase the level of production and productivity of crop there is a need of utilizing biofertilizers for partial supplementing chemical fertilizers. Biofertilizers *viz. Rhizobium leguminasarum* (N-fixation) and *Pseudomonas striata* (P-solubilizer) are inputs containing micro-organisms which mobilize nutrient elements from non-usable form to usable form through biological processes. They are cost-effective and renewable source of energy during their production, improve crop growth and quality of the products by producing plant harmons, help in sustainable crop production through maintenance of soil productivity. Inoculation with *Rhizobium* culture increases the nodulation and nitrogen fixation in peas resulting in higher yield of green pod. Thus apart from meeting its own requirement of nitrogen, peas are known to leave behind residual nitrogen in soil 50-60 kg/ha (Kanwar *et al.*, 1990).

The microorganism used as biofertilizers regulate the physiological processes in the ecosystem by involvement in the decomposition of organic matter, fixation of atmospheric N, secretion of growth promoting nutrients, immobilization of these assimiable nutrients and protection of plant. Soil microbial biomass facilitates nutrient transformation in soil and in vital for the soil fertility and plant availability.

Keeping in mind the above factors present study was undertaken:

1. To study the effects of biofertilizers on nodulation behaviour, growth, yield and nutritional quality
2. To study the effects of biofertilizers with nutrient sources on nodulation behaviour, growth yield and nutritional quality

Materials and Methods

The present investigation including ten different treatments of biofertilizers, lime and FYM with recommended doses of NPK either alone or in combination in pea was carried out under field condition at Vegetable Research Farm, Hill Campus, G.B. Pant University of Agriculture and Technology during 1999-2000. Biofertilizers (*Rhizobium + Pseudomonas*) used as a seed inoculation @250g/10kg, FYM applied as a soil treatment @ 20 Mg/ha, N P K @ 50: 50: 50 Kg/ha as basal doze and Lime @ 4 Mg/ha was broad cast in the experimental field at the time of field preparation. The experiment was design in FRBD with three replication.

Result and Discussion

Composite inoculation of both *Rhizobium leguminosarum* + *Pseudomonas striata* significantly increased the nodule number by 81.28 per cent. It has been proved that *Rhizobium* inoculation helped in increasing the nitrogen content especially in the form of NH_2^+ in the soil, which is preferred by the plants. The P-solubilizing organism they help solubilizing the native as well as applied phosphates thus increasing the phosphate content in soil and subsequently plant uptake. Similar findings were also reported by Jayasheel and Suvarna (2000). Further liming also increased the nodule number by 433.85 per cent at 140 days after sowing. Liming increased the Ca levels in the soil, which enhance the root penetration into the deeper layers and also induced the normal distribution of nodules. Similarly, Raychaudhari *et al.* (1998) reported 304.24 per cent increased in nodule number in soybean.

Biofertilizers *viz. Rhizobium leguminosarum* and *Pseudomonas straita* increased the shoot length by 71.43 per cent over no inoculation. The beneficial effect of inoculation of *Rhizobium leguminosarum* and *Pseudomonas straita* was also reported by Jain *et al.* (1999).

Table 15.2: Effect of Biofertilizers on Nodulation

Treatment	*Nodule Number at 140 DAS*
No biofertilizers	32.6
Rhizobium leguminosarum + *Pseudomonas striata*	59.1
S.Em +	0.71
C.D. at 5 per cent	2.09

Interaction Effect of BF x Source of Nodulation

Treatment	*BF_0*	*BF_1*
Control	3.4	17.8
FYM	11.4	27.8
NPK	8.6	120.9
FYM+NPK	19.9	45.2
Lime	42.5	62.3
Sem ±	0.82	
C.D. at 5 per cent	2.44	

The interaction effect between the biofertilizers *viz. Rhizobium leguminosarum* + *Pseudomonas straita*, nutrient sources and lime were found significant in case of shoot length. Seed co-inoculated with *Rhizobium leguminosarum* + *Pseudomonas straita* alone was superior to FYM and NPK application. While, showed comparable effect with combined application of 50 per cent recommended dose of FYM + NPK. Similarly, application of lime @ 4Mg ha^{-1} lime and NPK (25:25:25 kg/ha) with co-inoculation of biofertilizers was at par. Co-inoculation of *Rhizobium leguminosarum* + *Pseudomonas straita* along with FYM significantly increased the plant height over co-inoculation of *Rhizobium leguminosarum* + *Pseudomonas striata* alone, FYM, NPK, combined application of 50 per cent of recommended dose of FYM + NPK and NPK with biofertilizers. Maximum shoot length was obtained with lime, co-inoculated with *Rhizobium leguminosarum* + *Pseudomonas striata* followed by combined application of 50 per cent recommended dose of FYM + NPK + co-inoculation of *Rhizobium leguminosarum* and *Pseudomonas striata*. However, the response of FYM and NPK was at par.

Table 15.3: Effect of Biofertilizers on Shoot Length

Treatment	*Shoot Length (cm)*
No biofertilizer	15.4
R. leguminosarum + *P. striata*	26.4
Sem ±	0.47
C.D. at 5 per cent	1.41

Interaction effect of BF x nutritional sources on shoot length

Treatment	BF_0	BF_1
Control	11.2	16.4
FYM	13.9	27.3
NPK	13.1	21.6
FYM + NPK	16.8	30.7
Lime	21.9	36.1
Sem ±		1.06
C.D. at 5 per cent		3.16

Number of Grains/Pod

Inoculation of seed with *Rhizobium leguminosarum* and *Pseudomonas striata* significantly increased the number of grains per pod by 25 per cent. This is because grains *Rhizobium* inoculation increase the nodulation and nitro fixation capacity in pea which is positively correlated to grain yields. Kanaujia *et al.* (*1998*) also reported 10.13 per cent increase in grains with *Rhizobium* inoculation.

The interaction effect between the biofertilizers, NPK and NPK in the presence of biofertilizers *viz. Rhizobium leguminosarum* + *Pseudomonas striata*, lime and lime in the presence of co-inoculation exhibited significant increased in number of gains per pod. While other treatments showed non-significant. Lime in the presence of biofertilizers *viz. Rhizobium leguminosarum* + *Pseudomonas striata* increased the number of grain per pod *i.e.* 8.8 grains per pod. While, application of NPK in the presence of biofertilizers was at par with FYM + NPK 50:50 of the recommended doses.

Table 15.4: Effect of Biofertilizers on Grains/pod

Treatment	*No. of Grains/Pod*
No biofertilizer	6.0
R. leguminosarum + P. striata	7.5
Sem ±	0.25
C.D. at 5 per cent	0.73

Interaction ERffect of BF x Nutritional Sources and Amendments on Grains per Pod

Treatment	*BF_0*	*BF_1*
Control	5.1	6.9
FYM	6.0	7.0
NPK	5.6	7.3
FYM + NPK	6.3	7.3
Lime	7.1	8.8*
Sem ±		0.55
C.D. at 5 per cent		1.63

Table 15.5: Effect of Biofertilizers on Green Pod Yield

Treatment	*Green Pod Yield*
No biofertilizer	49.0
R. leguminosarum + P. striata	74.5
Sem ±	0.51
C.D. at 5 per cent	1.52

Interaction Effect of BF x Nutritional Sources and Amendments on Green Pod Yield

Treatment	*BF_0*	*BF_1*
Control	34.2	50.3
FYM	47.9	66.9*
NPK	45.7	63.3*
FYM + NPK	52.6	92.5
Lime	64.9	99.7
Sem ±		1.14
C.D. at 5 per cent		3.41

Yields

Green pod yield was significantly increased with co-inoculation of *Rhizobium leguminosarum + Pseudomonas striata* by 52.04 per cent. The increased in yield appears to be due to proper establishment and greater infection of *Rhizobium* strain, more number of nodules and their weight, which resulted in supply of N on larger quantity to plants. Similar results have been reported by Parashar *et al.* (1999) and Prakash *et al.* (2000) in broad bean and peanut and per cent increased was 88 and 94 per cent respectively. Biofertilizer as composite inoculants in presence of either FYM and NPK or mixture of the two as well as lime further accentuated the green pod yield of pod and highest pod yield was recorded with lime and biofertilizers *i.e.* 99.7 q/ha followed by NPK + FYM (92.5 q/ha).

The interaction effect between the treatments were also found significant among each other. It is further evident that the application of the lime applied @ 4.0 Mg ha^{-1} exhibited comparable effects with the FYM application @ 20 t/ha with seed inoculation with biofertilizers or application of NPK (25: 25:25) along with seed inoculation.

Similarly, seed inoculation with biofertilizers alone was superior to FYM and NPK while showed comparable effect with NPK and FYM combination whereas seed inoculation with *Rhizobium leguminosarum* + *Pseudomonas striata* in presence of either lime or in combination of FYM + NPK was superior to other treatments.

Table 15.6: Effect of Biofertilizers on Number of Pods per Plant

Treatment	*Number of Pods per plant*
No biofertilizer	8.7
R. leguminosarum+ P. striata	17.0
Sem ±	0.33
C.D. at 5 per cent	0.98

Interaction Effect of BF x Nutritional Sources

Treatment	BF_0	BF_1
Control	4.6	9.3
FYM	7.8	16.1
NPK	6.0	13.2
FYM + NPK	11.0	21.2
Lime	14.0	25.3
Sem ±		0.74
C.D. at 5 per cent		2.20

Number of pods per plant significantly increased by co inoculation of biofertilizers by 95.40 per cent over control. This may be due to the fact that with the increase of nodulation as influenced by biofertilizers resulted in an increase of nitrogenous activity of nodule in pea, which resulted in an higher nitrogen fixation in the plant and improves source and sink relation as well as nitrogen content of rhizophere.

Conclusion

It is quite evident that though we require more food to feed our huge population for which more and more crop production is vital but not at the cost of polluting the environment, which in turn will cause health hazard. It is true that crop production cannot be sustained with inorganic fertilizers used in indiscriminate way resulting in pollution of soil and water. Further; it is true that we cannot achieve the target of producing more and more crop year after year with the use of only organic manure. Therefore, integrated nutrient management, conservation of natural resources and emphasizing soil based practices are of substantial value.

It can be concluded that to boost up the yield of the crop on sustained basis without affecting environment, liberal application of organic manure, in addition to need based application of inorganic fertilizers is essential.

References

Jain, P.C., Nagarajan, P. and Jagdish, P., 1999., Response of chickpea (*Cicer arietinum* L.) to phosphorus and biofertilizer. *Legume Research*, 24: 241–244.

Jaysheel, N. and Suvarna, C.V., 2000. Effect of dual inoculation of Redgram with *Rhizobium* and *Glomus mosseae* at different levels of phosporus. Extended summary: *International Conference on Managing Natural Resources. IARI, New Delhi.* Vol. 2: 688–689.

Kanaujia, S.P., Sharma, S.K. and Rastogi, K.B., 1998. Effect of phosphorus, potassium and *Rhizobium* inoculation on growth and yield of pea (*Pisum sativum* L.). *Annals of Agricultural Research*, 19: 219–221.

Kanwar, J.S., Chadha, M.L. and Khurana, D.S., 1990. Effect of nitrogen and bacterial inoculation on the seed yield of pea. *The Pb. Veg. Grower*, 25: 12–15.

Parashar, A., Awasthi, C.P., Singh, A.K. and Gupta, M.K., 1999. Effect of *Rhizobium* inoculation and phosphorus application on yield and quality of broad bean (*Vicia faba* L.). *Legume Res.*, 23: 162–161.

Prakash, R. and Gupta, B.R., 2000. Effect of *Rhizobium* and co–inoculants as affected by sulphur on nodulation, yield and quality of pea nut (*Arachis hypogaea* L.). In: *International Conference on Managing Natural Resources for Sustainable Agriculture Production in the 21st Century*, New Delhi, India, pp. 677–678.

Raychaudhuri, M., Kumar, K. and Raychaudhuri, S., 1998. Response of soybean to lime phosphorus on an ultisol of Manipur. *J. of the Indian Soc. of Soil Science*, 46: 629–632.

Ecosystem Diversity and Carbon Sequestration: Climate Change Challenges and a Way Out for Ushering in a Sustainable Future *Pages 148–152*
Editors: **P.L. Gautam, Vir Singh and Uma Melkania**
Published by: **DAYA PUBLISHING HOUSE, 2009**

Chapter 16

Effect of Sugar Industry Effluents on Seeds Germination and Seedling Growth of *Linum usitatissimum* L.: The Green Revolution Bowl Reels Under Industrial Pollution

Neelam and Ila Prakash

Physico-chemical and biological characteristics of the effluent of Baghpat Co-operative Sugar Mill Ltd., Baghpat of western Uttar Pradesh were investigated. The effluent was found acidic in nature, rich in suspended organic and dissolved matter with higher amounts of nitrate, phosphate, chloride, calcium and high value of BOD and nil dissolved oxygen (DO).

Linseed (*Linum usitatissimum L.* var. *Padmini*) seeds were kept on adsorbent paper for germination placed in petridishes and treated with equal amount of different concentrations of effluent (25, 50, 75 and 100 per cent) and distilled water for control. The study revealed that seed germination and seedling growth was maximum, even better than control in lower concentration (25 per cent) and declined strikingly with higher concentration reaching minimum at 100 per cent concentration of the effluent. Thus sugar mill effluent can be used for irrigation purpose at lower concentration (up to 25 per cent) without use of additional fertilizers for better crop results.

Keywords: *Linseed seeds, Sugar mill effluent, Germination, Seedlings.*

Introduction

Western Uttar Pradesh is one of the most important Green Revolution bowls of India. Sugarcane development has been at very rapid pace during the Green

Revolution era. Sugar industry has proliferated thanks to rapid expansion of the cultivated area under the water-guzzling commercial crop of sugarcane. Pollution created by this industry has been on the increase and of serious consequences for the whole area, especially to agriculture sector. This paper is a work on one of the issues associated with the sugar industry pollution.

The untreated or partially treated effluents discharged from industries find their ways into water bodies through cultivated fields. The toxic nature of the untreated effluents is hazardous to soil productivity of the affected area. There is a growing concern to check the pollution created by waste water and re-use the industrial waste water after suitable treatment for the irrigation purpose (Pound and Crites 1973; Sopper and Kaidos 1974, Kannabiran and Pragasam 1993, 2001; Subramani *et al.*, 1995). Several studies carried out about the effect of effluents on various growth parameters of various crops, *viz.* seed germination, seedling growth, crop yield etc. showed that properly treated effluent can be used for irrigation with its fertilizers value (Pragasam and Kannabiran 2001; Mukherjee and Sahai 1988). The present study has been undertaken to study the sugar mill effluents and assess effect of its varying concentration on seed germination and seedling growth of *Linum usitatissimum* L. (CV. Padmini) for the possible utilization of effluent for irrigation of crop land.

Materials and Methods

Baghpat Co-operative Sugar Mill Ltd. is situated in Baghpat city of western Uttar Pradesh which discharges the mill effluents into open land. For our experimentation, the sugar mill effluent was collected from the source at the point of discharge. The physico-chemical analysis of the effluents was carried out according to APHA (1981). The collected effluent was brought to room temperature before it was used in experimentation. The collected effluent was considered as stock (100 per cent concentration) and diluted to 25 per cent, 50 per cent, 75 per cent with distilled water for experimentation. A set with distilled water was also maintained and taken as control.

Healthy seeds of Linseed (C.V. Padmini) were obtained from IARI, Pusa, New Delhi. The seeds were pre-treated with 0.2 per cent solution of sodium hypochloride for 2 min. to sterilize them, followed by a thorough washing with water and air drying before keeping them in Petri plates for germination. The seeds were soaked in respective dilutions and control for over night. In all the 25 Petri plates, layers of absorbent paper were used with five replicates for each dilution and control. Twenty five seeds were kept in each plate for germination at room temperature. Different concentrations (25 per cent, 50 per cent, 75 per cent and 100 per cent) of the effluents, 10 ml in each, were added to each plate on alternate days. The seed germination was recorded at 48 hrs after soaking the seeds. Then, the observations were recorded on third, fifth and seventh days for the parameters like length of radicle and plumule.

Results and Discussion

The Physicochemical characteristics of Baghpat Co-operative Sugar Mill effluent are given in Table 16.1. The effluent was alkaline in nature (Verma *et al.*, 1978). The large amounts of suspended, dissolved and volatile solids were responsible for the high BOD and COD. Considerable amounts of chlorides, sulphates and phosphorus were also present (Tables 16.2 and 16.3). There was significant decrease in percentage germination and seedling growth with concentrated effluent. The data indicate that 100 per cent germination was in 25 per cent concentration whereas it was recorded 91.20 per cent in control. The minimum germination recorded was 76.60 per cent in 100 per cent concentration. In seedling growth, the radicle and plumule length were recorded 0.25 cm and 0.14 cm respectively initially which later increased to 2.25 cm and 1.95 cm on the 7th day in 100 per cent concentration. In the remaining sets, the maximum average length reached up to 4.37 cm and 3.15 cm (25 per cent concentration), 3.25 cm and 2.75 cm (50 per cent concentration), 2.97 cm. and 2.67 cm (75 per cent concentration) and 3.40 cm and 2.87 cm (control).

Table 16.1: Physico-chemical Characteristics of Sugar Mill Effluent

Sl.No.	*Parameters*	*Minimum*	*Maximum*
1.	Odour	Organic	organic
2.	Appearance	Clear	Clear
3.	Temperature	29.5°C	36.0°C
4.	pH	6.8	7.3
5.	DO	Nil	Nil
6.	BOD	660	1648
7.	COD	680	2265
8.	Total Nitrogen	1.4	6.0
9.	Phosphate	0.53	1.2
10.	Total solid	2818	5000
11.	Suspended solid	1800	3235
12.	Dissolved solid	955	1375

All values are in mg/l except pH, appearance and odour.

Table 16.2: Effect of Different Concentration of Sugar Mill Effluent on Germination of Linseed Seeds (Germination percentage was noted after 48 hrs of soaking)

Concentration of Effluent (%)	*Percentage of Germination*
Control	91.20
25 per cent	100.00
50 per cent	90.30
75 per cent	85.30
100 per cent	76.60

It was noted from the study that the germination and seedling growth of linseed was severely affected by the sugar mill effluent at higher concentration. The high salt concentration in the effluent causes a high osmotic pressure, which might be responsible for the suppression of germination and seedling growth of seeds (Sahai *et al.*, 1988; Bahadur *et al.*, 1990).

Table 16.3: Effect of Sugar Mill Effluent on Seedling Growth of Linseed

Treatment	3rd Day per Seedling (cm)		5th Day per Seedling (cm)		7th Day per Seedling (cm)	
	Radicle	Plumule	Radicle	Plumule	Radicle	Plumule
C	0.42	0.38	2.74	2.52	3.40	2.87
25 per cent	0.57	0.52	3.08	2.95	4.37	3.15
50 per cent	0.37	0.31	2.65	2.47	3.25	2.75
75 per cent	0.28	0.19	1.93	1.87	2.97	2.67
100 per cent	0.25	0.14	1.20	1.12	2.25	1.95

The observations made in the present study show the toxic effects of high concentrations of the sugar mill effluent on one hand and the beneficial effects of the same at lower concentrations, on the other. The use of the effluent in low concentration not only checks it from being an environment hazard, but also serves as a potential source of liquid fertilizer for agricultural crops.

Acknowledgements

The authors are thankful to Dr. H. Kumar for providing laboratory and other facilities and to Prof. V. Singh for supporting our research activities.

References

APHA, 1981. *Standard Methods for the Examination of Water and Waste Water, 15th edn.*. Amer. Public Health Assoc., New York.

Bahadur, Brijesh and Sharma, B.K., 1990. Effect of industrial effluent on seed germination and early seedling growth of *Triticum aestivum* var. U.P. 115. *Acta Botanica Indica*, 18: 80–83.

Kannabiran, B. and Pragasam, A., 1993. Effect of distillery effluent on seed germination, seedling growth and pigment content of *Vigna mungo* L. Heper (CVT9). *Geobios*, 20: 108–112.

Neelam, Sahai, R., 1988. Effect of fertilizer factory effluent on seed germination, seedling growth, pigment content and biomass of *Sesamum indicum* L. *J. Environ. Biol.*, 9: 45–50.

Mukherjee, V. and Sahai, R., 1988. Effect of distillery waste on seed germination, seedling establishment and early seedling growth of *Cajanus cajan* L. (var. 5-16) *Acta Botanica Indica*, 16: 182–185.

Pound, C.E. and Crites, R.W., 1973. *Waste Water Treatment and Reuse by Land Application.* Vols. 1 and 2. Environ. Protection Agency Ad. Oklanoma. E.P.A., 660/2–73-006.

Pragasam, A. and Kannabiran, B, 2001. Effects of distillery waste on the photosynthetic pigments and nitrogenous compounds in *Vigna munga* L. Heper. *J. Samy. Bot. Cl.*, 18: 21–23.

Sopper, W.E. and Karados, L.T. (Eds.), 1974. *Recycling Treated Municipal Waste Water through Forests and Cropland.* Environ. Protection agency, University Park, Pa. E.P.A. 660/2- 73-006.

Subramanium, A., Sundaramonorthy, P. and Lakshmanachry, A.S., 1995. Effect of distillery effluent on *Vigna radiata*. *Poll. Res.*, 14: 477–481.

Verma, S.R., 1978. Studies on the sugar factories and their wastes in western U.P. *Ind. J. Environ. Health*, 20: 205–218.

Ecosystem Diversity and Carbon Sequestration: Climate Change Challenges and a Way Out for Ushering in a Sustainable Future *Pages 153–161*
Editors: P.L. Gautam, Vir Singh and Uma Melkania
Published by: DAYA PUBLISHING HOUSE, 2009

Chapter 17

Soil Carbon Sequestration: A Study in *Eucalyptus* Hybrid Plantations

Asha Upadhyay and Uma Melkania

The international panel of scientists by evaluation of evidences of climate change found that the anthropogenic interventions have been generating greenhouse gases that trap heat and produce runaway global warming. The Kyoto Agreement established a binding target for the green house gas emission (GHG) reduction of 5 per cent below 1990 level. A time period of 2008-2012 is available to the industrialized world to cut the GHG emission. Developing countries are asked to contribute to climate change mitigation through participation in clean development mechanism (CDM) as partners and host for renewable energy and plantation forestry projects.

Plantation forestry creates the sink for carbon-dioxide along with the restoration of degraded lands. By sequestration of the atmospheric carbon-dioxide the plantations transfer it to long lived pools and securely store it with lowering the degree of fear for remittance. Soils under these plantations can also sequester carbon in terms of soil organic carbon (SOC) and soil inorganic carbon (SIC) stocks. Storage of carbon in below ground systems is the best long-term option for carbon storage in terrestrial ecosystems because most soil organic matter has a long time than plant biomass.

The present study investigates the impact of *Eucalyptus* plantation on soil carbon dynamics. The soil organic matter content under this plantation was assessed along with the climatic variables and microbial activity. It is experienced during the study that capacity of soil as carbon sink is governed to a greater extent by climatic variables and conditions supporting microbial activity. There is a great possibility of increase in soil organic carbon sequestration by management of land and selected removal of litter. Thus through applying judicious land use and management practices the soil under such afforestation and reforestation plantations will result into a secure sink for sequestered carbon. Soil carbon sequestration improves soil health and sustains biomass productivity along with secure storage of greenhouse gases indirectly.

Introduction

Understanding the consequences of anthropogenic interventions in earth's climate system has advanced the knowledge base on climate change and response options to reduce the greenhouse gas levels. The interest is growing to find out measures aimed at capturing the atmospheric carbon dioxide levels. The reduction in greenhouse gas (GHG) emission associated with change in technological strategies and development of sink in green biomass to reduce the already accumulated gaseous carbon can be viewed as the best expected steps to mitigate climate change.

Owing the capacity to fix carbon dioxide as biomass forestry globally has taken central place to mitigate climate change. Carbon sequestration has become as an important value of forests other than as natural resource. Forest trees remove the carbon dioxide from the atmosphere by the natural process of photosynthesis and store it in the form of organic carbon in the cells. The part of this accumulated organic carbon is used up by the tree itself and re-emitted to the atmosphere during the respiration process and rest is built up as tree biomass. A portion of the fixed biomass carbon pool goes to pedosphere in the form of litter fall and starts the biogeo-chemical cycle in the soil. All the nutrients including carbon are released in soil by the decomposition process run by the microbes in soil environment. Carbon, initially in the atmosphere, through leaf fall and decomposition is now potentially transferred to soil carbon sequestration process. Soil carbon sequestration implies increase in soil organic carbon (SOC) and soil inorganic carbon (SIC) stocks through judicious land use and management practices.

After Russia's approval on Kyoto Protocol there is pressure on the developed countries to reduce the GHG levels to 5.2 per cent below the 1990 level in the first commitment period. Clean development mechanism (CDM) provides an additional opportunity to Annex I countries in achieving compliance with their Quantified Emission Limitation and reduction (QELR) commitments under Article 3 of Kyoto Protocol. The only two activities approved till present under CDM for the first commitment period are Afforestation and Reforestation (A and R). Although the definition of A and R is still being debated but its implication may be an attractive option for carbon mitigation as (1) the wood products generated as a result of project activity serve as substitute for the similar products unsustainably harvested from the natural forests (2) it sequesters the carbon in vegetation and soil (3) it increases income to the local people involved and grant carbon credits to the country running the project to offset GHG emission (4) it involves both developed and developing countries to play their role in mitigating climate change with sustainable development of humanity as the main goal to be achieved.

The soil carbon sequestration is not only important in view point of carbon price driven by mitigation projects but also crucial due to enhancement of soil quality, soil biodiversity, elemental cycling and nutrient reserves (Lal 2004). Soil contain 3.5 per cent of carbon reserves of the earth, compared with 1.7 per cent in the atmosphere, 8.9 per cent in fossil fuels, 1.0 per cent in biota and 84.9 per cent in the oceans (Lal *et al.*, 1995). Hence, there is need to improve SOC levels not only to maintain land productivity but also to provide a sink for atmospheric CO_2.

The only option available to increase SOC is to increase the plant biomass on the available lands which in turn will help to add the requisite the SOC level. Raising trees on degraded and barren lands may not result into natural forest-like situation but still can enhance SOC thus, actively participating in soil carbon sequestration. There are various factors involved in the soil carbon sequestration as photosynthetic efficiency of the tree, rate of litter addition to soil, rate of humification and decomposition, moisture status of soil, activity of soil biota and climatic variables (Singh and Rathod, 2002). Choice of tree for plantation is itself a significant factor affecting soil carbon sequestration. After the approval of A and R activities as CDM projects it is expected a large area to be covered under plantation forests in India. *Eucalyptus* hybrid is on the prime choice as extensively planted commercial tree in India by both the government and private sectors.

This paper represents the results of soil organic carbon storage in the soils of *Eucalyptus* hybrid plantation and its comparison with the adjacent grassland having no plantation tree.

Material and Methods

Site Conditions

The field study was conducted in plantation forest of *Tarai* region of Uttarakhand namely Gangapur Patia, located in northern part of west Himalayas in the *Tarai* belt of Uttarakhand. It includes part of district Udham Singh Nagar, Uttarakhand and covers an area of 26 ha total. Out of which 3.0 ha is covered in the present study. The *Eucalyptus* hybrid plantation was raised to establish seedling seed production area of *Eucalyptus* for improved quality seed production under tree improvement program. *Eucalyptus* hybrid is supposed to be a cross between *Eucalyptus tereticornis* and *Eucalyptus camaldulensis* developed under natural conditions (Bulgannawar 1992).This experiment was established in 2001. Plants were planted at 4 × 3 meter spacing. Adjacent grassland having no plantation was also selected as control to compare different parameters for study.

The climate of the study site is humid, sub-tropical with dry hot summer and cold winter. Generally, the monsoon starts in the third or fourth week of June and lasts up to September or first week of October. The mean daily maximum and mean daily minimum temperature varies widely in all the seasons. The mean daily maximum temperature for the study year (2005-06) ranged between 18.9°C and 41.9°C and mean daily minimum between 3.7°C and 26.8°C. The approximate variation in precipitation was from traces (0.01 cm) to 258.6 mm. The soil pH under *Eucalyptus* hybrid plantation was in acidic range (7.2-6.1) while of the control site was in neutral range (6.26-7.29). The five most dominant under storey species recorded under

Eucalyptus hybrid plantation were *Lantana camara, Ageratum conyzoides, Murraya koenigii, Rauwolfia serpentina* and *Abutilon indicum.*

Soil Sampling

Litter dynamics study was carried out by laying out the plot of one m^2 area randomly and in replication. Leaf litter fall was monitored at one-month interval starting from May 2005 to May 2006 in the randomly marked points under the plantation. Litter was collected and weighed for dry mass and cumulative litter mass was calculated. Soil samples were collected randomly at 0-15 cm and 15-30 cm depths from each study site, using a mechanical auger in each month of the study period. Samples from three different points were homogenized and the process in repeated to collect five representative samples. Soil samples were carried to laboratory and air dried and stored at 4°C till the analysis for selected parameters. Soil organic carbon for various samples was determined by partial oxidation method (Walkley and Black 1934), soil bulk density following Black (1965) and soil carbon stock following Joao Carlos *et al.* (2001).

Results

Litter Production

The amount and pattern of leaf litter fall was observed in the *Eucalyptus* hybrid plantation for one year at the interval of one month (Figure 17.1). Litter fall was observed throughout the year in the plantation trees with peaks in the months of November and May during the study period. Litter accumulated ranged from 19.92 to 61.14 kg ha^{-1} during the whole study period with a sum of 289.6 kg ha^{-1} for the

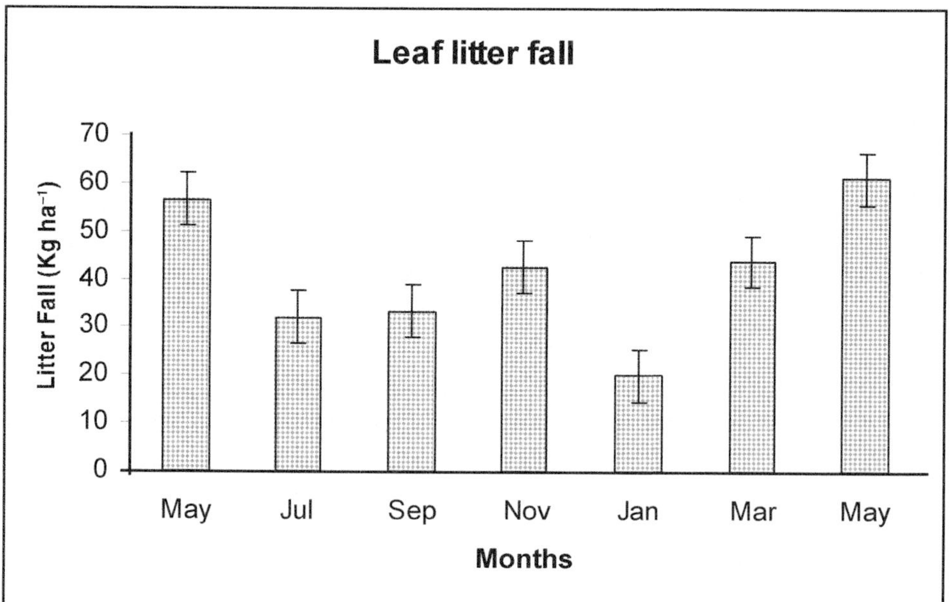

Figure 17.1: Leaf Litter Fall Dynamics (Kg ha^{-1}) in *Eucalyptus* Hybrid Plantation Systems

whole year. The variation in litter was due to seasonal changes and was found significant ($P < 0.01$).

Soil Organic Carbon

The soil organic carbon (SOC) content in the *Eucalyptus* hybrid plantation soil ranged from 1.81 to 3.06 per cent in the surface layer while from 1.68 to 2.68 per cent in the subsurface layer for the study period (Figure 17.2). In the grassland or control site it ranged from 1.39 to 2.87 per cent in the surface layer and from 1.01 to 2.41 per cent in the subsurface layer (Figure 17.3). A decreasing trend of SOC was observed with depths for both the study sites ($P < 0.01$). The values of SOC also differed significantly between the plantation and control site ($P < 0.01$). The values of SOC were low for the grassland as compared to the *Eucalyptus* hybrid plantation site.

Bulk Density and Soil Organic Carbon Stock

The bulk density of the *Eucalyptus* hybrid plantation site was recorded 1.43 g cm^{-3} for the surface soil and 1.57 g cm^{-3} for subsurface soil while, for the grassland 1.56 g cm^{-3} was observed for surface and 1.57 g cm^{-3} for the subsurface soil. The bulk density of subsurface soil was higher than the surface soil in both the study sites.

The initial soil organic carbon stock of *Eucalyptus* hybrid plantation observed in May 2005 was 38.82, 79.13 t ha^{-1}, which increased to 51.69, 102.68 t ha^{-1} in May 2006 in the surface and subsurface layers, respectively (Table 17.1). However, for the control site the initial soil organic carbon stock was 32.53, 51.81 t ha^{-1} in surface layer and at the end 67.16, 123.63 t ha^{-1} in the subsurface layer for the similar period of study.

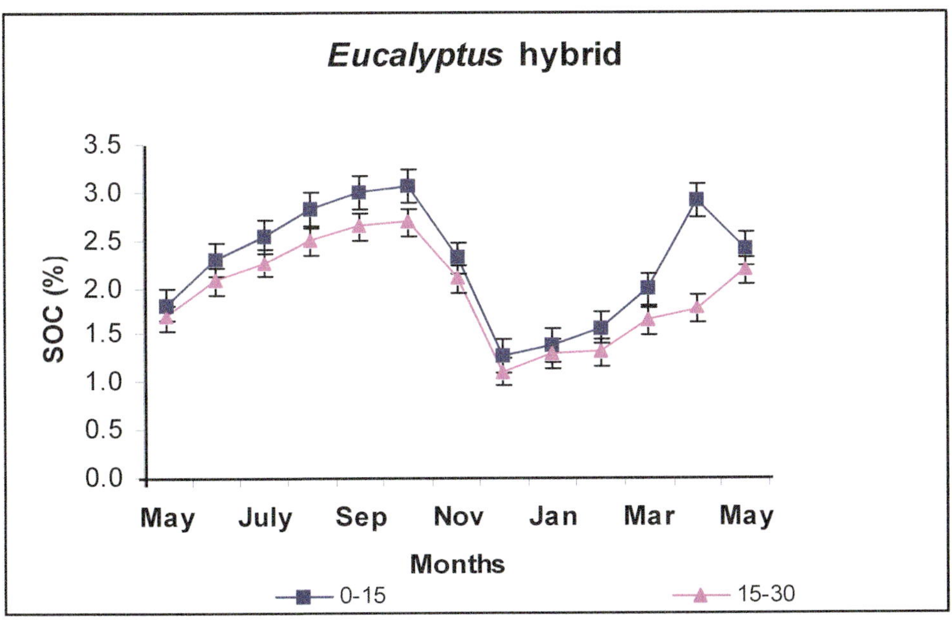

Figure 17.2: Temporal Changes in the Soil Organic Carbon (per cent) in *Eucalyptus* Hybrid Plantation System (Error bars are ± standard error)

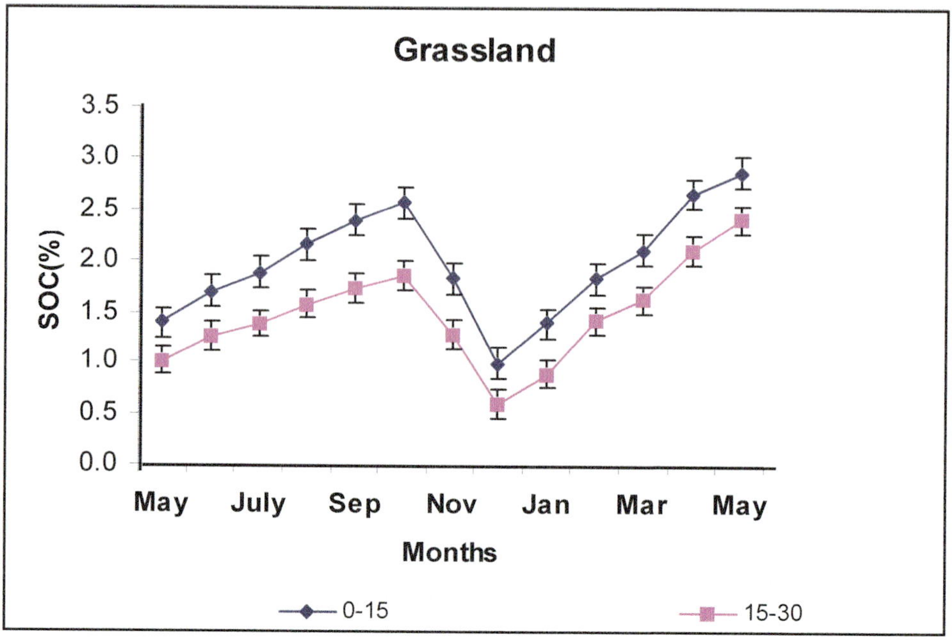

Figure 17.3: Temporal Changes in the Soil Organic Carbon (Per cent) in Grassland System (Error bars are ± standard error)

Overall the soil organic carbon stock of subsurface layer was higher for both the study sites and increasing trend was observed with the time of study. The results indicated that plantation of the *Eucalyptus* hybrid increased the soil organic carbon sequestration as compared to the grassland without any plantation.

Discussion

In the present study it has been inferred that in the same climatic and edaphic conditions the plantation activity helps to increase the soil organic carbon stock. The observed soil organic carbon was also higher in the plantation site. Vegetation was the major factor affecting the soil carbon sequestration and storage. The appropriate afforestation technique not only sequesters carbon in the plant biomass but also in the soil (Ramachandran 2007).

The carbon stocks of the grass land was increased more than the plantation site when the initial and final month of study were compared, but the carbon in the grass land is labile to leaching due to anthropogenic disturbances and the climatic factors. Due to the absence of vegetation cover the soil of grassland is expected to expose to external factors resulting into higher microbial decomposition and leaching in long time run (Singh *et al.*, 2000). This may be the probable reason of higher mean soil organic carbon stock in the plantation site as compared to the control or grassland site. This in turn indicates about the role of *Eucalyptus* hybrid plantation in the soil carbon storage.

Table 17.1: Temporal Changes in Soil Organic Carbon Stocks of the Plantation and Control Sites

Month	*Eucalyptus Hybrid*		*Grassland*	
	0-15 cm	*15-30 cm*	*0-15 cm*	*15-30 cm*
MAY	38.82±0.04	79.13±0.09	32.53±0.06	51.81±0.06
JUN	49.34±0.09	97.50±0.14	39.78±0.06	64.13±0.09
LULY	54.48±0.05	106.45±0.09	44.23±0.08	70.79±0.02
AUG	60.49±0.05	117.28 ±0.05	50.54±0.09	81.05± 0.04
SEP	64.35±0.06	124.34±0.04	56.16±0.09	89.26 ±0.09
OCT	65.64±0.08	126.23±0.07	60.14±0.03	95.42±0.008
NOV	49.55±0.08	98.44±0.07	43.06±0.07	65.15±0.009
DEC	27.24±0.01	51.81±0.01	23.17±0.11	30.27±0.005
JAN	29.60±0.01	60.76±0.01	32.53±0.08	45.66±0.11
FEB	33.46±0.01	61.70±0.02	43.06±0.09	72.33±0.06
MAR	42.47±0.01	77.24±0.01	49.37±0.05	83.62±0.09
APR	62.42±0.01	83.37±0.09	62.24±0.05	108.24±0.06
MAY	51.69±0.01	102.68±0.02	67.16±0.05	123.63±0.11
Mean	48.43	91.30	46.46	75.49

Values are mean of five replicates with standard error ±.

The litter fall added to the soil by the tree was the major source of carbon added to the soil which was absent in the grassland lacking tree plantation. The added litter (Lal 1989 and Lugo and Brown 1993) and the proliferated root systems (Blevines and Frye 1993) of the growing plantation tree influence the SOC store in the soil. A study by Singh and Singh (1993) in Karnal (India) indicated that raising tree on alkali soil increased the SOC from 0.12 to 0.58 per cent in the 20 years. *Eucalyptus* being the long term rotation plant as compared to the fast growing short rotation plants cultivated for pulp wood helps to sequester more carbon than in the grass land soil (Mendham *et al.*, 2003). Similar findings were reported by Balamurugan *et al.* (2000), Trouve *et al.* (1994), Gill and Abrol (1990) and Singh *et al.* (1981) in the different plantation and agroforestry systems.

Conclusion

This study clearly demonstrates the role of plantations in improving the soil carbon stock in the Tarai region of Uttarakhand. *Eucalyptus* hybrid tree system is proved to be efficient in soil carbon sequestration when employed in the plantation forestry for the CDM projects. Higher amount of litter addition in the plantation system showed positive impact on the soil carbon sequestration and accumulation. The climatic variables influence the soil carbon stock up to some extends but under similar climatic conditions the vegetation cover helps to improve the SOC stock. Thus judicial site specific afforestation on the barren land and degraded soils can help to mitigate the atmospheric carbon-dioxide.

The strategy must be employed to create vegetation barrier of the trees to control the movement of soil nutrients specially SOC. Raising *Eucalyptus* hybrid plantation, promising species for CO_2 sequestration and greenhouse gas mitigation the SOC in the top soil can be increased and may contribute to substantial soil conservation along with greenhouse gas mitigation. Site specific plantation of *Eucalyptus* hybrid can be a better climate change mitigation option because of the secondary environmental benefits associated with carbon sequestration in the context of mitigating global climate change. Such plantations with managed extraction of litter and wood can help to mitigate the demand for pulpwood and reduce pressure on natural forests. Hence, carbon inventories of popular plantation species in this area are required for the future sustainable management of the forest ecosystems. It is also prerequisite to claim carbon credits while proposing a mitigation CDM project.

References

Balamurugan, J., Kumaraswamy, K. and Rajarajan, A., 2000. Effects of *Eucalyptus citriodora* on the physical and chemical properties of soils. *Journal of the Indian Society of Soil Science*, 48(3): 491–495.

Black, C.A., 1965. *Methods of Soil Analysis*. Academic press Inc., New York, pp. 367.

Blevines, R.L. and Frye, W.W., 1993. Conservation tillage: An ecological approach in soil management. *Advances in Agronomy*, 51: 33–78.

Bulgannawar, G.M., Shantoppa, S. and Satish Chandra, K.M., 1992. Tree improvement in *Eucalyptus hybrid* through vegetative propagation. In: *Vegetative Propagation and Biotechnologies for Tree Improvement*. Natraj Publication, Dehradun, p. 31–43.

Gill, H.S. and Abrol, I.P., 1990. Evaluation of coastal sandy soils and their saline ground water for afforestation: a case study from India. *New Forests*, 4: 37–53.

Joao Carlos de, M.S., Carlos, C.C., Warren, A.D., Rattan, L., Solismar, P., Venske, F., Marisa, C.P. and Brigitte, E.F., 2001. Organic matter dynamics and carbon sequestration rates for a tillage chronosequence in a Brazilian Oxisol. *Soil Science Society of American Journal*, 65(5): 1486–1499.

Lal, R., 1989. Conservation tillage for sustainable agriculture: tropics vs. temperate environment. *Advances in Agronomy*, 42: 186–197.

Lal, R., 2004. Soil carbon sequestration impacts on global climate change and food security. *Science*, 304(5677): 1623–1627.

Lal, R., Kimble, J.M., Levines, E. and Whiteman, C., 1995. World soil and greenhouse effect: an overview. In: *Soil and Global Change*, (Eds.) Lal, R., R.M. Kimble, J. Levine and B.A. Stewart. CRC Press. Boca Raton, Fl. pp. 1–25.

Lugo, A.E. and Brown, S., 1993. Management of tropical soils as sinks or sources of atmospheric carbon. *Plant and Soil*, 149: 27–41.

Mendham, D.S., O'Connell, A.M. and Grove, T.S., 2003. Change in soil carbon after land clearing or afforestation in highly weathered lateritic and sandy soils of south-western Australia. *Agriculture Ecosystems and Environment*, 95(1): 143–156.

Ramachandran, R., 2007. The heat is on. *Frontline*, 24(4): 4–9.

Singh, G., Gupta, G.N. and Kuppusamy, V., 2000. Seasonal changes in organic carbon and nutrient availability in arid zone agroforestry systems. *Tropical Ecology*, 41(1):17–23.

Singh, G. and Rathod, T.R., 2002. Plant growth, biomass production and soil water dynamics in a shifting dune of India desert. *Forest Ecology and Management*, 171: 309–320.

Singh, G. and Singh, N.T., 1993. Mesquitefoe vegetation of salt-affected soils. *Research Bulletin* No. 18. Central Soil Salinity Research Institute, Karnal.

Singh, K., Chauhan, H.S., Rajput, D.K. and Singh, D.V., 1981. Report of a 60 month study on litter production, changes in soil chemical properties and productivity under Poplar *Eucalyptus* inter cropped with aromatic grasses agroforestry system. 9: 37–45.

Trouve, C., Mariotti, A., Schwartz, D. and Guillet, B., 1994. Soil organic carbon dynamics under Eucalyptus and Pinus planted on savannas in the Congo. *Soil Biology and Biochemistry*, 26(2): 287–295.

Walkley, A. and Black, C.A., 1934. An examination of Degtjareff methods for determining soil organic matter and a proposed modification of the chromic acid titration method. *Soil Science*, 37: 29–38.

Ecosystem Diversity and Carbon Sequestration: Climate Change Challenges and a Way Out for Ushering in a Sustainable Future *Pages 162–172*
Editors: **P.L. Gautam, Vir Singh and Uma Melkania**
Published by: **DAYA PUBLISHING HOUSE, 2009**

Chapter 18

Alternate Use of Biomass for Sustainable Development: Gasification Technology for Solving Energy Crisis in Rural Areas

Raj Narayan Pateriya and Sadachari Singh Tomar

The mankind has been interested in exploiting the natural resources for the fulfillment of needs without fixing the standard limit to depletion of natural resources or ecological/environmental costs. In fact, any economic activity that continues without a healthy balance between ecology and environment shall result in unsustainable nature of development. In most of the regions, due to over exploitation of forests and continuing application of wood as fuel, is blooming increasingly uncertain. Therefore, to meet the energy demand of the majority of the people is possible through the alternate use of the biomass which is available in the abundant quantity in our country. The potential of the biomass for the study area is calculated with the help of quadrat method. Realizing the importance of biomass resources in the energy scenario to meet the energy demand at the village level, the gasification was the most appropriate option due to the fact that the other alternatives have their own limitations, *e.g.* biogas can not replace directly the petroleum diesel due to its high carbon content, high amount of hydrogen sulphide gas and handling problem, besides its bulky structure. But such type of limitation does not exist with gasification technology. The density of the different types of trees varied from 0.48 to 3.68, abundance from 1.28 to 5.82 and frequency percentage from 28 to 84. There were 729 trees/shrubs per hectare. The dry biomass obtained was 8.4 kg/tree/year. Thus the estimated forest dry biomass available in the study area was 3130 tonnes. Out of this, 2190 tonnes forest biomass was surplus which can be available for producer gas technology. The potential of producer gas from crop residues and forest biomass worked out to be $1.48X10^3$ GJ and $5.69X10^3$ GJ.

Introduction

Sustainability ensures that development meets the needs of the present without compromising the ability of the future generations to meet their demands. The sustainable development does not mean blind opposition to progress, but opposition to blind progress. There is also the need to only cautiously interfere with natural process. A better future for all of us can be ensured only through sustainable development of environment. India is a large country endowed with variety of natural resources. The available natural resources are used to meet the needs of its ever-growing population. Energy is the prime mover of development of a country, and thus the extent of availability of appropriate quality and mix of energy to various sectors remains a major policy issue before the government. In one sense, civilization is the process whereby man strives to become master of his environment. The civilization process has been dependent upon man's ability to work far beyond the limitation of his own muscles. Even man's survival has always depended upon his ability to derive adequate energy in the form of food calories from his environment. According to Earl Cook, the human energy consumption through history can be represented by six stages of human development as given in Table 18.1.

Table 18.1: Human Energy Consumption Throughout History

Sl.No.	*Stage of Human Development*	*Daily Per Capita Consumption (1,000 kilo cal)*					
		Date	*Food*	*Residential and Commercial*	*Industry and Agricultural*	*Transportation*	*Total*
1.	Primitive Man	1000000 Years Ago	2	-	-	-	2
2.	Hunting Man	10000 Years Ago	3	2	-	-	5
3.	Primitive Agricultural Man	5500 B.C.	4	4	4	-	12
4.	Advanced Agricultural Man	1400 A.D.	6	12	7	1	26
5.	Industrial Man	19th Century	7	32	24	14	77
6.	Technological Man	Modern Age	10	66	91	63	230

Importance of Biomass and Gasification Technology

Biomass is a natural product of solar energy and therefore, a renewable source of carbon and hydrogen which are the basic constituents of energy and chemical products. Even through the total renewable biomass of resources for energy far exceeds the world's total energy requirement, its volume exploitation remains limited because of the present low cost of fossil fuels and the heterogeneous nature of biomass. The use of biomass, or materials derived from living organism, is as old as life itself. Plants which fix carbon and hydrogen utilizing solar energy through photosynthesis have thus always formed either directly or indirectly the source of energy. In fact, traditionally the poorer sections have continued to depend on the direct utilization of biomass in the form of agricultural residues, cow dung, firewood and variety of other biomass. The biomass gasification technology appears to be promising in this respect. The conversation efficiency of some of the fuels is given in Table 18.2.

Table 18.2: Heat Value and Conversion Efficiency of Selected Fuels

Sl.No.	*Fuel and Unit of Measurement*	*Calorific Value Kcal*	*Mode of Burning*	*Conversion Efficiency %*
1.	Biogas (m^3)	4713	Standard burner	60.0 per cent
2.	Producer gas	1200	Standard burner	60.0 per cent
3.	Butane (LPG) gas	10882	Standard burner	60.0 per cent
4.	Kerosene (lit)	9122	Pressure stove	50.0 per cent
5.	Agro-residue (kg.)	4000	Open chulha	15-20
6.	Fire wood	4700	Open chulha	17-20
7.	Cow dung cake	2092	Open chulha	9-11

Biomass and its effective utilization as an energy source through various conversion techniques is increasingly seen as an important contributor even in the relatively near term energy scenario of the rural areas. Its importance in this particular region is also due to unavailability of the electricity and the fact that the rural livings can not afford the expenditure of the engine power.

Alternate Uses of Agro Residues

The energy issues taking care of the environment development problems need to be tackled before any real progress can be made towards sustainable developments.

Of all the promising renewable sources of energy such as solar, wind, micro hydel, geothermal and ocean energy, biomass is unique because, in addition to energy, it is the only renewable sources of carbon which is the basic building block for chemical industry. In addition to their use as fuel, agro-residues are also used in other conventional utilization patterns, namely food, fodder, fiber applications and fertilizers. Therefore, their management strategies should be so developed that they do not basically disturb their traditional uses, such as for thatching roofs and inputs to cottage industries for making ropes and baskets etc.

In the last few years the interest and activity in method for obtaining energy from biomass has expanded dramatically. As a part of expanded interest the fuel from biomass and the charring of biomass is preferred in view of the following reasons:

1. Alternatives between biological routes and thermo-chemical route; thermo-chemical routes are generally more efficient.
2. The biological routes are generally more efficient in terms of materials and organic matter recycling.

The agroresidues and other biomass are being used for a variety of applications. These are competing requirements for food, fodder, fuel, fiber and feedstock. In view

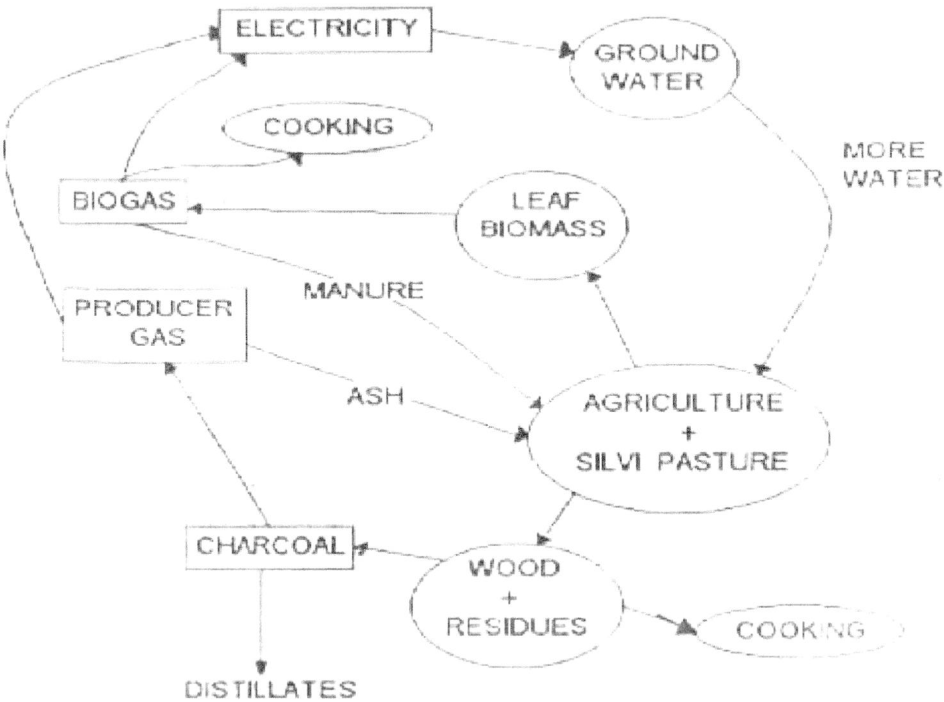

Figure 18.1: Land–Bio Energy–Water Nexus

of these competing uses of agro-residues there is need to develop energy and agro-residue utilisation planning both at micro and macro levels. As far as possible, the value addition of these rural resources should be carried out at the village level. This would also generate local employment and resources for rural population. One of the biggest fall outs of having this decentralized power generation cum agro-product processing approach is to retard the perpetual migration of rural population to cities. Therefore, the present study takes the close look to find out the potential of surplus biomass and its alternative use to meet the energy requirement of the study area.

Methodology

Determining of the Plant Biomass for the Study Area

Forest biomass and crop residues are the two major types of biomass available for energy purpose. In most of the villages, due to over exploitation of forests, continuing availability of wood as fuel is becoming increasingly uncertain, but with rising agricultural production, the production and availability of crop residues will continue to increase. For each tonne of main product of cereals, almost twice as much residue is produced. Although a variety of waste biomass is available but it is too scattered and availability is subject to seasonal variations.

Determining the Size of Quadrat

The biomass for the area is calculated with the help of quadrat. Quadrat is the name given to the different size of rectangular or it may be circular and spherical etc. frame. Determination of the size and number of quadrat is a pre requisite. The quadrats were thrown randomly in the different places and count of the trees/plants in the quadrat noted down. The size of the quadrat was fixed with the help of a string and 3 pegs fixed on the ground in L shape structure, then another string and pegs were taken to fix the area of 10 x 10m^2 and species of all kinds were counted. The size of the L shape structure was increased until and unless the total number of species remained constant. The first constant species size was taken as the size of quadrat.

Determining the Number of Quadrats

When the size of the quadrat is fixed we require fixing the number of quadrats for adequate sampling. The fix size of quadrat laid down on the forest area and number of species were noted down. The procedure was repeated until the number of species was not found constant. Then the first constant species number of quadrat will be the number of the quadrat.

To determine the ground plant biomass for the study area, we have followed by finding out the total biomass and then the surplus of biomass by quadrat method. The frequency, abundance and density of the plant species are also determined by quadrat method. This exercise was repeated by the requisite numbers of quadrats and the different species number of the particular species in the quadrat and the number of the quadrats which got that species were noted down. The frequency, abundance and density of the plant species were calculated with the help of following relationship:

(*i*) Frequency of the plant species
$$= \frac{\text{Total number of quadrates in which that species was present}}{\text{Numbers of the total quadrates which were studied}} \times 100$$

(*ii*) Abundance of the plant species
$$= \frac{\text{Total number of plant species}}{\text{Total numbers of quadrates in which all species were found}}$$

(*iii*) Density of the plant species
$$= \frac{\text{Total number of plant species}}{\text{Total number of quadrates studied}}$$

Determination of the Agricultural Crop Residues and Other Biomass

The rapid increase in the energy demand pushes to the alternative use of the biomass. The availability of agro-residues and useful product for different crops are presented in Table 18.3.

Table 18.3: Ratio of Crop Residue and Grain of the Main Crops for the Study Area

Sl.No.	*Crop Residue*	*Ratio*
1.	Wheat straw	1.47
2.	Gram straw	1.00
3.	Paddy straw	1.57
4.	Pigeon pea	1.32
5.	Jowar straw	2.00
6	Mustard sticks	1.85
7.	Bajra straw	2.00
8.	Groundnut haulms	2.05
9.	Maize stalk	1.56

The availability of agro-residues indicates that the potential exists to meet the demand. The traditional usage pattern and net surplus available either through adopting conservation measures or identifying surplus amount which should be ascertained before adapting the gasification technology.

Results and Discussion

Potential of Forest Biomass

The study area was located in Chitrakoot region of Kymore plateau of Satpura hills zone in M.P. The total forest area was about 503.46 ha, and the land which did not give enough economic return or was otherwise useless can broadly be termed as waste lands. The area of such type of land was reported to be 33 ha which was degraded and was presently lying unutilized. In the month of April the forest area exhibits typical summer aspect, with most of trees to be naked by shedding their leaves, while a few others develop some new leaves. The soil was covered with litter, but during the visit of forest it was observed that the litter was burnt away from several places. Some investigations were made to find out the frequency of trees and their abundance by Quadrat method. Samples were taken from different locations of the forest. Density and abundance of the different shrubs are presented in the Table 6. The size and number of the quadrat were found $100 \times 100m^2$ and 25 respectively (Tables 18.4 and 18.5).

The size of the quadrat was found 100x100 m^2, numbers of quadrates were 25, potential of forest biomass was observed $3.13x10^3$ tonnes and the total biomass (agriculture and forest) was $3.952x10^3$ tonnes. The estimated potential for producer gas was $7.17x10^3$ GJ. The density of different types of trees varied 0.48 to 3.68; abundance 1.28 to 5.82 and frequency percentage was observed between 28 and 84. Different types of trees/shrubs were observed to be 729 per hectare. Therefore the potential of producer gas from the crop residues and forest biomass was $1.48x10^3$ GJ and $5.69x10^3$ GJ.

Table 18.4: Determination of the Surplus Biomass Available by the Forest and Crop Residues

Sl.No.	Size of the Quadrat (m^2)	Total No. of Species Recorded
1.	10 × 10	08
2.	20 × 20	11
3.	30 × 30	14
4.	40 × 40	15
5.	50 × 50	18
6.	60 × 60	20
7.	70 × 70	28
8.	80 × 80	32
9.	90 × 90	45
10.	100 × 100	52
11.	110 × 110	52
12.	120 × 120	52

Table 18.5: Determination the Number of Quadrat

Sl.No.	Number of the Quadrates Studied	Number of Species Recorded
1.	10	18
2.	11	23
3.	12	25
4.	13	28
5.	14	29
6.	15	31
7.	16	32
8.	17	33
9.	18	34
10.	19	38
11.	20	39
12.	21	46
13.	22	42
14.	23	47
16.	24	49
17.	25	52
18.	27	52
19.	28	52
20.	29	52

Table 18.6: Determination of Plant Species, Plant Density and Abundance by Quadrat Method

Sl.No.	Plant Species	No. of the Plant Species in the Individual Quadrat																									Total No. of Plants	Total No. of Quadrates in which Species Found	Total No. of Quadrates Studied	Density	Abundance	Frequency %
		1	2	3	4	5	6	7	8	9	10	11	12	13	14	15	16	17	18	19	20	21	22	23	24	25						
1.	Tendu (*Diospyrous melanoxylon*)	1	-	1	3	-	1	3	6	5	-	2	-	1	1	1	2	2	-	-	-	3	-	-	-	1	40	16	25	1.6	2.50	64
2.	Jamun (*Syzizium cumini*)	-	-	-	-	-	1	2	2	1	-	2	2	-	-	-	-	2	-	-	-	-	1	-	1	1	15	10	25	0.6	1.50	40
3.	Mahua (*Madhuca indica*)	-	-	-	2	-	2	2	3	1	-	-	1	-	-	-	-	1	-	-	-	-	-	-	-	-	12	7	25	0.48	1.71	28
4.	Khair (*Acacia catechu*)	-	-	2	3	-	4	5	3	-	1	-	-	-	-	-	2	3	1	2	-	-	-	-	-	-	26	10	25	1.04	2.60	40
5.	Arjuna (*Terminallia arjuna*)	3	2	3	4	2	8	7	5	2	2	-	-	-	-	-	6	7	8	-	5	-	-	-	3	7	74	16	25	2.96	4.62	64
6.	Harsingar (*Nyctanthes arbortristis*)	2	-	3	5	7	8	7	7	7	8	-	-	5	-	-	3	7	8	-	-	-	1	-	1	-	87	15	25	3.48	5.32	60
7.	Saja (*Terminalia tomentosa*)	-	-	6	2	4	3	3	2	1	4	3	2	4	2	6	4	3	2	1	7	2	-	6	8	-	75	21	25	3.00	3.57	84
8.	Gurja (*Lamea coromandelica*)	5	8	5	9	3	4	6	3	3	2	2	-	-	-	-	2	2	3	-	1	5	-	3	-	-	66	17	25	3.64	3.88	68

Contd...

Table 18.6–Contd...

Sl.No.	Plant Species	No. of the Plant Species in the Individual Quadrat																									Total No. of Plants	Total No. of Quadrates in which Species Found	Total No. of Quadrates Studied	Density	Abundance	Frequency %
		1	2	3	4	5	6	7	8	9	10	11	12	13	14	15	16	17	18	19	20	21	22	23	24	25						
9.	Dudhi (*Wrightia tinctoria*)	-	-	8	9	8	9	8	4	5	6	2	1	-	-	-	2	3	1	-	-	-	-	-	-	-	66	13	25	2.64	5.07	5 2
10.	Achar (*Buchnania lanzan*)	4	5	7	8	9	6	4	5	3	5	-	1	1	-	-	1	-	-	2	-	-	-	-	-	-	61	14	25	2.44	4.35	56
11.	Mango (*Mangifera indica*)	3	2	2	2	1	2	1	1	1	-	-	-	-	-	-	2	1	-	-	-	-	-	-	-	-	18	11	25	0.72	1.63	44
12.	Neem (*Azadirachta indica*)	-	-	2	5	3	3	1	1	7	1	-	-	1	-	-	1	-	-	-	1	-	-	1	-	-	27	12	25	1.08	2.25	48
13.	Amala (*Emblica officinalis*)	1	2	3	7	8	9	5	4	3	2	1	1	-	-	-	-	4	9	1	-	-	-	2	-	-	62	16	25	2.48	3.87	64
14.	Salhai (*Boswelia robusta*)	1	2	1	5	3	4	2	3	1	2	1	1	2	1	-	-	3	5	-	-	1	-	-	-	-	38	17	25	1.52	2.23	68
15.	Babul (*Acacia nilotica*)	1	1	-	2	1	2	3	4	7	1	-	-	-	-	1	-	-	-	-	1	7	6	5	3	2	53	60	25	2.12	3.31	64
16.	Ankol (*Syzygium sanicifoniu*)	1	-	-	1	-	2	1	2	1	-	-	-	-	-	-	-	-	-	-	-	-	-	-	-	-	9	7	25	0.36	1.28	28
17.	Others	-	-	-	1	-	1	1	1	-	-	1	-	3	-	-	-	2	-	2	-	-	-	-	-	-	12	8	25			
	Total	= 729																														

Conversion Process of Biomass

The methods of converting biomass into useful form of energy can be broadly classified into biochemical and thermo chemical. The biochemical route of conversion of biomass to other useful forms is low energy process and relies upon the action of bacteria, which degrade complex molecules of biomass into similar ones. The production of biogas, a mixture of methane and carbon-dioxide, from animal dung by means of anaerobic digestion is an important example of this process. Enzymatic hydrolysis of wood and fermentation can produce ethanol, a liquid fuel that is comparatively cheaper to transport, easy to liquefy and relatively less polluting to burn. But the proposed process can be feasible only if forest biomass is used as the basic raw material for integrated process.

Table 18.7: Suitability of Different Biomass

Sl.No.	*Biomass*	*Group Characteristics*	*Applications*
1.	Most species of fuel wood	High density, CV and FC low ash content	Fuel for downdraft gasifiers and combustion
2.	Stalks of Arhar	Medium density and CV	Fuel for downdraft gasifiers and domestic use
3.	Paddy husk and groundnut shells	Low, to medium density and CV, High ash content	Fuel for throat less gasifiers
4.	Cattle dung, straw	Low density, lignin and physical strength	Biogas and alcohol production
5.	Maize cobs, paddy husk	High penthosans	Fuel for up draft and down draft gasifire
6.	Paddy husk and straw	High silica	Crystalline and amorphous silica production
7.	Wheat and paddy straws	High to medium celluloses	Paper industry

The gasification is a thermo chemical process in which biomass is subjected to a high temperature and depends on the quantity of oxygen supplied. Preparation of charcoal from wood and incineration of municipal solid wastes are examples of pyrolysis. Under certain conditions of temperature and oxygen supply, a gaseous mixture rich in carbon monoxide and hydrogen sulphide, referred to as producer gas, is formed.

References

Anonymous, 1994. *Agricultural Statistics*. Directorate of Agriculture, Govt. of M.P., Bhopal.

Chaturvedi, A.N., 1984. Assessment of biomass production. *Indian Forester*, 110(8). 726–737.

Doughlas, M.C., 1997. *Energy Technology Hand Book*. McGraw Hill Book Comp. New York. pp. 444–449.

Grover, P.D. *et al.*, 1998. Thermo-chemical characterization of biomass gasification. *Report on Physico-chemical Parameters of Biomass Residues*. Biomass Conversion Laboratory, Chemical Engineering Department, IIT Delhi, Hauzkhas, New Delhi.

Pandey, P.K., Bhattacharya, D. and Pateriya R.N., 2003. A case study of rural energy resources in Jakhi village. In: *Proceeding of International Conference Energy and Environmental Technologies for Sustainable Development*, MNIT, Jaipur, India

Singh, S., Verma, S.R. and Mittal, J.P., 1997. *Energy Requirements for Production of Major Crops in India*. AMA, New Delhi.

Ecosystem Diversity and Carbon Sequestration: Climate Change Challenges and a Way Out for Ushering in a Sustainable Future *Pages 173–186*
Editors: P.L. Gautam, Vir Singh and Uma Melkania
Published by: DAYA PUBLISHING HOUSE, 2009

Chapter 19

Arbuscular Mycorrhizal Fungi: A Unique Organism of Potential Implications for Carbon Sequestration

Rashmi Srivastava, Shruti Chaturvedi, Preeti Chaturvedi and A.K. Sharma

Soil carbon sequestration is the process of transferring carbon dioxide from the atmosphere into the soil through crop residues and other organic solids, and in a form that is not immediately re-emitted. This transfer or "sequestering" of carbon helps off-set emissions from fossil fuel combustion and other carbon-emitting activities while enhancing soil quality and long-term agronomic productivity. Soil carbon sequestration can be accomplished by management systems that add high amounts of biomass to the soil, cause minimal soil disturbance, conserve soil and water, improve soil structure, and enhance soil fauna activity. Continuous no-till crop production is a prime example.

Agriculture Management Practices and Carbon Sequestration

Organic matter has been lost from our cropland mainly through ploughing, which makes soil more likely to erode. Restoring soil carbon is essential to enhancing soil quality, sustaining and improving food production, maintaining clean water, and reducing increases in atmospheric CO_2. Short-sighted farming practices have resulted in loss of an estimated 4±1 gigatons (Gt) of carbon from soils of the United States, and 78±12 Gt from the world's soils, a large fraction of which ended up in the atmosphere (Lal 1999). Soil carbon loss has come principally from ploughing that turns over the soil, making it susceptible to accelerated erosion (Lal *et al.*, 1999). This is exemplified by the Dust Bowl era in the United States and is a serious issue in most developing countries. Although some carbon is sequestered (Smith *et al.*, 2001), accelerated water erosion is responsible for net emission of about 1 Gt C/year (Lal, 2003). Leaving crop residues after harvest increases the carbon content of soil and controls erosion, but the benefits are lost if the biomass is ploughed under, because microorganisms quickly degrade residue C to CO_2 (Reicosky *et al.*, 1997). Essential nutrients that adhere to soil organic carbon (SOC) disappear with its depletion. Thus, farmers require more fertilizer, irrigation, and pesticides to preserve yield. Water quality can deteriorate when less SOC is available for natural filtering. No-till agriculture (in which seeds are implanted without turning the soil with a plough) reduces the loss of the SOC pool (Dalal *et al.*, 1995; West and Post 2002), while conserving soil water and inhibiting weeds. Soil C enhancement would improve agronomic productivity (Bauer and Black 1994) and resource-use efficiency of impoverished soils. The beneficial effects of enhanced SOC cannot be fully replaced by increased levels of fertilizer, especially in soils of the tropics (Kanchikerimath and Singh 2001). No-till, in combination with mulching and crop rotation to enhance the SOC pool (Jenkinson 1991; Smith and Powlson 2000), is also a viable strategy for sustainable management of soils of the tropics in general. No-till would decrease silt in rivers and lakes, which would lower transport of SOC. Of all cultivated land (1379 M ha globally), no-till is currently practiced on only 5 per cent of the world's cropland. Rapid adoption of no-till farming in South America is attributed to cooperative agricultural extension education efforts (in which university staff work with the farming community), use of crop residue for mulch rather than for fodder or fuel, and development of systems to make no-till farming work. The success of no-till sowing of wheat after rice in the South Asian rice-wheat belt is encouraging. However, intense ploughing of water-saturated soil (puddling) for the rice crop and lack of residue mulch because of prior removal or burning at the time of sowing wheat minimize benefits. Furthermore, expansion into Africa and Asia remains a challenge, because crop residues are removed from the land, and animal waste is primarily used as fuel not as fertilizer. Identifying economic, clean, and healthy sources of household cooking fuel remains a challenge in developing countries. Topsoil is even used for bricks to meet the demand for housing. Farmers in India, for example, sell topsoil to 1-m depth for up to Rs 60,000/acre. Identifying alternate material for brick making is a high priority, but finding agriculturally marginal lands from which soil can be mined to deeper depths may also be needed. No-till agriculture, together with leaving crop residue in fields, does have costs. The yield may be lower in poorly drained and compacted soils and in places where springtime soil warming is slow. Initially, more fertilizer may be required, but, as SOC increases, the soil becomes more productive, requiring the same or even less fertilizer. Crop residue left in the fields would not be available for animal feed, energy production, biofuels (ethanol or hydrogen), or other uses and may increase incidence of pests and pathogens.

Soil Organic Matter and its Management

The sequestration of carbon in soils used for agriculture, forestry, and land reclamation has been recognized as a potential option to mitigate global change (Batjes 1996; Smith *et al.*, 1997; Lal 2003). Soils are the largest carbon reservoirs of the terrestrial carbon cycle. Soils contain about three times more C than vegetation and twice as much as that present in the atmosphere. Soils contain much more

C (1500 Pg of C to 1 m depth and 2500 Pg of C to 2 m) than is contained in vegetation (650 Pg of C) and twice as much C as the atmosphere (750 Pg of C). Carbon in the form of organic matter is a key element to healthy soil. It is estimated that each tonne of soil organic matter releases 3.667 tonnes of CO_2, which is lost into the atmosphere. Similarly, the build-up of each tonne of soil organic matter removes 3.667 tonnes of CO_2 from the atmosphere. The conversion of natural habitats to cropland and pasture, and unsustainable land practices such as excessive tillage frees carbon from organic matter, releasing it to the atmosphere as CO_2. Due to depletion of organic carbon, soils develop a carbon deficit and which could be regained by reabsorbing it from the atmosphere.

Soil organic matter (SOM) comprises the sum of all organic substances in the soil, and is defined as a *"mixture of plant and animal residues at various stages of decomposition, of substances synthesised microbiologically and/or chemically from the breakdown products, and of the bodies of micro-organisms and small animals and their decomposing products"*. It has a stabilising effect on soil structure, improves moisture retention, and protects soil against erosion. A wide range of factors affects levels of SOM, including moisture status, temperature, oxygen supply, drainage, soil acidity, nutrient supply, clay content and mineralogy. SOM accumulates best at low temperatures in acid parent materials and in anaerobic conditions. Large amounts of SOM and carbon are lost from the soil because of present intensive agriculture practices and to rectify the problem, the following measures have been suggested.

- Reduce ploughing through conservation and zero-tillage
- Use mixed rotations using cover crops and green manures
- Minimize summer fallows and periods with no ground cover
- Apply composts and manures to soil
- Improve pasture and rangelands through grazing and vegetation management
- Use perennial rather than annual grasses, as perennials have 60-80 per cent of biomass below ground compared with 20 per cent for annuals
- Restore and protect wetlands (provided carbon sequestration is greater than methane production)
- Convert agricultural land to woodlands
- Adopt agroforestry in cropping systems
- Cultivate crops for bio-fuels (grasses and coppiced trees)

However, increasing population and industrialization has reduced the land availability and especially in India, most of the population is vegetarian and hence dependent on the cereals and grains and this requires a continuous effort to maximize the productivity. It means that we can compromise only to certain extent with the present agricultural practices. Hence, the above listed measures are practically not possible as to meet the requirement of practices to be followed for effective carbon sequestration. It is, therefore, necessary to look for an alternate strategy, which could fit in the present agriculture scenario considering the measures mentioned and the practical problems of feeding growing population. Use of arbuscular mycorrhizal fungi, an essential component of terrestrial plants, could necessarily be the integral part of agriculture/forestry/horticulture system.

Arbuscular Mycorrhizal Fungi and Carbon Sequestration

Arbuscular mycorrhizal (AM) fungi form symbiotic relationships within roots of approximately 80 per cent of all plant taxa. The AM fungi colonize fine roots behind the area of active cell elongation. Hyphae radiate out from the roots, effectively performing the functions of uptake of nutrients, particularly phosphorus (P), and of water in exchange for photosynthetically derived carbon (C) from their host. They form a significant pathway for the transfer of photosynthetic C to soils. For example, AM fungi are estimated to utilize 5–25 per cent of photosynthetically fixed C in temperate herbaceous plant species and up to 45 per cent in temperate trees. Studies in tropical forests have indicated that 20–80 per cent of fine root length is colonized by AM fungi, and that spore production can be substantial. Many tree species are highly dependent on AM fungi, being unable to grow beyond the seed reserves if they are not inoculated with AM fungi. The C allocated to arbuscular mycorrhizas and thus their contribution to soil C could be particularly high in the tropics because of the low nutrient levels in highly weathered tropical soils. One of the compounds produced by AM fungi is a recalcitrant glycoprotein, glomalin. Concentrations of glomalin range from 2 to 15 mg g^{-1} of soil in temperate climates, and over 60 mg cm^{-3} were found in a chronosequence of Hawaiian soils. In addition to containing substantial carbon (and up to 5 per cent iron), glomalin enhances soil aggregation, thereby protecting carbonaceous material from rapid degradation in soils. Ecosystem productivity is thus enhanced due to improved levels of soil aeration, drainage and microbial activity. How glomalin concentrations vary across landscapes, and the influence of soil fertility on glomalin concentrations may provide insights into the importance of AM fungi and their products in soil C sequestration across tropical forest landscapes.

Recent research suggests that mycorrhizal fungi might be an important component of the SOC pool, in addition to facilitating carbon sequestration by stabilizing soil aggregates. Symbiotic associations between plant roots and Arbuscular mycorrhizal fungi (AMF) are ubiquitous in terrestrial ecosystems (Smith and Read 1997). The role of AMF in mediating the ecosystem response to global change has been reviewed recently (Rillig *et al.*, 2002; Stadon *et al.*, 2002). In view of the importance of AMF in ecosystem processes, these reviews highlight a significant void in research addressing the contributions of AMF to terrestrial carbon cycling. Recent studies using ^{14}C labeling indicated that photosynthate is transferred from host plants to AMF hyphae within hours after labeling (Johnson *et al.*, 2002). It is also generally accepted that AMF receive all their carbohydrate from the host plant and that the association of AMF with roots could create a sink demand for carbohydrate (increasing sink strength), which could result in a 4–20 per cent drain of carbon from the host plant and could indirectly influence carbon storage in soils (Graham 2000). Furthermore, up-regulation of photosynthesis by AMF is indicated where the amount of fungus in the root system is related directly to net carbon gain of the host

(Miller *et al.*, 2002). Such fungus mediated effects on plant growth can potentially improve carbon sequestration by increasing net primary production, especially in nutrient-limited environments.

The mycorrhizal symbiosis provides a number of benefits to the plant host. Rapid growth of fine, ephemeral hyphae in microsites containing high concentrations of nutrients such as P, N, Fe, Cu and Zn (Clark and Zeto 1996; Douds and Millner 1999; Pawlowska *et al.*, 2000) provides an efficient mechanism for nutrient uptake. The mycorrhizal relationship reduces the growth of plant pathogens, especially fungal pathogens, by increasing host resistance (*i.e.* triggering a defense response), altering root exudations to stimulate the growth of microbes antagonistic to pathogens, competing for photosynthetic carbon, and reducing the number of infection sites (Hooker and Black 1995; Borowicz 2001). The type of pathogen (nematode or fungal), pathogen species, mode of action (necrotrophic or wilt for fungal pathogens and migratory or sedentary for nematodes), and pathogen density help to determine the severity of disease (Borowicz 2001). As with other benefits in the mycorrhizal relationship, the magnitude and direction of effects on disease resistance are dependent upon host genotype, AM species and isolate, timing of AM colonization, other soil organisms and abiotic factors. Mycorrhizal fungi may hyper-accumulate toxic heavy metals, such as Cd and Pb, (Diaz *et al.*, 1996; Gonzalez-Chavez *et al.*, 2002). Metal uptake depends upon soil fertility, metal concentration, soil pH, host plant, and AM species and may interfere with P nutrition in the host plant (Diaz *et al.*, 1996; Gonzalez-Chavez *et al.*, 2002).

In addition to improving plant health, mycorrhizal fungi also contribute to soil health. Fungal hyphae improve soil structure by helping to form water-stable soil aggregates (Miller and Jastrow 1990; Tisdall *et al.*, 1997; Rillig and Steinberg 2002). Mycorrhizal fungi also improve rhizosphere health by stimulating root exudation, which promotes the growth of other soil microbes (Paul and Clark 1996; Borowicz 2001). Arbuscular mycorrhizal fungi contribute directly to SOM by accounting for 5 to 50 per cent of the total microbial biomass in soil (Olsson *et al.*, 1999).

Glomalin: A Glycoprotein Produced by AM Fungi

The identification of glomalin, a glycoprotein produced by arbuscular mycorrhizal fungi, has lead to a reevaluation of fungal contributions to SOM and aggregate stability. Glomalin was identified at the USDA in the early 1990's during work to produce monoclonal antibodies reactive with AM fungi. One of these antibodies reacted with a substance on hyphae of a number of AM species (Wright *et al.*, 1996). This substance was named glomalin after Glomales, the order to which AM fungi belong. Several other typical soil fungi, such as Rhizoctonia, Gaeumannomyces, Endogone, Mucor, and Phytophthora were tested for cross-reactivity with the antibody against glomalin but were not immunoreactive (Wright *et al.*, 1996). The glomalin fraction is operationally defined by its extraction procedure but is further characterized by total and immunoreactive protein assays (Wright *et al.*, 1996). Glomalin has been found in abundance (typically, 2 to 15 mg g^{-1} and up to >60 mg g^{-1}) in a wide range of soil environments (acidic, calcareous, grassland and cropland) (Wright and Upadhyaya 1998; Wright *et al.*, 1999) and appears to be as ubiquitous as AM fungi themselves (Carlile and Watkinson 1996; Wright and Upadhyaya 1998; Olsson *et al.*, 1999).

Glomalin was revealed on AM fungal hyphae using an indirect immunofluorescence procedure that employs the antibody against glomalin and a second antibody tagged with fluorescein isothiocyanate (FITC) molecule (Wright 2000). Evidence that glomalin is produced by AM fungi, not plant roots, was obtained early in the investigation of the reaction of the monoclonal antibody against glomalin. In a blind experiment, immunofluorescence correctly identified glomalin only on roots that were later described as having AM colonization by J.B. Morton (West Virginia University). In more recent work with an axenic culture of transformed carrot roots, glomalin was extracted from hyphae in a root free zone (Rillig and Steinberg 2002). Glomalin also is routinely extracted from hyphae up to 7 cm away from roots in pot cultures where hyphae are separated from roots by a 38-µm nylon mesh bag (Wright and Upadhyaya 1999). Immunofluorescence assays show that glomalin coats AM fungal hyphae, sloughs from hyphae onto colonized roots, organic matter, soil particles, horticultural or nylon mesh, and glass beads, and is found on arbuscules within root cells (Wright *et al.*, 1996; Wright and Upadhyaya 1999; Wright 2000).

Typically, glomalin concentration in these pools is measured by a Bradford total protein assay (*i.e.* TG and EEG) and immuno-reactive protein (*i.e.* IRTG and IREEG) assays (Wright *et al.*, 1996). The Bradford protein assay is non-specific and will detect any proteinaceous material. Bradford concentrations are based on comparison with a bovine serum albumin (BSA) standard curve. The immunoreactive protein assay, or enzyme-linked immunosorbent assay (ELISA), uses the monoclonal antibody specific for glomalin, but certain artificial conditions may reduce immunoreactivity. ELISA values are determined by comparison to 100 per cent immunoreactive glomalin extracted from hyphae or soil (Wright *et al.*, 1996). The total protein assay measures concentrations ranging between 1.25 to 5.0 µg while ELISA numbers range from 0.005 to 0.04 µg (Wright and Upadhyaya 1999). Since the range of Bradford values is 100 times greater than ELISA numbers, values of over 100 per cent could be supported.

Comparisons of the total and immunoreactive pools of glomalin extracted from soil or pot culture show that not all of the extracted material is immunoreactive. Reduction in immunoreactivity may be due to exposure to conditions that affect the site of binding of the antibody. The reactive site for a monoclonal antibody is very specific (Goding 1986). In this case, the monoclonal antibody was generated against glomalin on spores of AM fungi and not against extracted glomalin. Therefore, some reactivity is lost probably because of conformational changes during the high temperature (121°C) and long time period (at least 0.5 to 1.0 h) used in the extraction procedure (unpublished data). In the soil, organic matter, metals (such as iron), clay minerals, and other substances may bind to glomalin causing conformational changes or masking the reactive site. Degradation is another factor in soil extracts and may result in a decline in immunoreactivity (Wright and Upadhyaya 1999). In addition, conformational changes may occur in the molecule when it sloughed from the hyphae and is in the scum pool due to self-aggregation via hydrophobic interactions.

Glomalin is present on extramatrical hyphae of all AMF tested to date (*i.e.* representatives from all known genera, except *Sclerosystis*) (Wright *et al.*, 1996). As hyphae degrade, this hydrophobic, highly stable glycoprotein sloughs off to coat organic matter and other soil particles. Wright *et al.* (1996) hypothesize that glomalin forms a conglomeration with root fragments and organic matter, thus protecting it from degradation by microorganisms. As a result, Wright and Upadhyaya (1998) found a strong correlation between glomalin concentration and soil aggregation.

The correlation between glomalin concentration and soil aggregation also may be influenced by iron. Iron- and Al-(hydr)oxides are speculated to be involved in aggregate formation by bridging organic matter to clay minerals and to contribute to the persistence of aggregates (Wright and Upadhyaya 1998; Bird *et al.*, 2002). Kemper and Chepil (1965) found that aggregate stability is a function of organic matter, clay and free iron oxides.

Evolution of Land Plants and the Environment of Early Earth

In the early Devonian, at least 400 million years ago, plants moved from the oceans onto land (Taylor *et al.*, 1995). The environment of this early earth was very harsh with few plant-available nutrients and little structure to the soil (Lewis 1987). Weathering of silicates in an abiotic environment was very slow, mostly due to mechanical and chemical weathering, which in turn were hampered by rapid wind and rain erosion in the absence of crust stabilizing algae, lichens, and more advanced biota (Schwartzman and Volk 1989). Devonian land plants were morphologically very simple with few leaves, branches, or roots. Evolutionary progress was expressed in variant genotypes with greater nutrient-extraction ability and habits promoting water retention (Pirozynski and Malloch 1975; Schwartzman and Volk 1989; Taylor *et al.*, 1995). In this evolutionary process, plants did not readily evolve root hairs to improve efficiency of adsorption. Instead, they exploited their relationship with organisms, such as fungi and algae, to combat nutrient deprivation (Pirozynski and Malloch 1975; Taylor *et al.*, 1995).

Recent evidence of similar sequences in the SSU rRNA region between several *Glomus* species and *Geosiphon pyriforme* indicates that this may be the precursor for the AM symbiosis in early land plants (Gehrig *et al.*, 1996). *Geosiphon pyriforme* forms a structure for nutrient exchange in its *Cyanobacterium* host, *Nostoc puntiforme*, which is analogous to the arbuscule found in AM fungi (Gehrig *et al.*, 1996).

Ecological, geological, and cytological evidence has shown that there were explosive periods of morphological innovation during the Devonian Period (Gensel and Andrews 1987; Schwartzman and Volk 1989). During this time, photosynthetic free oxygen became more abundant creating the ozone layer to protect land plants from UV radiation (Pirozynski and Malloch 1975). This allowed plants to flourish on land and the number of genera of plants increased from 1 to 28 with plants changing from tiny creeping structures to a diverse array of 0.5-9 m high structures with leaves, roots, reproductive systems, and secondary growth (Gensel and Andrews 1987). Algae and fungi have been implicated in the formation of calcareous crusts and biomineralization that made nutrients available to plant and increased the rate of soil development (Stubblefield and Taylor 1988). Soil and plant development would lower CO_2 and surface temperatures for the evolution of more complex life forms (Schwartzman and Volk 1989).

In mutualistic relationships, both plant host and fungal invader obtain benefits that outweigh the inherent costs of the symbiosis. The mycorrhizal symbiosis that subsequently evolved optimizes Liebig's "Law of the Minimum" (Read 1991). Carbon, the limiting nutrient for AM fungi, was supplied in abundance by plants whose growth was limited by access to soil minerals, which the fungal hyphae could readily supply (Read 1991).

As protected root inhabitants, endomycorrhizal fungi have changed little from their ancient relatives, because they have not been subjected to the selective pressures that have caused morphological changes among the plants (Lewis 1987; Stubblefield and Taylor 1988; Morton 1990). Indeed, the fossil record shows that the fungal structures found in the roots of Devonian plants are almost exactly the same as the structures found in the roots of modern plants (Stubblefield and Taylor 1988; Taylor *et al.*, 1995). The intimate relationship between plant and fungi is best appreciated if it is realized that these organisms belong to different kingdoms that have evolved separately but in complete interdependence (Wessels 1997).

In satisfying the need in soils for a persistent material capable of performing a number of functions necessary for plant growth, nature has adopted a simple and elegant solution whereby plants benefit vitally from the ubiquitous, immediately surrounding organic medium resulting from the decay of the plants' predecessors (MacCarthy 2001). The general functions of humic substances could, in principle, be satisfied by a suite of biomolecules that possess the requisite combination of hydrophilic, acidic, complexing, and sorptive properties, but many of these molecules do not survive for long periods in the environment (Burdon 2001; MacCarthy 2001). Therefore, the formation of humic substances or, more likely, glomalin would provide the organic environment that plants need for productive growth. Glomalin would have evolved this function, because supposedly it was first formed when there was no other organic matter in the soil.

Soil Organic Matter

In soil, plant- and microbially-produced organic carbon is found in two pools: (1) the labile, "light" or particulate organic matter (POM) fraction and (2) the recalcitrant, "heavy" or humic fraction. The POM fraction represents fresh or partially decomposed plant material, while the humic fraction is more completely decomposed material. The POM fraction is similar in chemical composition to plant material (Cambardella and Elliott 1992) and may be separated by floatation in high-density liquids, such as NaCl or sodium polytungstate (Wolf *et al.*, 1994; Gale and Cambardella 2000). Changes in POM concentration are correlated with changes in soil fertility due to tillage practices or environmental factors (Cambardella and Elliott 1992). As POM degrades further, it is transformed into humic substances.

The "heavy" fraction contains three types of humic substances: (1) humic acid (HA), (2) fulvic acid (FA), and (3) humin. Historically, solubility characteristics and color have separated humic substances into the acid- and alkaline-soluble, yellow to light brown FA fraction; the alkaline-soluble, dark brown to black HA fraction; and the insoluble, black humin. The low turnover rate, especially in the HA and humin fractions, makes these components important in sustainable agriculture and C sequestration. Humic substances are considered important in sustainable agriculture because they enhance water-holding capacity, permeability, soil aggregation, buffering capacity, and cation exchange capacity.

Typically, humic substances account for the majority of SOM [about 70 to 80 per cent of the soil organic carbon (SOC)] (Hayes and Graham 2000; Hayes and Clapp 2001). These substances are derived from biological and chemical transformations of organic debris (*i.e.* POM) and are higher in C but lower in O and H than POM (Swift 1996; Hayes and Graham 2000). The composition of humic substances varies slightly from sample to sample according to types of plant material and microbial populations. Frequently, because biochemical compounds such as amino acids, carbohydrates, and lipids, often are complexed to and extracted with humic substances, the extract is operationally but not structurally defined (Clapp and Hayes 1999; Hayes and Graham 2000). Low molecular weight humic substances can be bridged together via the introduction of iron salts to form high-molecular weight substances (Hatcher *et al.*, 2001). Despite being chemically extracted in alkaline solution and separated by acidification, neither HAs nor FAs are pure compounds. Some residues or impurities in HAs may be removed by increasing the ionic strength of the extraction solution or by using XAD resins (Hayes and Clapp 2001). These impurities contribute to the high concentrations for HAs reported for most soils.

The recent discovery of glomalin has led to a re-examination of SOM components (Wright *et al.*, 1996; Wright and Upadhyaya 1996). Glomalin is a ubiquitous and abundant glycoproteinaceous molecule (Wright and Upadhyaya 1999; Rillig *et al.*, 2001b). However, unlike POM or humic substances, glomalin is not derived from the decomposition of plant- or microbially-produced material. Glomalin forms a hydrophobic sheath on hyphae that may keep material from being lost from across the hyphal membrane and/or may protect the hyphae from microbial attack (Wright and Upadhyaya 1998). Its presence in soil helps to stabilize aggregates, which in turn help to protect the molecule from degradation (Wright and Upadhyaya 1998). Glomalin appears to be highly correlated with aggregate stability (Wright and Upadhyaya 1998) and with carbon sequestration in the soil by helping to physically protect organic matter within aggregates (Rillig *et al.*, 1999; Rillig *et al.*, 2001a).

Soil Aggregation

Loss of topsoil due to erosion is a serious problem in agro-ecosystems. Pimentel *et al.* (1995) estimated that during the last 40 years nearly one-third of the world's arable land was lost to erosion with a current rate of 10 million hectares per year. Soil aggregates are important for: (1) maintaining soil porosity, which provides aeration and water infiltration rates favourable for plant and microbial growth, (2) increasing stability against wind and water erosion, and (3) storing carbon by protecting organic matter from microbial decomposition (Hassink and Whitmore 1997; Rillig *et al.*, 1999; Bird *et al.*, 2002). Since both aggregate stability and SOM decline upon cultivation, it may be that SOM (*i.e.* POM, humic substances, microbially-produced molecules, and fungal hyphae) plays a role in aggregate formation, but the exact mechanism is not understood. Aggregate formation is a complex process of physical and chemical interactions (Kemper and Chepil 1965; Tisdall and Oades 1982; Miller and Jastrow 1990).

Electron microscopy shows that aggregates are a conglomeration of soil minerals (clay particles, fine sand and silt), small plant or microbial debris, bacteria, free amorphous organic matter and organic matter strongly associated with clay coatings (Chenu *et al.*, 2000; Six *et al.*, 2001). Fungal hyphae may initiate aggregate formation by providing the framework upon which organic mater collects (Miller and Jastrow 1990; Tisdall *et al.*, 1997). Chemical processes then contribute to aggregate formation and stability by coating with hydrophobic polymers, binding mineral particles with organic polymers, bridging organic matter to clay particles by polyvalent cations, and gluing with polysaccharides (Degens 1997; Piccolo and Mbagwu 1999; Chenu *et al.*, 2000). Drying and wetting actions, shrinking and swelling of clays, freeze-thaw cycles, compaction, and enmeshing by fungal hyphae and/or fine roots physically stabilize aggregates (Chaney and Swift 1986; Degens 1997).

In reduced or no-till systems, Chaney and Swift (1986) found that the stubble and mulch litter promote aggregate formation because fungal decomposition of organic matter produces gluing agents, such as polysaccharides and mucigels. Caesar-TonThat and Cochran (2002) found that ligninolytic basidiomycetes produce large quantities of hydrophobic polysaccharides, glycolipids or glycoproteins that bind to and stabilize soil particles in water-stable aggregates. However, many of the polysaccharides produced by microbial degradation will glue aggregates together quickly but are water-soluble and ephemeral and do not to contribute to the long-term stability of aggregates (Chaney and Swift 1986; Six *et al.*, 2001).

Without the protection of hydrophobic coatings, soil aggregates may be disrupted by rainfall because of slaking, the differential swelling of clays, mechanical dispersion by the kinetic energy of raindrops, and physio-chemical dispersion. The aliphatic molecules involved in aggregate formation increase water stability and long-term survival of aggregates, because attractive forces between these molecules are much stronger internally than externally (Degens 1997; Piccolo and Mbagwu 1999; Chenu *et al.*, 2000). Soil organic matter containing high concentrations of aliphatic groups, such as HA, may increase aggregate stability and the long-term stabilization of organic materials (Piccolo and Mbagwu 1999). These aliphatic, hydrophobic groups and polymers are the major contributors to the water-stability of aggregates. They increase the contact angle for water penetration, which restricts infiltration and slaking, lowers wettability and increases the internal cohesion of aggregates (Chenu *et al.*, 2000).

Glomalin contributes to the stabilization of aggregates by sloughing off hyphae onto the surrounding organic matter, binding to clays (probably via cation bridging by iron), and providing a hydrophobic coating (Wright and Upadhyaya 1999). This is demonstrated in a number of experiments, where total and, especially, immunoreactive concentration of glomalin are positively correlated with per cent water-stable soil aggregates in both agricultural and native soils (Wright and Upadhyaya 1998; Wright *et al.*, 1999; Wright and Anderson 2000; Bird *et al.*, 2002; Rillig *et al.*, 2003).

Glomalin Under Elevated CO_2

Several studies were conducted to compare glomalin concentrations to aggregate stability under elevated CO_2 conditions. In a native grassland ecosystem in northern California, TG and IRTG concentrations increased with higher CO_2 concentrations, along with hyphal length at one site, and aggregate stability in 1-2 and 0.25-1 mm aggregate size fractions (Rillig *et al.*, 1999). Long-term exposure to elevated atmospheric CO_2 conditions from a natural CO_2 spring in New Zealand formed by a thermal vent resulted in a linear increase in per cent root colonization by AM fungi, soil hyphal length, TG and EEG along a CO_2 gradient (Rillig *et al.*, 2000). In a Sorghum field, aggregate stability, hyphal length and EEG increased with elevated CO_2 (Rillig *et al.*, 2001a). In both the grasslands (Rillig *et al.*, 1999) and Sorghum field (Rillig *et al.*, 2001a), aggregate stability was correlated with glomalin concentrations. These studies show that under elevated CO_2 conditions, photosynthetic carbon is allocated belowground and glomalin may provide a significant sink to trap carbon in the soil. Using ^{14}C data, Rillig *et al.* (1999) calculated a turnover time for glomalin of 7 to 42 y. However, the incubation studies suggest that a long-lived, recalcitrant glomalin fraction exists with a much longer turnover time.

Conclusion

Role of microorganisms in the mineralization process is well documented in literature and hence, the carbon cycling is also one important process of microbial activity contributing to carbon enrichment either in soil or in environment. An Arbuscular mycorrhizal fungus is one of the significant components of rhizosphere which has got the potential in sequestering carbon. However, research is needed in this direction to utilize the potential of this phenomenon of glomalin production through AMF.

References

Batjes, N.H., 1996. *Eur. J. Soil Sci.*, 47: 151–163

Bauer, A. and Black, A.L., 1994. *Soil Sci. Soc. Am. J.*, 58: 185.

Bethlenfalvay, G.J. *et al.*., 1982. *Soil Sci. Soc. Am. J.*, 49: 1164–1168.

Bird, S.B. *et al.*, 2002. *Pollut.*, 116: 445–455.

Bolan, N.S., 1991. *Plant Soil*, 134: 189–207.

Borowicz, V.A. 2001. *Ecology*, 82: 3057–3068.

Brundrett, M., 1991. *Adv. Ecol. Res.*, 2: 171–220.

Burdon, J., 2001. *Soil Sci.*, 166: 75–769.

Caesar-TonThat, T.-C. and Cochran, V.L., 2000. *Biol. Fertil. Soils*, 32: 374–380.

Cambardella, C.A. and Elliott, E.T., 1992. *Soil Sci. Soc. Am. J.*, 56: 777–783.

Carlile, M.J. and Watkinson, S.C., 1996. In: *The Fungi*, (Eds.) M.J. Carlile and S.C. Watkinson. Academic Press, New York, p. 307–371.

Chaney, K. and Swift, R.S., 1986. *J. Soil Sci.*, 37: 329–335.

Chenu, C. *et al.*, 2000. *Soil Sci. Soc. Am. J.*, 64: 1479–1486.

Clapp, C.E. and Hayes, M.H.B., 1999. *Soil Sci.*, 164: 899–913.

Clark, R.B. and Zeto, S.K., 1996. *J. Plant Nutrit.*, 19: 247–264.

Dalal, R.C., *et al.*, 1995. *Aust. J. Exp. Agric*, 37: 903.

Degens, B.P., 1997. *Aust. J. Soil Res.*, 35: 431–459.

Diaz, G. *et al.*, 1996. *Plant Soil*, 180: 241–249.

Douds, D.D. and Millner, P.D., 1999. *Agri., Eco. Environ.*, 74: 77–93.

Driver, J.D., *et al.*, 2005. *Soil Biol. Biochem.*, 37: 101–106.

Eviner, V.T. and Chapin, F.S., 2002. *Plant Soil*, 246: 211–219.

Gale, W.J. and Cambardella, C.A., 2000. *Soil Sci. Soc. Am. J.*, 64: 190–195.

Gehrig, H.A. *et al.*, 1996. *J. Mole. Evol.*, 43: 71–81.

Gensel, P.G. and Andrew, H.N., 1987. *American Scientist*, 75: 478–489.

George, E.K. *et al.*, 1992. In: *Mycorrhizas in Ecosystems*, (Eds.) D.J. Read *et al.* CAB International. Cambridge, UK, p. 42–47.

Gianinazzi, S. *et al.*, 1995. *Crit. Rev. Biotechn.*, 15: 305–311.

Goding, J.W., 1983. *Monoclonal Antibodies: Principles and Practice.* Academic Press, Inc. New York, USA.

Gonzalez-Chavez, C. *et al.*, 2002. *Plant Soil*, 240: 287–297.

Graham, J.H., 2000. In: *Current Advances in Mycorrhizal Research*, (Eds.) Podila, G.K. and D.D. Douds. pp. 127–140.

Hassink, J. and Whitmore, A.P., 1997. *Soil Sci. Soc. Am. J.*, 61: 131–139.

Hatcher, P.G. *et al.*, 2001. *Soil Sci.*, 166: 770–794.

Hayes, M.H.B. and Clapp, C.E., 2001. *Soil Sci.*, 166: 723–737.

Hayes, M.H.B. and Graham, C.L., 2000. Procedures for the isolation and fractionation of humic substances. In: *4th Seminar on Humic Substances*, (Eds.) G. Davies and E.A. Ghabbour. The Royal Society of Chemistry, Cambridge, p. 91–109.

Hooker, J.E. and Black, K.E., 1995. *Crit. Rev. Biotechnol.*, 15: 201–212.

Jackson, M.L., 1958. *Chemical Analysis of Soils.* Prentice Hall Inc., New Delhi.

Jenkinson, D.S., 2001. *Agron. J.*, 83: 2.

Johnson, D. *et al.*, 2002. *Soil Biol. Biochem.*, 34: 1521–1524

Kanchikerimath, M. and Singh, M., 2001. *Agric. Ecosys. Environ.*, 86: 155.

Kemper, W.D. and Chepil, W.S., 1965. Size distribution of aggregates. In: *Methods of Soil Analysis, Part I*, (Ed.) C.A. Black. Agronomy No. 9. American Society of Agronomy, p. 499–509.

Lal, R., 1999. *Prog. Environ. Sci.*, 1: 307.

Lal, R., 2003. *Crit. Rev. Plant Sci.*, 22: 151–184.

Lal, R., 2003. *Environ. Int.*, 29: 437.

Lal, R. *et al.*, 1999. *J. Soil Water Conserv.*, 54: 374.

Lewis, D.H., 1987. In: *Evolutionary Biology of the Fungi*, (Eds.) A.D.M. Rayner *et al.* Cambridge University Press, Cambridge, UK, p. 161–178.

MacCarthy, P., 2001. *Soil Sci.*, 166: 738–751.

Miller, R.M. and Jastrow, J.D., 1990. *Soil Biol. Biochem.*, 22: 579–584.

Miller, R.M. *et al.*, 2002. *New Phytol.*, 155: 149–162

Millner, P.D. and Wright, S.F., 2002. *Symbiosis*, 33: 101–123.

Morton, J.B., 1990. *Mycotaxon*, 37: 493–515.

Olsson, P.A., *et al.*, 1999. *Soil Biol. Biochem.*, 31: 1879–1887.

Page, A.L., 1982. *Methods of Soil Analysis, Part I.* ASA and SSSA, Madison, Wisconsin.

Paul, E.A. and Clark, F.E., 1996. *Soil Microbiology and Biochemistry*, 2nd Edition. Academic Press: New York.

Pawlowska, T.E., *et al.*, 2000. *App. Environ. Micro.*, 66: 2526–2530.

Piccolo, A. and Mbagwu, J.S.C., 1999. *Soil Sci. Soc. Am. J.*, 63: 1801–1810.

Pimentel, D. *et al.*, 1995. *Science*, 267: 1117–1123.

Pirozynski, K.A. and Malloch, D.W., 1975. *BioSystems*, 6: 153–164.

Read, D.J., 1991. Mycorrhizas in ecosystems–Nature's response to the "Law of the Minimum". *In* D.L. Hawksworth (ed.) Frontiers in Mycology. p. 101–130. CAB International. Cambridge, UK.

Reicosky, D.C., *et al.*, 1997. *Soil Tillage Res.*, 41: 105.

Rillig, M.C. and Mummey, D.L., 2006. *New Phytol.*, 171: 41– 53.

Rillig, M.C. and Steinberg, P.D., 2002. *Soil Biol. Biochem.*, 34: 1371–1374.

Rillig, M.C. *et al.*, 2000. *Ecol. Lett.*, 3: 475–478.

Rillig, M.C. *et al.*, 2002. In: *Mycorrhizal Ecology (Ecological Studies Series) (Vol. 157)*, (Eds.) van der Heijden, M., Sanders, I., *et al.*. Springer, pp. 135–160.

Rillig, M.C. *et al.*, 2003. *Plant Soil*, 253: 293–299.

Rillig, M.C. *et al.*, 2001b. *Plant Soil*, 233: 167–177.

Rillig, M.C. *et al.*, 1999. *Nature*, 400: 628.

Rillig, M.C. *et al.*, 2001a. *Glob. Change Biol.*, 7: 333–337.

Rosier, C.L. *et al.*, 2006. *Soil Biol. Biochemistry* (In press).

Schwartzman, D.W. and Volk, T., 1989. *Nature*, 340: 457–460.

Simon, L.*et al.*, 1993. *Nature*, 363: 67–69

Six, J., A. *et al.*, 2001. *Plant Soil*, 234: 27–36.

Smith, P. and Powlson, D.S., 2000. *Science*, 287, 428.

Smith, P. *et al.*, 1997. *Glob. Change Biol.*, 3: 67–79

Smith, S. V. *et al.*, 2001. *Global Biogeochem. Cycles*, 15: 697.

Smith, S.E. and Read, D.J., 1997. *Mycorrhizal symbiosis*, 2nd Edition. Academic Press, Sand Diego.

Staddon, P.L. *et al.*, 2002. *Plant Soil*, 244: 253–261.

Stubblefield, S.P. and Taylor, T.N., 1988. *New Phytol.*, 108: 3–25.

Swift, R.S., 1996. In: *Methods of Soil Analysis, Part 3: Chemical Methods*, (Eds.) D.L. Sparks *et al.* SSSA Book Series, No. 5 Madison, WI, p. 1018–1020.

Taylor, T.N. *et al.*, 1995. *Mycologia*, 87: 560–573.

Tinker, P.B. *et al.*, 1994. *New Phytol.*, 128: 115–122.

Tisdall, J.M. *et al.*, 1997. Aggregation of soil by fungal hyphae. *Aust. J. Soil Res.*, 35: 55–60.

Tisdall, J.M., and Oades, J.M., 1982. *J. Soil Sci.*, 33: 141–163.

Varma, A., 1995. *Crit. Rev. Biotechn.*, 15: 179–199.

Wessels, J.G.H., 1997. *Adv. Microb. Physiol.*, 38: 1–45.

West, T.O. and Post, W.M., 2002. *Soil Sci. Soc. Am. J.*, 66: 1930.

Wolf, D.C., *et al.*, 1994. In: *Methods of Soil Analysis, Part 2: Microbiological and Biochemical Properties*, (Eds.) R.W. Weaver *et al.* SSSA Book Series, no. 5 Madison, WI, p. 879.

Wright, S.F., 2000. *Plant Soil*, 226: 171–177.

Wright, S.F. and Anderson, R.L., 2000. *Biol. Fertil. Soils*, 31: 249–253.

Wright, S.F. and Millner, P.D., 1994. In: *Advances in Soil Science; Soil Biology: Effects on Soil Quality*, (Eds.) J.L. Hatfeld and B.A. Stewart. Lewis Publishers, Boca Raton, FL, USA, p. 29–59.

Wright, S.F. and Upadhyaya, A., 1996. *Soil Sci.*, 161: 575–585.

Wright, S.F. *et al.*, 1996. *Plant Soil*, 181: 193–203.

Wright, S.F. *et al.*, 1999. *Soil Sci. Soc. Am. J.*, 63: 1825–1829.

Wright, S.F. and Upadhyaya, A., 1998. *Plant Soil*, 198: 97–107.

Wright, S.F. and Upadhyaya, A., 1999. *Mycorrhiza*, 8: 283–285.

Ecosystem Diversity and Carbon Sequestration: Climate Change Challenges and a Way Out for Ushering in a Sustainable Future *Pages 187–197*
Editors: P.L. Gautam, Vir Singh and Uma Melkania
Published by: DAYA PUBLISHING HOUSE, 2009

Chapter 20

Role of Plant Transcription Factor-DOF in Enhancing Nitrogen Use Efficiency: Molecular Means for Promoting Organic Farming

Dinesh Yadav, Nidhi Gupta, Anil Kumar, Pushpa Lohani, Munna Singh and U.S. Singh

Nitrogen use efficiency (NUE) at the plant level is its ability to utilize the available nitrogen (N) resources to optimize its productivity. This includes nitrogen uptake and assimilatory processes, redistribution within the cell and balance between storage and current use at the cellular and whole plant level. In terms of agriculture, it is the optimal utilization of nitrogenous manures or fertilizers for plant growth, yield and protein content, as atmospheric nitrogen gas is not utilized by higher plants, except symbiotic legumes. Worldwide the NUE for cereal production accounts for only 33 per cent, the remaining 67 per cent represents a $15.9 billion annual loss of nitrogen fertilizer. A number of management strategies to improve the efficiency of utilization of nitrogenous fertilizers include application of different types of fertilizers, their mode of application, avoiding runoff, mitigation of losses from soil and plants, use of slow-release fertilizers, nitrification inhibitors, use of organic manures, green manuring, use of legumes in cropping systems, correction in their imbalanced use and integrated nutrient management. At molecular level the identification of novel nitrogen responsive genes and their *cis* and *trans* acting gene elements can be a strategy to improve nitrogen use efficiency of crops. The Dof (DNA binding with one finger) transcription factor acts as master regulator in the expression of photosynthetic genes and thereby improving nitrogen assimilation of crop plants and hence a target for improving the nitrogen use efficiency of crops.

Plant Transcription Factor

The control of eukaryotic gene expression is highly coordinated operation involving many DNA-binding proteins referred as transcription factors (TFs) along with cis-elements sequences present in promoter region. Many transcription factors have been reported, each recognizing specific regulatory sequences and capable of activating transcription via the possession of two functionally separate protein domains. TFs can be classified as i) Ubiquitous TFs, required for the expression of all structural genes transcribed by RNA polymerase II and ii) Specific TFs, present in a restricted number of cell types and are responsible for transcription of genes in a cell- or tissue-specific manner. Efficiency of transcription depends on the various interactions between different sets of TFs and their corresponding DNA regulatory sequences. Since each cell comprises of same set of genes and their associated regulatory sequences, the cell- or tissue specific regulation of gene expression depends on the availability of specific transcription factor in any given time. Gene expression during development, differentiation and cellular responses to environmental signals largely depends on the presence and absence of specific transcription factors.

Plant transcription factor comprises of several domains like a DNA-binding domain, an oligomerisation domain and a transcription regulation domain. Most of the transcription factors have only one type of DNA-binding region and one type of oligomerisation region, though more than one domain have also been reported in some plants. The DNA-binding region is quite crucial as it contains amino acid (mostly basic) that interacts with DNA at the cis-sequence elements and hence determines the specificity of the transcription factor. The oligomerisation region is responsible for determining protein-protein interactions, as most transcription factor can form hetero-and/or homodimers, which ultimately affect sequence-element specificity, transactivation efficiency or DNA-binding affinity.

The transcription regulation domain is responsible for transcription regulation either repressing or activating. Transcription factors are often classified on the basis of their structural features. The main classes of transcription factors are: a) Helix-turn helix (HTH), b) Leucine zipper, c) Zinc finger binds DNA, d) Helix-loop-helix (HLH), and e) High-mobility group (HMG)-box motif. Helix-turn helix, Leucine zipper, Helix-loop-helix binds DNA as dimer while zinc finger can bind DNA both as dimer or monomers. The Transcriptional regulation of a gene is often governed by number of conserved sequence. The promoter is one of major player in gene regulation by affecting the efficiency of transcription by binding of RNA polymerase to different domains of conserved sequences. These sequence elements are termed as "cis-elements" as they are on the same DNA strand as the coding region of the gene. Some of the sequence elements like TATA box, CAAT box, GC box etc. are crucial in regulation of plant gene expression. Besides the cis-elements sequences, other conserved sequences are also found in promoter region, which are involved with binding of specific proteins, termed as "*transcription factors*". These transcription factors (or trans-acting factors) are involved in linking various signals (both external and internal) to gene expression and are responsible for determining the level, place and timing of expression.

Zinc-Finger Transcription Factors in Plants

Plants exhibits number of unique biological processes like photosynthesis, nitrogen fixation, the reproductive process, development and responses to environmental signals and hence assumed to have some transcription factors that are unique to plants only besides common TFs found in eukaryotes. The term "*Zinc finger*" represents the sequence motifs in which cysteines and/or histidines coordinate a zinc atom(s) to form local peptide structures that are required for their specific functions. The Zinc-finger motifs, often classified based on the arrangement of the zinc-binding amino-acids, are frequently observed in many transcription factors performing a critical roles in interactions with other molecules involved with gene expression. Some of the commonly found Zinc-finger transcription factors reported in plants are:

TF111A Type

It comprises of two cysteines and two histidines in conserved sequence motifs (CX2-4FX5LX2HX3-5H) tetrahedrally coordinate a zinc atom to form a compact structure that interacts with the major groove of DNA in a sequence-specific manner. Example EPF family of petunia regulating the petal-specific expression of the gene for enolpyruvylshikimate-3-phosphate synthase (EPSPS), an enzyme involved in shikimate pathway, leading to the flower pigment, anthocyanins.

SUPERMAN

It contains only one TFIIIA-type zinc finger and two overlapping leucine-zipper-like motifs in the C-terminal region, which might be associated with protein-protein interactions. Example SUPERMAN protein in Arabidoposis involved with flower development.

WRKY-Family

It represents a new type of Cys2/His2-type zinc finger (CX4-CX22-23HX1H) present in the conserved regions of WRKY family of proteins involved with activation of PR1 (pathogen-related1) genes in response to fungal pathogens.

GATA1-Like

It forms one of the major families of the Cys2/Cys2-type zinc-finger transcription factors in eukaryotes. The DNA-binding domain consensus is CX2CX17CX2C and contains a zinc atom coordinated by the conserved four cysteines. The gene for the first GATA1 type protein in plants was isolated from tobacco and referred as NTL1, which is identical to NIT2, a transcription factor in the fungus *Neurospora crassa* (involved with activation of genes for nitrogen metabolic enzymes during nitrogen-limiting conditions. The role of NTL1 is yet not clear in plants. GATA1 like proteins have been found with CONSTANS protein of Arabidopsis involved with floral induction.

RING-finger Type

It is commonly found regulatory protein of plants, animals, fungus, virus and protozoa and characterized by a bipartite asymmetric motif, C-X2-C-loop I-C-X-H-

X2-C-X2-C-loop II-C-X-C (loops II and I are variable in length). These are involved in some aspects of transcriptional regulation like COP1 in plants are associated with regulation of light inducible genes.

PHD-finger Type

The PHD (plant homeodomain finger), first reported in two plant proteins containing homeodomains, Arabidopsis HAT3.1 and maize Zmhox1a, is similar to RING-finger in the arrangement of putative zinc-binding amino acids. It is a cysteine-rich sequence motif with Cys_4-His-Cys_3 structure as compared to RING finger with Cys_3-His-Cys_4. All these protein exhibits either sequence–specific or nonspecific DNA-binding activity.

LIM–Family

It is also a cysteine-rich motif of CX_2-$CX_{17\text{-}19}HX_2CX_2CX_{16\text{-}20}CX_{2\text{-}3}C$ and based on its association with other functional domains. They are further classified as: (a) LIM-HD proteins–containing homeodomain, (b) LIM-only proteins–no homeodomains, and c) Two LIM domains linked to protein kinase domains. Lim domains serve as the site for protein-protein interactions with itself (homeodimerisation), with helix-loop-helix type transcription factor or with protein kinases. In plants LIM-domain containing proteins has been reported from sunflower as PLIM-1, which is involved in the formation and growth of pollen tube by organizing the actin cytoskeleton at its adhesion sites with plasma membrane.

Dof–family

The Dof (DNA binding with one finger) family comprises of sequence specific DNA-binding proteins that interact with the promoter sequences of several genes involved with tissue specific and stress-specific expression. The proteins of the Dof family have a conserved domain (Dof domain) including a zinc-finger-like motif, $CX_2CX_{21}CX_2C$, followed by a basic region similar to that of GATA1 motif. Still the Dof family forms a very distinct class because the spacing between the second and the third conserved cysteines is longer than those in the GATA family and it is quite unique based on their ability to bind to DNA with only one finger. The Dof-domain like sequences are unique to plants as it has not been reported in yeasts and animals. These family of proteins are involved in the transcriptional regulations of various biological processes unique to plants like flower development, light-regulated morphogenesis, pathogen response etc. The first member of this protein family, Dof1, from maize is involved in the light-regulated activation of the gene for C4–photosynthetic phosphoenolpyruvate carboxylase (C4 PEPC). PBF (prolamin-box binding factor), another Dof-family protein of maize is involved with endosperm-specific expression of genes for zein and other seed storage proteins. Dof-family proteins are also associated with promoters of pathogen-responsive genes mediating expression in the presence of signals such as auxin, salicylic acid and hydrogen peroxide. Dof proteins have been associated with diverse promoters of plant specific genes, suggesting multifarious roles in plants.

Role of DOF Proteins in Plants

Dof proteins being plant specific are involved with number of processes unique to plant system like expression of a photosynthetic gene, influencing genes for seed-storage proteins, a plant oncogene, genes for pathogen responsiveness and genes responsive to a plant hormone. Dof proteins regulate these diverse functions based on their interactions with other transcription factors and proteins. Some of the studied roles of Dof proteins are:

Tissue Specific Expression

Dof expression is tissue specific; it is not expressed in all plant cells at any given time, suggesting its role in development of plants. Dof1 of maize is constitutively expressed in leaves, stems and roots, whereas the closely related Dof2 is expressed mainly in stems and roots. In leaves, Dof1 responds to light and activates the transcription of a photosynthetic gene, which is essential for fixing atmospheric carbon into sugars, named C4PEPC (Phosphoenol Pryuvate Carboxylase) gene. The differential expression of Dof1 and Dof2 may permit leaf-specific gene expression.

Biotic Stress

The Dof is involved in expression of genes related to biotic stresses, like in barley seeds. Dof is known to be involved in biotic stress by regulating cystatin gene mainly by influencing the promoters of cystatin.

Plant Hormone Response

Dof proteins are regulators of plant hormones responsive genes in plants, and mediate the response of Gibberellin and Auxin by influencing the phytohormone signaling pathway with a coordinated interaction of many proteins, *i.e.* transcription factors and their coordinated binding with cis-acting elements present in the promoter of phyto-hormones responsive genes.

Accumulation of Seed Storage Proteins

One of the roles of Dof proteins is the accumulation of seed storage proteins like prolamins in cereals. The genes encoding prolamine are coordinately expressed in the developing endosperm where they are under spatial and temporal transcription control, involving cis-acting and trans-acting motifs in their promoters and tans-acting transcription factors. A number of consensus sequences in gene promoters have been shown to be involved in conferring endosperm specificity in cereals like prolamin box (P-box), which has the recognition core for Dof proteins *i.e.* (A/T) AAAG. The Dof protein binds with this P-box is identified as PBF (Prolamin Box Binding Factor) and found to be activating the expression of Zein protein, a prolamin of Maize seeds, in co-ordination with another basic leucine zipper transcription factor Opaque 2.

Assimilation and Reassimilation of Nitrogen in Plants.

Dof proteins are involved with the assimilation and reassimilation of nitrogen in plants by influencing the genes involved in the pathway. The process of nitrogen assimilation in plants is a complex mechanism involving several enzymes. The main sources of nitrogen available to plants are in inorganic form NO^-_3 and NH_4^+. In all higher plants, inorganic nitrogen is first reduced to ammonia prior to its incorporation into organic form. The assimilation of inorganic nitrogen into organic form has marked effect on plant productivity, biomass and crop yield. Ammonia is assimilated into organic form as glutamine and glutamate, which serve as the nitrogen donors in the biosynthesis of essentially all amino-acids, nucleic acids and other nitrogen containing compounds such as chlorophyll.

Glutamine and glutamate are first amino-acids synthesized in plants from ammonia, which can further be used to form aspartate and asparagines and all of these four amino-acids are finally used to translocate organic nitrogen from sources to sinks. The enzymes involved in the primary assimilation of ammonium into these four N-transport amino-acids glutamine, glutamate, aspartate and asparagines are glutamine synthetase (GS), glutamate synthase (GOGAT), aspartate amino transferase (AAT) and asparagines synthetase respectively.

Nitrogen Remobilization during Grain Refilling

The nitrogen remobilization during grain refilling is a complex process involving many enzymes, their respective genes and trans-acting transcription factors regulating the process. This catabolic process is well organized on the whole plant level, on the organ level, on the cellular level and sub cellular level. Up to 75 per cent of nitrogen present in mesophyll cells is located in chloroplast. Stromal enzymes mainly RUBISCO, represent the major fraction of chloroplast nitrogen. The rate of senescence and remobilization of leaf nitrogen are related to the nitrogen nutrition status of the plant and as source/sink relationship. The chloroplast enzyme ribulose-1, 5-bisphosphate carboxylase is responsible for the photosynthetic fixation of CO_2 in plants. It is also the major protein of the leaf and plays a fundamental role in nitrogen economy of plant, since the protein content of developing seed in most annual plant is derived from the degradation of protein in vegetative organs.

Carbon and Nitrogen Metabolism

It now becomes evident that for assimilation of nitrogen, in the form of ammonia into amino acids needs carbon skeletons in form of 2–oxoglutarate, oxalo-acetic acid and citrate. These carbon skeletons are produced through sequential reactions from photo-assimilated carbohydrates. The levels of carbon and nitrogen metabolites mutually influence each other, implying the intimate link between carbon and nitrogen metabolisms. Therefore, modulation of carbon skeleton production might be an alternative approach to improve nitrogen assimilation in plants and finally the nutritional quality of grains. However, a number of enzymes are involved in carbon skeleton production; it is not practical to intensify the pathway supplying carbon skeleton by the transfer of individual genes for respective enzymes. Due to this reason, a master regulator has been identified in plants, which regulate the expression of photosynthetic genes, and helps to modulate this biological process.

Dof proteins suits this master regulator as it is found to be an activator for multiple gene expression associated with the organic acid metabolism including phosphoenolphruvate carboxylase (PEPC) gene expression. Dof1 of maize found to be an activator of PEPC gene expression and enhances transcription from the promoters of Orthophosphate dikinase and Pyruvate Kinase, as expression of Dof1-specific antisense RNA or the DNA binding domain of Dof1 alone reduced the expression of these genes.

Orthophosphate dikinase catalyzes the formation of phosphoenolpyruvate in the reaction:

$$\text{Pyruvate} + \text{ATP} + \text{P}_i \longrightarrow \text{Phosphoenol pyruvate (PEP)} + \text{AMP} + \text{PPi} + 2\text{H}^+$$

PEP is the substrate for PEPC, which catalyzes the reaction:

$$\text{PEP} + \text{HCO}_3 \longrightarrow \text{Oxaloacetate} + \text{P}_2$$

Oxaloacetate serves as carbon skeleton to synthesize glutamate by enzyme Glutamate synthase.

Nitrogen Use Efficiency

Nitrogen assimilation is one of the factors influencing the growth and development of plants, which directly affects the crop productivity. Nitrogen deficiency in agricultural systems is a worldwide problem India being no exception. To increase crop productivity per unit area and agricultural production to meet the demand of the ever-increasing population, which may touch the 1.3 billion mark by 2025 AD, millions of tones of nitrogenous fertilizers are applied to the soil. These are accompanied by other major and minor nutrients depending on the health status of the soils. Two crops, *viz.* wheat and rice, which form the staple diet of teeming millions, consume 70 per cent of the fertilizers. There is likelihood of change in this pattern of utilization due to diversification of crops, export orientation and emphasis on agro-processing. For wheat and rice, the utilization efficiency of nitrogenous fertilizer under field conditions is around 50 and 25–30 per cent, respectively. This poor nitrogen use efficiency is of great concern. Burgeoning population of the world needs crop genotypes responding to higher nitrogen and showing direct relationship to yield with use of nitrogen inputs. However, for fulfilling the high global demand of organic produce, the low nitrogen responsive genotypes with greater nitrogen use efficiency and grain yields are required.

Different factors are responsible for low nitrogen use efficiency especially in cereals like gaseous plant emission, soil denitrification, surface runoffs, volatilization and leaching, which varies with the crops. Nitrogen use efficiency (NUE) at the plant level is its ability to utilize the available nitrogen (N) resources to optimize its productivity. This includes nitrogen uptake and assimilatory processes, redistribution within the cell and balance between storage and current use at the cellular and whole plant level. In terms of agriculture, it is the optimal utilization of nitrogenous manures or fertilizers for plant growth, yield and protein content, as atmospheric nitrogen gas is not utilized by higher plants, except symbiotic legumes. There are number of standard practices for increasing the NUE of a crop:

1. *Rotations*: Number of crops rotations like soybean-corn, wheat legumes have proved to improve the nitrogen use efficiency than the monoculture of

soybean or wheat. The more intensive systems, requires greater fertilizer N inputs but results in higher total yield and thus economically more viable.

2. *Forage production systems*: This results in lower plant gaseous N emission and improved NUE, as the plants is not allowed to approach flowering, where maximum nitrogen losses is reported.
3. *Hybrids or cultivars*: Ample evidence is available, where selected hybrids or cultivars have proven better for NUE. Wheat varieties with a high harvest index and low forage yield have low plant nitrogen loss and increased NUE. Similarly high NUE has been observed with cultivars having high harvest index.
4. Conservation tillage is also responsible for increasing the NUE, though the direct effect needs to be investigated.
5. *Irrigation*: Supplying fertilizers in irrigation water, applying fertilizers to plants rather than the soils, and use of slow-release fertilizer can mitigate the loss of N and thus increasing the NUE. Similarly application of urease and nitrification inhibitors can directly influence the yield of the crop by influencing the NUE.
6. *Precision agriculture*: It allows precise and timely application of nitrogenous fertilizers to meet the need of plants based on their position in the landscape, rather than applying to the whole land.
7. *NH_4-N Source*: This particular source is generally not subjected to leaching or denitrification losses, as the N in the form of NH_4 in the soil is available for late-season uptake. It has been reported that N supplied as NH_4 as compared to NO_3^-, increases nitrogen uptake in wheat by 35 per cent. It is also evident that NO_3^- assimilation requires high energy than the $NH_{4,}$ which may be responsible for greater dry weight production of plants supplied solely with NH_4.
8. *In-season and foliar applied Nitrogen*: It has been observed that NUE decreases with increasing N level supplied by fertilizers, especially under dry soil conditions. In–season and foliar applied nitrogen generally increases the NUE.

A number of management strategies to improve the efficiency of utilization of nitrogenous fertilizers include application of different types of fertilizers, their mode of application, avoiding runoff, mitigation of losses from soil and plants, use of slow-release fertilizers, nitrification inhibitors, use of organic manures, green manuring, use of legumes in cropping systems, correction in their imbalanced use and integrated nutrient management. While the amount of N available to the plant can be improved by using sustained-release fertilizers, split applications, minimizing fertilizer losses and other nutrient management and crop management strategies, the inherent efficiency of the plant to utilize available N-inputs for higher productivity needs to be tackled biologically. Since overuse of inorganic nitrogenous fertilizers is hazardous to the environment, unused fertilizer is washing off fields into rivers, poisoning coastal waters and causing acid rain. Therefore, the creation of crops with nitrogen efficiency is agriculturally important and might pose a big challenge for molecular breeding. Marker assisted selection of genotypes responding to low and high nitrogen inputs will enable us to enhance agricultural productivity and production of organic produce. This, in turn, led to the identification of novel nitrogen responsive genes and their *cis* and *trans* acting gene elements.

Nitrogen assimilation in plants requires not only inorganic nitrogen (nitrate and ammonia) but also the carbon skeleton (2-oxoglutarate, oxaloacetic acid and citrate) that is produced through sequential reactions from photo-assimilated carbohydrates. The levels of carbon and nitrogen metabolites mutually influence each other, implying the intimate link between carbon and nitrogen metabolism. Therefore, modulation of carbon skeleton production might be an alternate approach to improve nitrogen assimilation in plants. However, because a number of enzymes are involved in carbon skeleton production, it is not practical to intensify the pathway supplying carbon skeletons by the transfer of individual genes for respective enzymes. Thus utilization of transcription factors might be a powerful approach to modification of metabolism for generation of crops having superior characteristics with improved nitrogen utilization efficiency and improved nutritional quality of grains because a single transcription factor frequently regulates coordinated expression of a set of genes for respective pathways. In terms of finding a global target for manipulation of Nitrogen Use Efficiency (NUE), recent studies revealed that Dof (DNA binding with one finger) transcription factor acts as master regulator in the expression of photosynthetic genes and thereby improving nitrogen assimilation of crop plants. Therefore, Dof could be utilized for development of crop varieties having high nitrogen use efficiency and increased protein content of grains under both high nitrogen and organic farming conditions. Thus transgenic crops having Dof gene will promote organic farming.

References

Abrol, Y.P. (ed), 1993. *Nitrogen: Soils, Physiology, Biochemistry, Microbiology, Genetics.* Indian National Science Academy, New Delhi, Spl. Vol., p. 285.

Baumann, K. and Gualberti, G., 1999. The DNA binding site of the Dof protein NtBBF1 is essential for tissue-specific and auxin-regulated expression of the rolB oncogene in plants. *Plant Cell*, 11: 323–333.

Diaz, I. *et al.*, 2002. The GAMYB protein from barley interacts with the Dof transcription factor BPBF and activates endosperm-specific genes during seed development. *Plant Journal*, 29(4): 453–464.

Edwards, J.W. and Coruzzi, G.M., 1990. Cell-specific gene expression in plants. *Annu Rev. Genet.*, 24: 275–303.

Gualberti, G. *et al.*, 2002. Mutations in the Dof Zinc finger genes DAG2 and DAG1 Influence with opposite Effects the Germination of *Arabidopsis* seeds. *The Plant Cell*, 14:1253–1263.

Lea, P.J. and Miflin, B.J., 1980. Transport and metabolism of asparagines and other nitrogen compounds within the plant. In: *The Biochemistry of Plants: Amino-Acids and Derivatives*, (Ed.) B.J. Miflin. Academic Press, New York, 5: 569–607.

Lea, P.J. and Thurman, D.A., 1972. Intracellular Location and Properties of Plant L-Glutamate Dehydrogenases. *Journal of Experimental Botany*, 23: 440–449.

Lea, P.J., Robinson, S.A. and Stewart, G.R., 1990. The enzymology and metabolism of glutamine, glutamate, and asparagines: In: *The Biochemistry of Plants*, (Eds.) B.J. Miflin and P.J. Lea. Academic Press, New York, pp. 121–159.

Mena, M. and Carbonero, P., 2002. A role for the Dof transcription factor BPBF in the regulation of gibberellin-responsive genes in barley aleuronic cells. *Plant Physiology*, 130: 111–119.

Mena, M. *et al.*, 1998. An endosperm Specific Dof protein from barley, highly conserved in wheat, binds to and activates transcription from the prolamin-box of a native β-hordein promoter in barley endosperm. *Plant Journal*, 16: 53–62.

Miflin, B.J. and Lea, P.J. 1976. The Pathway of nitrogen assimilation in plants. *Phytochemistry*, 15: 873–885.

Miflin, B.J. and Lea, P.J. 1980. Ammonia Assimilation: In: *The Biochemistry of Plants: Vol. 5: Amino Acids and Derivatives*, (Eds.) B.J. Miflin, E.E. Con. Academic Press, New York, pp. 169–202.

Nelson, T. and Langdole, J.A., 1993. C4 photosynthetic genes and their expression patterns during leaf development. In: *Control of Plant Gene Expression*, (Ed.) D.P.S. Verma. CRC Press, Boca Raton, FL, pp. 259–274.

Raun, W.R. and Johnson, G.V., 2007. Improving nitrogen use efficiency for cereal production. *Agron. J.*, 91:357–363.

Takatsuji, H., 1998. Zinc-finger transcription factors in plants. *Cell. Mol. Life. Sci.*, 54: 582–596.

Tong Zhu, *et al.*, 2003. Transcriptional control of nutrient partitioning during rice grain filling. *Plant Biotechnology Journal*, 1: 59–70.

Yanagisawa, S., 1995. A novel DNA-binding domain that may form a single zinc finger motif. *Nucleic Acids Research*, 213: 3403–3410.

Yanagisawa, S., 1996. Dof DNA-binding proteins contains a novel zinc finger motif. *Trends in Plant Science*, 1: 213–214.

Yanagisawa, S., 1997. Dof DNA binding domains of plant transcription factors contribute to multiple protein-protein interactions. *European Journal of Biochemistry*, 250: 403–410.

Yanagisawa, S., 2000. Dof1 and Dof2 transcription factors are associated with expression of multiple genes involved in carbon metabolism in maize. *Plant Journal*, 21: 281–288.

Yanagisawa, S., 2002. The Dof family of plant transcription factors. In: *Trends in Plant Science*, pp. 555–560.

Yanagisawa, S. and Schmidt, R.J., 1999. Diversity and similarity among recognition sequences of Dof transcription factors. *The Plant Journal*, 17(2): 209–214.

Yanagisawa, S. and Sheen, J., 1998. Involvement of maize Dof zinc finger proteins in tissue specific and light-regulated gene expression. *Plant Cell*, 10: 75–89.

Yanagisawa, S. and Tetuya, M., 2004. Metabolic engineering with Dof 1 transcription factor in plants: Improved nitrogen assimilation and growth under low-nitrogen conditions. *Proceedings of National Academy of Science*, 101: 7833–7838.

Ecosystem Diversity and Carbon Sequestration: Climate Change Challenges and a Way Out for Ushering in a Sustainable Future *Pages 198–202*
Editors: **P.L. Gautam, Vir Singh and Uma Melkania**
Published by: **DAYA PUBLISHING HOUSE, 2009**

Chapter 21

Fibre Yielding Plants and Carbon Sequestration: Banking on Ecological Attributes of Economic Plants

Sapna Gautam and Uma Melkania

World is experiencing global warming. Many officials and policy-makers have made the topic of global warming known for years. Global warming is an increase in the average temperature of the Earth's near-surface air and oceans. When used without further qualification, the term is understood as the observed warming in recent decades and its projected continuation. An increase in global temperatures can, in turn, cause other changes, including a rising sea level and changes in the amount and pattern of precipitation. There may also be increases in the frequency and intensity of extreme weather events, though it is difficult to connect specific events to global warming.

Other consequences include changes in agricultural yields, glacier retreat, reduced summer stream flows, species extinction and increases in the ranges of disease vectors. Remaining scientific uncertainties include the exact degree of climate change expected in the future, and especially how changes will vary from region to region across the globe. To survive, a certain level of the greenhouse effect needs to take place naturally, otherwise the Earth's temperature would be below freezing and life as we know it would be impossible. However, when the greenhouse effect becomes too intense, according to scientists, the Earth's temperature grows warmer and a domino effect of circumstances may arise. These may include everything from melting of polar ice caps to a rise in sea level to above-average temperatures across the board. As a result, many people are continually looking for ways to reduce this damaging situation. Soil carbon sequestration may help. Most national governments have signed and ratified the Kyoto Protocol aimed at combating greenhouse gas emissions.

Over the past century, human activities such as the burning of fossil fuels, deforestation, grassland conversion, and other land-use changes have contributed to a large increase in the amounts of carbon dioxide (CO_2) and other greenhouse gases in the atmosphere. Most scientists believe that this rapid increase in atmospheric CO_2 has resulted in some human-induced global warming. Most experts agree that if the warming trend continues, climate change is inevitable. According to the U.S. Environmental Protection Agency (EPA), terrestrial carbon sequestration is the process through which carbon dioxide (CO_2) from the atmosphere is absorbed by trees, plants and crops through photosynthesis and stored as carbon in biomass (tree trunks, branches, foliage and roots) and soils. Farms and other agricultural businesses contribute to greenhouse gases in two ways. First, certain agricultural applications such as harvesting forestry for wood and paper production removes acres upon acres of trees that would normally "consume" carbon dioxide (CO_2). Second, equipment, harvesting and fertilizing processes, and other means of operation can contribute to greenhouse gas production in itself. Through soil sequestration, agriculture can offset the production of CO_2 by taking measures to balance or reduce it in the hope of preventing climate change.

Economic models as of today are undoubtedly the main culprits behind the current environmental mess. Industry should come forward to link itself with environmentally safe and C-sequestration-enhancing processes. This paper underlines the scope of fibre-yielding plants that, in addition to conferring enormous benefits to the textile industry, would play crucial role in removing CO_2 from the atmosphere and thus contributing to averting the processes of global warming and consequent climate changes.

How is Sequestration Achieved?

According to the EPA, there are three general means by which agricultural and forestry practices can reduce greenhouse gases:

1. Avoiding emissions by maintaining existing carbon storage in trees and soils.
2. Increasing carbon storage by tree planting or converting from conventional to conservation tillage practices on agricultural lands.
3. Substituting bio-based fuels and products for fossil fuels, such as coal and oil, and energy-intensive products that generate greater quantities of carbon dioxide (CO_2) when used.

Sequestration results vary among plant species and soil. Additionally, sequestration can eventually reach a saturation point in soil that has regained the levels of carbon and organic matter or when trees or plants reach maturity. However, regular maintenance of plants can help ensure that carbon is not lost to the atmosphere once again.

One possible mechanism for climate change mitigation is carbon sequestration, the facilitated redistribution of carbon from the air to soils, terrestrial biomass, geologic formations, and the oceans. For semi-arid and sub-humid regions of the world, carbon sequestration in soils represents the most promising option for climate change mitigation.

What are the Other Benefits?

In addition to moderating climate changes, carbon sequestration can improve general quality of life in the soil, air, and water and with regard to wildlife. It improves the quality of both the habitats and the creatures that sustain in these habitats.

The earth desperately needs the attention and action of us all or our children's children will surely not have a world fit to live in. There is no one solution but amazingly, the simple fibre yielding plant can make a dramatic positive impact in many areas.

Bast and Other Plant Fibres

Environmental concerns have regenerated interest in the use of natural fibers for a much wider variety of products, including high-tech applications such as geotextiles, and composite materials for automotive and light industry use. Bast and Other Plant Fibers analyzes flax, hemp, jute, kenaf, ramie, sisal, coir, and nettle, and provides an index of fiber-yielding plants. Each chapter examining chemical and physical structure, fibre, yarn and fabric production, dying, handle and wear characteristics, economics, and environmental, health and safety issues.

'**Jute**' is the best of the coarser bast fibres and one of the most important commercial fibre crops of the world, particularly in India. The fibre is obtained from the stems (secondary phloem) of the genus chorchorus, which belongs to the family 'Tiliaceae'. The plant is a slender annual. It is tall and attains a height of 10'. It bears alternate simple leaves. The jute fibres are used for bags; the poorer fabrics are used in the making of gunny bags while better ones are used for the manufacture of twine upholstery rugs, carpets etc.

Pine Apple is a very familiar tropical fruit plant belonging to the family 'Bromeliaceae'. It furnishes fibres of great strength and fine quality. They are white, shiny, very durable, flexible and are not injured by water. The plants are cultivated for fibre, and are planted close to each other. The fibres are collected from the leaves. Usually leaves are cut after two years and the leaves are scraped out by hand, dried and combed out. The fibres are tied end to end and can be woven. In Philippines Islands the 'pina cloth', is the most delicate and highly prized fabric. In Taiwan, the fibre is used in making stronger fabrics.

Hemp is one of the most environmentally friendly crops to be found, as it actually improves soil condition. Hemp grows tall and thick, shading and mulching the ground, while its deep taproots break up and aerate the soil. This contributes to healthy microbial life and nutrient content in the soil, and the shading eliminates competing weeds. It is also naturally resistant to most insects, moulds, and other pests. Hemp requires little fertiliser and few if any pesticides, which means it is cheap and requires minimal effort to grow. It also means the environment is not contaminated. Hemp fibre is a popular choice for making textiles because it is strong, comfortable and colourfast. Hemp is a versatile fibre and can be mixed well with cotton, silk, linen and wool. The high cellulose content of the plant means that it can be used to make synthetic textiles such as rayon, too. Hemp clothing is strong and holds its shape, stretching less than any other natural fibre. It is durable, comfortable, and resistant to mould and ultraviolet light. As a renewable resource from living plants, hemp does not contribute to the greenhouse effect. The growing plants absorb as much carbon dioxide (CO_2) as will later be released when oil or other plant matter is burnt. Unlike fossil fuels (oil, coal and gas) or nuclear fuels, hemp could supply raw materials for thousands of years, without ever changing our climate and without producing hazardous waste.

Bamboo is the fastest growing canopy for the regreening of degraded lands, and its stands release 35 per cent more oxygen than equivalent stands of trees. Some bamboo even sequester up to 12 tons of carbon dioxide (CO_2) from the air per hectare. Bamboo can also lower light intensity and protects against ultraviolet rays. Traditional belief holds that being in a bamboo grove–the favorite dwelling place of Buddha–restores calmness to emotions and stimulates creativity.

Banana is in huge demand for the fibre, which is (blended with other natural fibres) used in making a wide range of goods as cordage, yarns, paper and paper cups, tea cups and tea bags, attractively patterned cloth, handbags/purses and shoes. Banana stores 5.7 t C ha^{-1}.

Opportunities

- ☆ Global demand for fibre yielding plants is forecast at more than 338,000 tons by 2008.
- ☆ Market demand for fibre yielding plants is being driven by:
    1. Widening of "fit-for-use" applications
    2. More competitive price structure
    3. Degradability standards

4. Government/legislative traction
5. Consumer education and adoption.

Agriculture is one of the major producers of the greenhouse gases that are causing climate change, but it can also contribute significantly to the reduction of such gases in the atmosphere. Agricultural soils are one of the planet's largest reservoirs of carbon and they have the potential to expand carbon sequestration–that is, to remove carbon from the atmosphere and fix it in the soil–thus mitigating climate change. Fibre-yielding plants managed in agro-ecosystems and in uncultivated areas would be instrumental towards enhancing carbon sequestration in nature and thus helping in mitigating ominous changes leading to global warming and climate change. The additional benefit would include a phenomenal increase in the economic returns to private and public sectors in the country.

References

Colourage. LIII(II) 57-61.

http://edcintl.cr.usgs.gov/carbonsequestration.html.

http://www.chennaimuseum.org/draft/gallery/05/02/ecnomic1.html.

http://www.epa.gov/sequestration/

http://www.netcomposites.com/netcommerce_features.asp?1011.

http://www.theorangecountynews.com/news/2007/0228/Local_news/050.html.

http://www.tve.org/ho/doc.cfm?aid=895.

Lal, A.R. 2006. *Impacts of Rapid Economic Growth on Environment.*

Storage, http://www.fao.org/DOCREP/004/y1318e/y1318e10.html.

Why Bamboo? http://bamboocentral.org/whybamboo. html.

HIMALAYAN MOUNTAINS
Rejuvenated Fragile Ecosystems can Give Appropriate Response to Global Warming

Ecosystem Diversity and Carbon Sequestration: Climate Change Challenges and a Way Out for Ushering in a Sustainable Future *Pages 205–214*
Editors: P.L. Gautam, Vir Singh and Uma Melkania
Published by: DAYA PUBLISHING HOUSE, 2009

Chapter 22

Sustainable Sloping Land Management Options: Potential Effects on Carbon Sequestration in Upland Soils in the Himalayas

Isabelle Providoli, Sanjeev Bhuchar, Keshar Man Sthapit, Madhav Dhakal and Eklabya Sharma

In recent decades unsustainable land management practices due to increasing population pressure in the Hindu Kush-Himalayas (HKH), especially in the "Middle Mountains", are leading to rapidly degrading catchments. These land use changes and degradations are affecting greenhouse gas emissions influencing climatic change. The Kyoto Protocol of the United Nations Framework Convention on Climate Change represents a major international effort to decrease greenhouse gas emissions to the atmosphere. In this context the Clean Development Mechanism (CDM) is an arrangement under the Kyoto Protocol allowing industrialized countries with a greenhouse gas reduction commitment to invest in emission reducing projects in developing countries as an alternative. The CDM is currently offered for afforestation and reforestation projects, but in the future it is expected to extend to carbon sequestration in agricultural soils. In this context sustainable sloping land management, in particular low-cost methods to sequester carbon in soils are emerging as a major international policy goal. However, due to the complex matter of carbon sequestration current applicable numbers of C stock in agricultural soils and other land use are missing. In order to address this C sequestration potential, promoted sustainable land management options of the ICIMOD coordinated "People and Resource Dynamics Project–PARDYP" (1996–2006) will be analyzed through the C sequestration lens. PARDYP was a research for development project which operated in five middle mountain watersheds across the HKH (China, India, Nepal and Pakistan). This perspective paper highlights the goal of our future research to analyze these promoted options of conservation agriculture, in particular the soil conservation practices on the C sequestration capacity. The aim of this study will be to contribute to the understanding of C sequestration in sustainable land management system emphasizing the importance of soil conservation practices. These findings of the short and longer term C sequestration will contribute to new rules for the CDM.

Keywords: *Soil carbon, Soil organic matter, Sequestration, Land management, Himalayam Upland soils.*

Introduction

During the last few decades, human activities have dramatically altered natural ecosystems due to the increasing demand for food, clean water, energy and other resources (Galloway *et al.*, 2003, Daily 2000). Activities such as clearing forests, practicing subsistence agriculture, intensifying farmland production, or expanding urban centers are having large impacts on ecosystems (Foley *et al.*, 2005). Several decades of research have revealed the environmental impact of land use throughout the globe, ranging from changes in atmospheric composition to the extensive modification of ecosystems.

Recent scientific opinion, led by the Inter-governmental Panel on Climate Change (IPCC), is that Global Climate Change is happening; it is inevitable and will present practical challenges to local ecosystems (IPCC 2001 2007). Furthermore, the IPPC report highlights that global atmospheric concentrations of carbon dioxide, methane and nitrous oxide have increased markedly as a result of human activities since 1750 and now far exceed pre-industrial values. The global atmospheric concentration of carbon dioxide, considered as the most important anthropogenic greenhouse gas (GHG), has increased from a pre-industrial value of about 280 ppm to 379 ppm in 2005. This increase in carbon dioxide concentration is primarily due to fossil fuel use and land-use change, while those of methane and nitrous oxide are primarily due to agriculture.

A major international effort to decrease greenhouse gas emissions to the atmosphere is the Kyoto Protocol, an agreement made under the United Nations Framework Convention on Climate Change (UNFCCC). Countries that ratify this protocol commit to reduce their emissions of carbon dioxide and five other greenhouse gases, or engage in emissions trading if they maintain or increase emissions of these gases. To help mitigate global warming, a variety of means of artificially capturing and storing carbon–as well as of enhancing natural sequestration processes–are being explored. In this context the Clean Development Mechanism (CDM) is an arrangement under the Kyoto Protocol allowing industrialized countries with a greenhouse gas reduction commitment to invest in emission reducing projects in developing countries as an alternative. The CDM is currently offered for afforestation and reforestation projects, but in the future it is expected to extend to carbon sequestration in agricultural soils. However, at present not enough is known and the potential of agriculture systems to sequester carbon has not yet been fully recognized (Montagnini and Nair 2004).

Carbon sequestration is the process that removes carbon from the atmosphere. In this process several natural carbon dioxide (CO_2) sinks can be distinguished such as oceans, soils, vegetation, and geologic formations. The soil carbon is a major component of the terrestrial carbon pool making up approximately 75 per cent of it (Swift 2001; FAO 2001; FAO 2004; Lal 2004). Consequently, soils are a major reservoir of carbon and an important sink. The primary way that carbon is stored in the soil is as soil organic matter (SOM). SOM is a complex mixture of carbon compounds, consisting of decomposing plant and animal tissue, microbes and carbon associated with soil minerals. It improves soil quality and nutrient supply, making the soil more productive. SOM also increases structure stability, resistance to rainfall impact, rate of infiltration and faunal activities (Roose and Barthes 2001). Therefore, SOM is closely linked with soil fertility and also seen as an index of sustainable land management (Nandwa 2001).

The carbon in soils comprises two components, the soil organic carbon (SOC) and the soil inorganic carbon (SIC) pool. Carbon can remain stored in soils for millennia, or be quickly released back into the atmosphere. Climatic conditions, natural vegetation, soil texture, and drainage all affect the amount and length of time carbon is stored. Microbial biomass C in soils is a good indicator of soil fertility and also relates to soil organic matter. Studies in a watershed in Sikkim show that soil microbial biomass C is higher in forests and agroforestry systems compared to agriculture land (Sharma *et al.*, 2004).

Especially land use and soil management play an important role in the carbon budget. Good management of agricultural soils can reduce carbon emission rates. Lately, soil organic carbon (humus) levels in many agricultural areas have been severely depleted. Recent studies from Perez *et al.* (2007) and Bationo *et al.* (2007) show that prevailing land management practices, such as deep plowing, crop residue removal, biomass burning, low-input farming, reduced fallow periods and overgrazing of vegetation aggravate the depletion of the soil organic C pool. In recent decades such unsustainable land management is also happening in the upland catchments of the Hindu Kush-Himalayas (HKH), especially in the "Middle Mountains". Therefore, the rehabilitation and sustainable management of these upland catchments is seen as not only important but also challenging for all mountain societies and governments in the region and associated downstream communities. It demands promotion of good land management practices (traditional and scientific), including conversion of marginal lands into restorative land use, adoption of conservation tillage with cover crops and crop residue mulch, better nutrient cycling including using compost and manure, and other systems of sustainable management of soil and water resources.

Sustainable sloping land management are emerging as a major international policy goal. Particular emphasis is on low-cost methods to sequester carbon in soils. Many observers believe that the CDM in farming systems could offer a great potential to reduce rural poverty by extending payments to low-income farmers who provide carbon storage through sustainable land-use systems. However, due to the complex matter of carbon sequestration current CDM-applicable numbers of C stock in agricultural soils and other land use are missing. Sustainable sloping land management options such as low cost soil conservation options can take an important role as carbon sequestration strategies.

The aim of this perspective paper is to contribute to the understanding of C sequestration in sustainable land management system emphasizing the importance of soil conservation practices. The paper looks at the sustainable land management options promoted by the ICIMOD-coordinated "People and Resource Dynamics Project–PARDYP" (1996-2006). The goal of our future research will be to analyze these promoted options of conservation agriculture and watershed management, in particular the soil conservation practices on the C sequestration capacity–*e.g.* the effect of erosion control through terracing, increased soil fertility (soil organic matter, SOM) through better compost application, effects of rehabilitation activities of degraded lands (terracing, planting, hedgerow planting) on soil C sequestration. In addition, the global network WOCAT (World Overview of Conservation, Approaches and Technologies) database offers tools and methods to capture and disseminate good options in sustainable soil management and to promote them. The better understanding of C sequestration in sustainable land management will contribute to development of new rules for the CDM.

The People and Resource Dynamics Project (PARDYP)

PARDYP was a research for development project and operated in five middle mountain watersheds across the HKH–two in Nepal, and one each in China, India and Pakistan (Figure 22.1) (White 2005). While the first phase (1996-1999) had a strong biophysical research dimension, the two following phases (1999–2002 and 2003-2006) shifted to stronger socio-economic and community-based approaches, focusing on development through community-based decision making processes and the development of relevant methodologies. The main objectives were (1) to find options and approaches to support sustainable and equitable access to water, land, and forests; (2) to improve productivity of farming systems; (3) to increase productivity of agricultural land; and (4) to test and disseminate improved water management options. The basic idea was to test, develop and promote simple options for sustainable land management, while building on local knowledge and practices involving local peoples.

PARDYP used participatory rural appraisals and informal meetings with villager and farmer groups to identify the major problems in the watersheds. The PARDYP team then suggested and tested potential interventions. The basic idea was to test, develop and promote simple but effective options, *e.g.* techniques for sustainable land management, while building on local knowledge and practices and involving local people. The key approach was social empowerment through awareness raising, involvement of local institutions and users, encouraging women involvement, user committee formulation and technical backstopping.

Promoted Sustainable Sloping Land Management Options and their Potential to Increase the Carbon Sequestration Status of the Soil

Options to increase soil fertility and to address water scarcity by increasing water availability and improving water use efficiency were promoted by PARDYP.

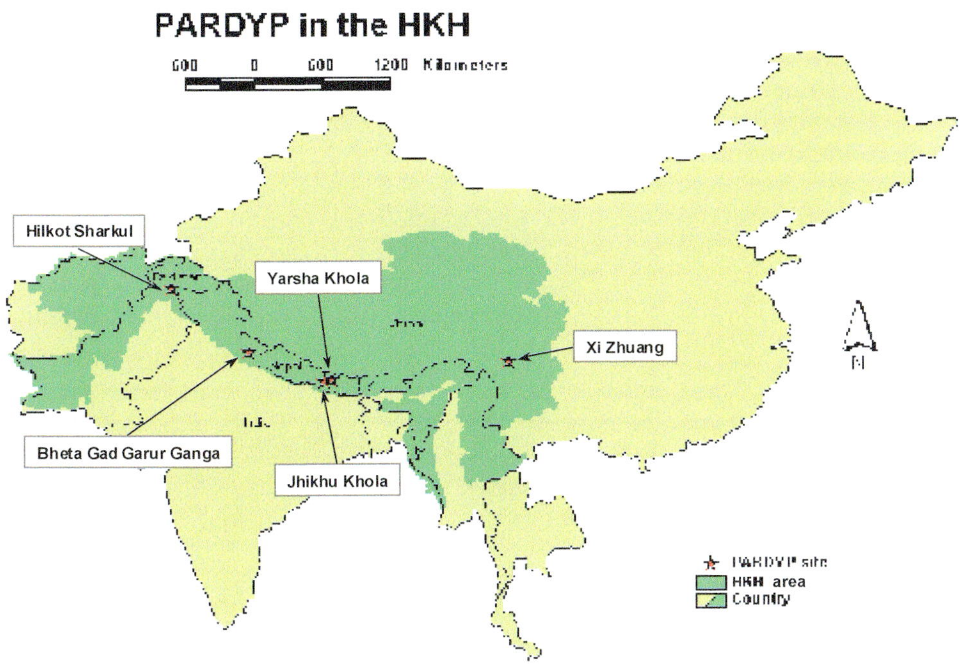

Figure 22.1: Location of PARDYP Watersheds in the HKH

Detailed information can be found in White (2005) and Providoli *et al.* (2007). In this paper we will mainly highlight the promoted sustainable sloping land management options which have a potential to contribute to carbon sequestration in the soil by enhancing soil fertility and increasing water availability by also controlling soil erosion (Table 22.1).

Options which increased soil fertility and had at the same time a high potential in sequestering carbon were the use of biofertilisers and improved composting.

1. Biofertilisers are microorganisms that can improve the availability of key soil nutrients either by fixing nitrogen or releasing important nutrients. Biofertilisers, such as vesicular arbuscular mycorrhiza (VAM), were promoted by PARDYP. They are cost effective, environmentally friendly, and help to maintain soil fertility naturally. A study on the influence of bacterial inoculation (*Azotobacter chroococcum* and *Azospirillum brasilense*) on maize crop in upland farming system in Sikkim increased yield by more than double showing great potential (Pandey *et al.*, 1998).
2. Improved compost a) Black plastic composting: PARDYP promoted a simple and cheap on-farm composting technology to improve the quality of traditional farm compost, which commonly consists of a mixture of cattle dung, urine, and plant material. Most of the farmers dump this heap in an open space for decomposition. In the black composting method, the

traditional compost heap is covered with a black polythene sheet which protects the compost from nutrient leaching during rainy days, and provides a favourable environment for the growth of microbes by increasing temperature and reducing evaporation. The method significantly accelerated compost decomposition (45-60 days) when compared to the traditional method (4-6 months) and also produces compost of much better quality (higher nutrient contents and improved decomposition). A disadvantage perceived by farmers, however, was the relatively high cost and low availability of the thick (durable) black plastic in local markets. Biocomposting of weeds in field conditions can also increase nutrient availability of carbon, phosphorus and potassium. The technology provides opportunity of farm weeds in accelerating nutrient cycling and availability, including carbon build-up in the soil (Sharma *et al.*, 2001).

Table 22.1: Options Contributing to Increasing Carbon Sequestration in Soil

Sl.No.	*Options*	*Effect on Soil*	*C Seq. Status*	*Country*
		SOIL FERTILITY		
1.	Biofertilisers	Increase nutrients	Improve SOM status	India, Pakistan, Nepal, China
2.	Improved composting (*a*) Black plastic composting (*b*) Vermi-composting	Better compost quality (increase nutrients)	Improve SOM status	Nepal, India
		EROSION CONTROL (Increasing water infiltration)		
3.	Terracing	Reducing erosion/C	Improve C content	Nepal
4.	Contour hedgerows	output	by reducing soil erosion	India, Nepal
5.	(*a*) Rehabilitation through eye brow terraces		increasing SOM input	Nepal
	(*b*) Rehabilitation with conservation pond and pits			India

3. Improved compost b) Vermi-composting is produced by using earthworms to convert biodegradable wastes into compost. The nutrient composition of vermi-composts is much better compared to compost made in pits or by bacterial decomposition. In PARDYP India, Red-worm or Tiger-worm (*Eisenia fetida*) and a local variety were tested: The *E. fetida* results were much better due to the species tolerance of a wide range of environmental conditions. In addition, it can feed on a variety of organic wastes (*e.g.* cattle dung, sheep dung, poultry dropping, weeds, used flowers, waste paper, waste from fruit, pulp, etc.) and grow and multiply very fast.

   Furthermore, in PARDYP effective options to increase water availability were achieved by increasing infiltration, water harvesting, and improving drinking water supply. In the context of carbon sequestration the options increasing infiltration and simultaneously controlling erosion were particularly interesting (Table 1).

4. By terracing and improving terraces the slope was modified and direct runoff and soil erosion was reduced. These options included also planting of vegetative measures, such as grasses. Through this intervention the soil moisture status was increased and a favourable micro-environment for plant growth was created.
5. By planting contour hedgerows erosion could be reduced by stabilising the soil. In addition nitrogen-fixing hedgerows improved soil nutrients and provided fodder availability.
6. Degraded lands were rehabilitated through eye brow terraces and conservations ponds/pits. These options were an easy way of rainwater harvesting and created a favourable micro-environment for plants to grow. The soil moisture status could be increased and erosion could be reduced.

In one of the study sites in Uttaranchal (India), the change in soil characteristics as a result of land rehabilitation activities was monitored (Kothyari *et al.*, 2005). The average of soil moisture increased from 12.3 to 21 per cent, total soil organic carbon increased from 1.0 to 1.5 per cent, and the soil pH increased from 5.9 to 6.3. The gradual increase in organic carbon was a positive indication of improvement in soil fertility as well as the increase in total nitrogen (N), and available phosphorus (P) and potassium (K) during the same period. In addition, the soil moisture content had a positive correlation with the soil microbial biomass. Conducted studies showed a gradual increase in micro-organisms density which also contributed to the increasing soil fertility. Furthermore, the site also recorded a significant increase in grass production as well as a gradual reduction in overland flow and sediment losses due to the improvement in vegetation cover.

Discussion

Sustainable soil management practices have a high potential to increase the carbon inputs and to reduce the outputs of carbon in soils. The findings of PARDYP show that low-cost sustainable sloping land management options had an effect on carbon sequestration, especially on soil fertility. The options improving soil fertility, the application of biofertiliser and the use of improved compost were effective for increasing soil quality and crop production in the project areas. By applying biofertiliser the SOM status of the soil was improved contributing to a general increase in carbon sequestration. Likewise, the options addressing soil erosion control had a high potential to increase carbon sequestration by improving the carbon content by reducing erosion and increasing SOM input. These options being low-cost have a great potential to be applied by low income farmers. Low cost soil conservation measures have in general the following characteristics: they are affordable for farmers, local resources including materials and human are used, they are simple and easy to use with minimum backstopping from the technicians, and are easy to maintain.

However, while implementing PARDYP the main focus was not on the assessment of the carbon status of the soil. Therefore at this stage not enough data are available to clearly highlight the effect of sustainable sloping land management. Our future research plans are to analyze these promoted options of conservation agriculture and watershed management in more detail. In particular the effect of erosion control through terracing, increasing soil fertility (soil organic matter, SOM) through better compost application, and effects of rehabilitation activities of degraded lands (terracing, planting, hedgerow planting) on soil C sequestration. The further understanding of the C sequestration in sustainable land management system will emphasize the importance of soil conservation practices.

In addition, the global network WOCAT (World Overview of Conservation, Approaches and Technologies) database offers tools and methods to capture and disseminate good options in sustainable agriculture and to promote them (www.wocat.net). In this WOCAT network also efforts were taken to highlight the relevance of carbon sequestration in sustainable agriculture and to capture carbon sinks potentials. Studies from the Philippines presented the role and importance of soil conservation on carbon sequestration. Natural vegetative strips could reduce erosion and carbon emission by 90 per cent. This database has a subnet called the Himalayan Conservation Approaches and Technologies (HIMCAT), including an extranet platform for the Himalayan region (http://extranet.icimod.org.np/himcat/), in which good soil and water conservation practices can be documented and shared. In the case of Nepal, technologies and approaches from the PARDYP experience will be shared through the WOCAT database.

Globally, good examples especially from West-Africa, are known (Bationo *et al.*, 2007 and Perez *et al.*, 2007). These studies especially highlight the potential of agroforestry systems for carbon sequestration, in particular if they were well managed and if soil conservation practices were included (Montagnini and Nair 2004). One study from the Himalayas in a cardamom-based agroforestry system in the Mamlay watershed in the southern part of Sikkim shows a higher floor litter carbon concentration than in other land use stands (Sharma *et al.*, 1997a, 1997b; Sharma 2003)). However, the number of studies in the HKH region on sustainable sloping land agriculture and the link to carbon sequestration is still very small. Therefore, more research on the potential of such options is needed.

All the examples presented in the paper highlight the potential of short and longer term carbon sequestration and carbon trading. Globally, markets for buying and selling carbon credits already exist and are getting increasingly available. Although very few of those credits include provisions for terrestrial carbon sequestration, soil carbon sequestration and broader sink activities could become eligible during subsequent Kyoto Protocol commitment. Hopefully, all these findings will contribute and lead to new rules for the CDM for terrestrial carbon sequestration, which represents a huge potential for poor farmers in developing countries. Therefore there is a strong need for further research because currently very few studies are available. With this perspective paper we wanted to highlight the need of further studies which we hopefully will be able to perform in the future.

Acknowledgements

This article takes into acknowledgment all the people who have worked with PARDYP. Furthermore, we acknowledge the Swiss Agency for Development and Cooperation (SDC), the International Development Research Centre (IDRC) of Canada, and the International Centre for Integrated Mountain Development (ICIMOD) for funding.

References

Bationo, A., Kihara, J., Vanlauwe, B., Waswa, B. and Kimetu, J., 2007. Soil organic carbon dynamics, functions and management in West African agro-ecosystems. *Agricultural Systems*, 94: 13–25.

Daily, G.C., 2000. Management objectives for the protection of ecosystem services. *Environmental Science and Policy*, 3: 333–339.

FAO, 2001. *Soil Carbon Sequestration for Improved Land Management*. Food and Agriculture Organization of the United Nations, Rome.

FAO, 2004. *Carbon Sequestration in Dry Land Soils*. Food and Agriculture Organization, Rome.

Foley, J.A., DeFries, R., Asner, G.P., Barford, C., Bonan, G., Carpenter, S.R., Chapin, F.S., Coe, M.T., Daily, G.C., Gibbs, H.K., Helkowski, J.H., Holloway, T., Howard, E.A., Kucharik, C.J., Monfreda, C., Patz, J.A., Prentice, I.C., Ramankutty, N. and Snyder, P.K., 2005. Global consequences of land use. *Science*, 309: 570–574.

Galloway, J.N., Aber, J.D, Erisman, J.W., Seitzinger, S.P., Howarth, R.W., Cowling, E.B. and Cosby, B.J., 2003. The nitrogen cascade. *BioScience* 53, 341–356.

IPCC, 2001. *Climate Change: Synthesis Report. A Contribution of Working Groups I, II and III to the Third Assessment Report of the Inter-governmental Panel on Climate Change*. Cambridge University Press, Cambridge, UK, and New York, USA.

IPCC, 2007. *IPCC Summary for Policymakers: Climate Change 2007: Climate Change Impacts, Adaptation and Vulnerability*. IPCC WGII Fourth Assessment Report.

Kothyari, B.P., Bhuchar, S.K., Bisht, S.S., Joshi, B.K., Bisht, B.S. and Mishra, A.K., 2005. Participatory rehabilitation of degraded lands for rural livelihood sustainability and improved biodiversity: A case study from the Indian Central Himalayas. In: *Renewable Natural Resources Management for Mountain Communities*, (Eds.) M. Stocking, H. Helleman and R. White. ICIMOD, Kathmandu, Nepal.

Lal, R., 2004. Soil carbon sequestration to mitigate climate change. *Geoderma*, 123: 1–22.

Montagnini, F. and Nair, P.K.R., 2004. Carbon sequestration: an underexploited environmental benefit of agroforestry systems. *Agroforestry Systems*, 61: 281–295.

Nandwa, S.M., 2001. Soil organic carbon (SOC) management for sustainable productivity of cropping and agroforestry systems in Eastern and Southern Africa. *Nutrient Cycling in Agro-ecosystems*, 61: 143–158.

Pandey, A., Sharma, E. and Palni, L.M.S., 1998. Influence of bacterial inoculation on maize in upland farming systems of the Sikkim Himalaya. *Soil Biology and Biochemistry*, 30(3): 379–384.

Perez, C., Roncoli, C., Neely, C. and Steiner, J.L., 2007. Can carbon sequestration markets benefit low-income producers in semi-arid Africa? Potentials and challenges. *Agricultural Systems*, 94: 2–12.

Providoli, I., Bhuchar, S., White, R., Sthapit, K.M. and Dhakal, M., 2007. People and Resource Dynamics Project (PARDYP): Participatory Options for Sustainable Sloping Land Management Promoted in Five Watersheds in the Himalayas. In: *International Conference on Sustainable Sloping Lands and Watershed Management*, December 12–15, 2006, Luang Prabang, Laos (Proceeding In Press).

Roose, E. and Barthes, B., 2001. Organic matter management for soil conservation and productivity restoration in Africa: A contribution from francophone research. *Nutrient Cycling in Agro-ecosystems*, 61: 159–170.

Sharma, P., 2003. Ecological linkages of carbon dynamics in relation to land-use/ cover change in a Himalayan Watershed. *Ph.D. Thesis*, University of North Bengal, India.

Sharma, P., Rai, S.C., Sharma, R. and Sharma, E., 2004. Effects of land-use change on soil microbial C, N and P in a Himalayan watershed. *Pedobiologia*, 48: 83–92.

Sharma, E., Rai, S.C. and Sharma, R., 2001. Soil, water and nutrient conservation in mountain farming systems: Case-study from the Sikkim Himalaya. *Journal of Environmental Management*, 61: 123–135.

Sharma, R., Sharma, E. and Purohit, A.N., 1997a. Cardamom, mandarin, and nitrogen–fixing trees in agroforestry systems in India's Himalayan region. I. Litterfall and decomposition. *Agroforestry Systems*, 35: 239–253.

Sharma, R., Sharma, E. and Purohit, A.N., 1997b. Cardamom, mandarin, and nitrogen–fixing trees in agroforestry systems in India's Himalayan region. II. Soil nutrient dynamics. *Agroforestry Systems*, 35: 255–268.

Swift, R.S., 2001. Sequestration of carbon by soil. *Soil Science*, 166(11): 858–871.

White, R., 2005. Overview of PARDYP Research: Promising Approaches to Watersheds Management. In: *Resource Constraints and Management Options in Mountain Watersheds of the Himalayas: Proceedings of a Regional Workshop,* (Eds.) R. White and S.K. Bhuchar. held 8–9 December, 2003 in Kathmandu, Nepal. Kathmandu, ICIMOD.

Ecosystem Diversity and Carbon Sequestration: Climate Change Challenges and a Way Out for Ushering in a Sustainable Future *Pages 215–221*
Editors: P.L. Gautam, Vir Singh and Uma Melkania
Published by: DAYA PUBLISHING HOUSE, 2009

Chapter 23

Rangelands Resources in the Mountains: Management Objective Should Focus on Carbon Sequestration Enhancement

R.D. Gaur, Vir Singh and Babita Bohra

Rangelands are more crucial in terms of their contributions to environment and society than often perceived by planners and policy makers. Rangeland science has assumed many dimensions of rangeland ecology. Like other disciplines, the rangeland ecology is slowly but steadily attracting attention of scholars to unravel complexity of the range ecosystems. Though relatively a young discipline, it is posing many challenges before the scholars, who include geographers, ecologists, environmentalist, sociologists, anthropologists, economists and agriculturists. One of the most crucial roles rangelands can play is carbon sequestration. Occupying the largest chunk of the geographical area in the Himalayan mountains, the rangelands can serve as the dumping grounds for carbon.

Defining Rangelands

Grasslands or grazing lands are often used as synonym of rangelands. Though these terms can be used interchangeably, the latter carry a broader dimension in terms of their floristic composition and utilisation. Rangelands can be dominated by diverse vegetation and are often associated with livelihood systems of the communities who manage or have access to them. Rangelands sometimes are also perceived as wastelands or degraded lands and are equated with desert of arid land.

Rangelands are the ecosystems or units of nature playing pivotal role in agriculture and livelihoods. Several agencies and workers have attempted to define rangelands in their own way. In 1973, National Committee on Rangeland

Management defined rangelands as "uncultivated areas (although sometimes disturbed by unthoughtful cultivation) that support natural or seeded herbaceous or shrubby vegetation with or without trees". The Society of Rangeland Management Website defines rangelands as the lands on which the native vegetation–predominantly grasses and grass like plants, forage or shrubs–is suitable for use by grazing and browsing. Earlier, Stoddart *et al.* (1975) described rangelands as those areas of the world which by reason of physical limitations, low and erratic precipitation, rough topography, poor drainage and extreme temperatures are unsuited to cultivation and which are a source of forage for free ranging native and domestic animals, as well as a source of wood products, water and wildlife.

There are several other loose definitions of the rangeland ecosystems. In the Himalayan context, Singh (2007) has proposed the following holistic definition:

Rangelands are the uncultivated areas or abandoned cultivated areas and occasionally fallow lands that harbour natural or seeded/planted vegetation of herbaceous and woody species and that serve as habitats of a variety of wild animals, support domesticated animals, provide unique products and ecosystem services especially vital for the livelihoods of livestock-dependent communities.

Extent of the Rangelands

No exact figures of rangeland area on earth are available. As the waters on the Earth, so are the rangelands on the land portion of the Earth. At global level, grasslands and rangelands occur in polar, temperate, subtropical and tropical latitudes, from low to high elevations. They vary from subtropical savannas to alpine meadows in the eastern, central, and western Himalayas and steppe formations on the plateau (Richard *et al.*, 2000).

Globally, rangelands cover 45 km^2 or one-quarter of the Earth's surface. In the Hindu Kush-Himalayan Tibet-Quinghai Plateau, rangelands and pastures cover some 60 per cent of the total area (Richard *et al.*, 2000). According to Zhaoli (2004), some estimates place 70 per cent of world's total area as rangelands. According to Sidahmed and Rota (2004), the rangelands encompass approximately half of the earth's land surface. More than 60 per cent of the Himalayan region consists of rangeland ecosystems. In many areas of the Himalayan mountains, more than 80 per cent of the geographical area is under rangelands.

Rangelands are the major ecosystems within the Hindu Kush-Himalayan region and, for a large number of local people, they provide the main or the only livelihood resource. At the same time, these rangelands also provide high value ecosystem functions and services (Zhaoli 2004). Rangelands being the core components of the farming systems in the Himalayan region hold key to ecological integrity, a vital prerequisite of agricultural sustainability.

Rangelands as Source of Fodder

Gaur (1999) has brought out comprehensive account of the biodiversity present in the Garhwal district of Uttarakhand. Rangelands serve as rich repositories of fodder plants. As such, these areas are crucial for livestock production, which is the major occupation of mountain farming communities and of a quite sizeable population of nomadic or transhumant people in the Himalayan mountains.

Table 23.1: Number of Fodder Species Belonging to Different Families in the Himalayan Rangelands

Sl.No.	*Family*	*Number of Species*	*Sl.No.*	*Family*	*Number of Species*
1.	Acanthaceae	7	35.	Fabaceae	69
2.	Amaranthaceae	8	36.	Fagaceae	5
3.	Anacardiaceae	4	37.	Flacourtiaceae	1
4.	Apiaceae	4	38.	Geraniaceae	1
5.	Apocynaceae	1	39.	Gesneriaceae	1
6.	Aquifoliaceae	3	40.	Hippocastanaceae	1
7.	Araliaceae	1	41.	Hydrangeaceae	1
8.	Asteraceae	19	42.	Hypericaceae	8
9.	Balsaminaceae	1	43.	Lamiaceae	6
10.	Basellaceae	1	44.	Laminaceae	1
11.	Berberidaceae	2	45.	Lauraceae	10
12.	Betulaceae	1	46.	Loranthaceae	2
13.	Burseraceae	2	47.	Magnoliaceae	1
14.	Caeselpiniaceae	11	48.	Malvaceae	3
15.	Campanulaceae	1	49.	Meliaceae	3
16.	Capparaceae	3	50.	Menispermaceae	2
17.	Caprifoliaceae	2	51.	Mimosaceae	9
18.	Caryophyllaceae	2	52.	Moraceae	21
19.	Celastraceae	5	53.	Myrsinaceae	2
20.	Chenopodiaceae	5	54.	Oleaceae	4
21.	Combretaceae	3	55.	Orobanchaceae	1
22.	Commelinaceae	1	56.	Oxalidaceae	2
23.	Convolvulaceae	2	57.	Passifloraceae	1
24.	Coriariaceae	1	58.	Plantaginaceae	1
25.	Cornaceae	3	59.	Poaceae	164
26.	Corylaceae	1	60.	Polygonaceae	7
27.	Cucurbitaceae	2	61.	Ranunculaceae	6
28.	Cyperaceae	5	62.	Rhamnaceae	6
29.	Dilleniaceae	1	63.	Rosaceae	8
30.	Dioscoreaceae	8	64.	Rubiaceae	13
31.	Ebenaceae	2	65.	Rutaceae	1
32.	Ehretiaceae	2	66.	Sabiaceae	1
33.	Elaeagnaceae	2	67.	Salicaceae	6
34.	Euphorbiaceae	7	68.	Sapindaceae	2

Contd...

Table 23.1–Contd...

Sl.No.	Family	Number of Species	Sl.No.	Family	Number of Species
69.	Sapotaceae	1	77.	Tiliaceae	6
70.	Saurauiaceae	1	78.	Ulmaceae	4
71.	Saxifragaceae	1	79.	Urticaceae	13
72.	Scrophulariaceae	2	80.	Verbenaceae	3
73.	Smilacaceae	2	81.	Violaceae	1
74.	Solanaceae	2	82.	Vitaceae	7
75.	Symplocaceae	3	83.	Zingiberaceae	1
76.	Theaceae	1			

Source: Based on Singh (2007) and Singh and Gaur (2008).

As many as 541 fodder species belonging to 83 families of the plants have been found occurring in the mid-altitude Himalayan rangelands (Table 1). Family Poaceae alone offers as many as 164 species of fodder value for livestock.

Biodiversity is influenced considerably by altitude and slope orientation. Grasslands at lower and mid-altitudes are often interspersed by trees and shrubs. Amatya (1990), Bhatt and Rawat (1993), Bana and Singh (2004), Singh and Gaur (2005), Singh and Bohra (2005) and Bohra (2006) have identified and documented an impressively large number of fodder species occurring in the rangeland ecosystems in parts of the HKH region. Prevalence of individual species is mainly according to altitude and slope orientation.

Alpine Ranges

Alpine zone ranges between 3100m and 5400m altitudes and comprises the highest vegetational zone in the Himalaya. This zone harbours the vegetation which is the climatic climax vegetation. Alpine meadows are the only "grasslands" in India which are composed of the climatic climax vegetation. Grasslands occurring in the regions other than in the Himalayan alpines have come into being due to anthropogenic activities or as a result of edaphic factors operating on certain sites.

Alpine meadows near timberline ecotone have well developed thick soil substratum enriched with organic matter content. Alpine ranges support unique plants (Rawat and Gaur 2007).

Alpine plants are distributed in accordance with the alpine habitats. Rawat and Gaur (2007) have classifies alpine habitats into 12 categories, *viz.*,

1. Meadows
2. Rock Associated Habitats
3. Timberline Forests and Shrubs
4. Marshes
5. Fellfields

6. Moraines
7. Boulder Fields and Block Fields
8. Talus
9. Scree
10. Stream Banks
11. Lakes and Ponds

Each habitat has distinct environmental and phytosociological characteristics. There are certain plant species which are restricted solely to meadows, while some species are restricted solely to rock associated habitats and so on. Some species are found overlapping other habitats in the alpine ranges. Species diversity generally decreases with the increase in altitude and at a point, with the occurrence of permanent glaciers, vegetation losses its existence. Contained in the alpine zone, all habitats, thus contribute to the uniqueness of the Himalayas.

In a comprehensive study of 24 alpine zones of the Uttarakhand Himalaya within the altitude range of 3100-5400m, Rawat and Gaur (2007) found a total of 424 species of angiosperms growing in 11 different habitats. Out of these species, 192 (45.25 per cent) are restricted to a single habitat only and thus are very habitat-specific. The remaining 232 (54.72 per cent) are distributed in more than one habitat. Of these, 159 (37.5 per cent) are distributed in two habitats, 60 (14.15 per cent) in three habitats and 13 (3.07 per cent) only are recorded in more than three habitats (Table 23.2). None of the alpine species grows in all types of habitats, obviously the ubiquitous nature in alpine zones is distinctly excluded (Rawat and Gaur 2007).

Table 23.2: Distribution of Species in Different Habitats

Alpine Habitat	*Altitude, m asl*	*Total Species*	*Restricted Species*	*Species Common in Other Habitats*
Meadows	3100-4600	301	133 (44.19%)	168 (55.81%)
Timberline Forests	3100-3600	77	7 (9.09%)	70 (90.91%)
Rock Associated Habitats	3100-5000	107	34 (31.77%)	73 (68.23%)
Marshes	3100-4500	30	2 (6.67%)	28 (93.33%)
Fellfields	4500-5000	28	-	28 (100%)
Moraines	3100-5000	16	1 (6.25%)	15 (93.75%)
Block Fields and Boulder Fields	3100-5000	33	2 (6.06%)	31 (93.94%)
Talus	3500-4500	56	-	56 (100%)
Scree	4500-5000	36	8 (22.22%)	28 (77.78%)
Stream Banks	3100-5000	31	-	31 (100%)
Ponds	3100-5000	5	5 (100%)	-

Source: Rawat and Gaur (2007).

Amongst the different habitats of the alpine ranges, meadows form the largest component. A total of 301 species have been found to occur in this range in Uttarakhand Himalaya of which 133 (44.19 per cent) are restricted to meadows only. The rest 168 species (55.81 per cent) are also found occurring in other alpine habitats. Maximum number of species sharing is with timberline forests and shrubs (58), followed by rock associated habitats (49), talus slopes (30), marshes (24), and fell fields (10) (Rawat and Gaur, 2007).

Rangeland Management

Rangeland management is the science and art of optimizing the return from rangeland through the manipulation of range ecosystems (Stoddart *et al.*, 1975). The American Society for Range Management (1964) defined range management as the art and science of planning and directing range use to obtain sustained maximum animal production consistent with perpetuation of natural resources. Viewed within the system framework, rangeland management has been defined as management of renewable natural resource composed of one or more range ecosystems for the optimum sustained yield by the combination of goods and services (Vallentine 1971).

These rangelands accommodate important watershed functions, support livestock industry and provide valuable as well as biologically diverse resources. Not only this, the rangelands also provide livelihood base for some primitive nomadic cultures as well as for modern farming cultures. Rangeland management involves all the natural factors, socio-economic parameters and policies.

Carbon sequestration has not been a focused objective of range management strategies so far. With the emerging issues of global warming and ominous climate changes, enhancement of carbon sequestration needs to become one of the major objectives of our range management strategies.

Since rangelands in the mountains are managed with the primacy of livestock production, conservation of the natural diversity of the fodder species–trees, shrubs and herbaceous plants–according to site specificities and their appropriate utilization pattern might be of potential implications for carbon sequestration. Several of the fodder trees prevail in agroforestry systems. Conversion of croplands into agroforestry systems would have enormous bearing on the rates of carbon sequestration in agricultural lands. Building up a synergy between uncultivated rabgelands areas and croplands would be phenomenal for carbon sequestration in the fragile mountains.

References

Amatya, S.M., 1990. *Fodder Trees and Their Lopping Cycle in Nepal*. Technical Report, Kathmandu.

Bana, O.P.S. and Singh, V., 2004. Silvipasture systems for livestock production in the Hindu Kush–Himalayan region: issues and options. In: *Livestock Production Systems for Sustainable Food Security and Livelihoods in Mountain Areas*, (Eds.) Singh, V. and P.L. Gautam. GB Pant University of Agriculture and Technology, Pantnagar.

Bhatt, A.B. and Rawat, N., 1993. Fodder resources of Garhwal: A search for non conventional fodder species. In: *Garhwal Himalaya: Ecology and Environment*, (Ed.) G.S. Rajwar. Ashish Publishing House, New Delhi, pp. 227–239.

Bohra, Babita, 2006. Dairy farming and rangeland resources in mountain agro-ecosystems in Uttarakhand. *Ph.D. Thesis*, G.B. Pant University of Agriculture and Technology, Pantnagar.

Camille, R., Basnet, K., Sah, J.P. and Raut, Y., (Eds.). 2000. *Grassland Ecology and Management in Protected Areas of Nepal. Kathmandu*: ICIMOD, WWF and Department of National Parks and Wildlife Conservation, HMG/Nepal.

Gaur, R.D., 1999. *Flora of the District Garhwal, North West Himalaya with Ethnobotanical Notes*. Transmedia, Srinagar, Garhwal.

Rawat, D.S. and Gaur, R.D., 2007. Habitat diversity and distribution of alpine plants. In: *Plant Morphology and Biotechnology*, (Ed.) P.C. Trivedi. Avishkar Publication, Jaipur, p. 37–52.

Sidahmed, A.E. and Rota, A., 2004. IFAD's Experience in supporting commmunity-based livestock and rangeland development. In: *Strategic Innovations for Improving Pastoral Livelihoods in the Hindu-Kush-Himalayan Highlands, Vol. II: Technical Papers*, (Eds.) C. Richard and Kate Haffman. ICIMOD, Kathmandu.

Singh, V., 2007. Studies on the rangeland management of selected areas in Kumaun Himalaya. *D.Phil. Thesis*, H.N.B. Garhwal University, Srinagar.

Singh, V. and Bohra, Babita, 2005. Livestock feed resources and feeding practices in hill framing systems: A review. *Indian Journal of Animal Sci.*, 75(1): 121–127.

Singh, V. and Gaur, R.D., 2005. The Himalayan rangelands: Ecosystem services and ecotourism opportunities. *National Seminar on Tourism and Himalayan Biodiversity*. Govt. Post Graduate College, Uttarkashi (UA), March 4–5.

Stoddart, L.A., Smith, A.D. and Box, T.W., 1975. *Range Management*. McGraw Hill Book Company, New Delhi.

The American Society of Range Management, 1964. *Glossary of Terms of Range Management*. American Society of Range Management, USA.

Vallentine, J., 1971. *Range Improvement and Development*. Young University Press, Proso, Utah, USA.

Zhaoli, Yan, 2004. Co-management of Rangelands: An Approach for Enhanced Livelihoods and Conservation. *ICIMOD Newsletter*, 45: 15–16.

Ecosystem Diversity and Carbon Sequestration: Climate Change Challenges and a Way Out for Ushering in a Sustainable Future *Pages 222–230*
Editors: P.L. Gautam, Vir Singh and Uma Melkania
Published by: DAYA PUBLISHING HOUSE, 2009

Chapter 24

Himalayan Conservation and Development: The Mighty Mountains can put the Earth's Climate Systems in Order

M.L. Dewan

Introduction

The Himalayan ecosystem is one of the most important and most threatened of life support systems on earth. In the shadow of the Himalayas live more than 150 million people, some of them the poorest in the world. The rivers which arise in the Himalayas and flow down into the wide Indo-Gangetic plain, support agriculture that sustains these people. The population pressure and land and its mounting demand for fuel and fodder has already denuded areas of the Himalayan ranges–the lifeline of India. The rate at which the forests are disappearing is much faster that the rate of afforestation.

My love for the Himalayas dates back to the year 1938 when I went to foot to the famed Amarnath Caves. My earliest contact with the mountain system in north–western India goes back even earlier for I was born in a foothill area surrounded by hills and mountains. My professional contact, however, began in the fifties when I was working in the Soil Conservation Department of the Damodar Valley Corporation (DVC).

During my FAO tenure, I had more occasions to work in the mountainous areas in Iran, Bulgaria and many other countries.

In 1975, However, I first started working on Himalayan Programme as FAO Chief of the Regional Asia Bureau. I felt that here is a need for close co-operation between the Himalayan countries as the Himalayas serve as source of floods and indirectly of droughts in the sub-continent. In October 1980, I visited Himachal Pradesh and brought to the notice of the Planning Commission, the state of the Himalayas as really crumbling and bleeding profusely. The Planning Commission in the person of Dr. M.S. Swaminathan organized the Himalayan Eco-development Task Force which discussed, for over one year, what to do about it and brought out a report of the Task Force in March 1982. This report which was submitted to the Prime Minister highlighted the action programmes which included some of the activities of the Department of Environment and others and especially the proposal for a Himalayan Environment and Development Institute to be started in the Seventh Five Year Plan.

My association with the National Land Resource Conservation and Development Commission (1984-85) gave me more opportunities to work in this direction. I emphasized that all these activities especially in the Himalayas and elsewhere in the country for conservation and development of the soil resources cannot be undertaken by the Government alone. The need for greater involvement of people also through non-government organizations then becomes evident. It is against this background that the programme for Himalayan Eco-System Development was initiated at the Centre for Policy Research which hosted this programme and through which contacts were made with many NGOs in western, central and north–eastern Himalayas. An operational plan was prepared by the end of December 1984 which suggested that an NGO meeting be organized March 1985 to work out a programme, and develop a strategy for large-scale involvement of NGOs, youth groups, universities, colleges and schools, the Territorial Army and other non–governmental and voluntary agencies for concerted efforts in saving the Himalayas.

The Himalayan mountain area is approximately 523,000 sq. km (16.13 per cent of the land surface of India), the Himalayan sediment areas, that is the areas which are directly fed by the Himalayan rivers and which form the Indo-Gangetic and Brahmaputra Plains, including 726,000 sq. km or 22.26 per cent of the total area. Thus the grand total of the Himalayan region is 1,249,000 sq. km or 38.4 per cent of the land resources area of India. In the Himalayas and foot hills live about 150 million people, and about 300 million people live in Himalayan mountain area and Himalayan Sediment Area. It is important that a comprehensive effort to protect the production base in the Indian Himalayas and plains by reducing soil losses and by preserving the ecological balance be made urgently. The most important task of the programme is to develop and apply or stimulate application of a technical package of practices for watershed management to develop or strengthen an appropriate infrastructure (Government has since established a Himalayan Institute for Conservation and Development in the Seventh Five Year Plan) to further improve the social, political and public policy on environmental issues in the Himalayas.

The Technical measures for conservation and development are essentially:

1. Tree planting and reforestation
2. Water harvesting
3. Terraces and other engineering structures
4. Agronomic, grassland and pasture development and other socio-economic measures for conservation and management of the soil.

The youth groups, the academic institutions, the People's Organisations and interested and qualified NGOs are essential as an additionally of effort to what government is already doing for the conservation and development of Himalayas.

The Plight of Himalayas

Why Himalayas?

The Himalayas have exercised a great influence on the environmental conditions of northern India and the people living in the Indo–Gangetic Plains. It has prevented the monsoon winds from crossing over to Tibet and forced them to precipitate most of the moisture on the Indian side in the form of rain and snow. It has also acted as a formidable barrier range and presented free migration of the fauna and flora as well as human culture across it. The complex geological history of the Himalayas and the tectonic stress which heaved up this range–geologically the youngest in the earth's history–have contributed to the aggravation of the earth's dynamic process of weathering, erosion, mass wasting, seismicity, and so on, which affect and modify the natural environment. The developmental activities of man in the area, like the construction of high dams, exploration for minerals and mining activity in conjunction with the quest for arable land have intensified these dynamic processes–commonly referred to as geological hazards. The human interference often gives these dynamic processes catastrophic proportions, leading to disasters and irreparable damage to the natural balance of the eco-system. It is therefore, necessary to acquire a better understanding and appreciation of the natural dynamic processes of the earth and a better integration of man and his activities into his environment so that the natural balances of the ecosystem of the Himalayas is not destroyed.

Brief Description of Himalayas and Statistical Overview

The Himalayas represent the youngest mountain range, having been uplifted about 60-70 million years ago. It is also geologically a very complex range, and its origin is not yet clearly understood. With the development of the means of communications in the region, many of its remote parts have now become accessible and are being systematically investigated and habituated due to pressure of population. These developmental activities have interfered with the natural eco-system of this range and caused deterioration in the environment by aggravating the natural dynamic processes of the earth. These dynamic processes are already in a more active state in the region because of the certain geological, geo-dynamical, weathering, erosion and seismic processes.

The developmental activities in the Himalayas have to face the challenge of these geological hazards and in some cases the developmental activities, such as the impounding of large reservoirs of water, tunneling operations and the addition of silt load of rivers by road cutting lead to the intensification of the natural dynamic process.

Though the knowledge about the fragility of the Himalayas is well known, the developmental activities did not take into account this aspect. The vast Himalayan region that extends from Jammu and Kashmir in the west to Arunachal Pradesh in the east has been under severe deforestation due to various developmental activities. The following tables provide Himalayan area statistics, extent of sedimentation of major reservoir in the country and the extent of damage caused by floods during the years 1953–1981.

Table 241: Himalayan Land resources

Sl.No.	*Name of Land Resource Region*	*Area Square Km*	*Percentage of the Land Resource Area (India)*
	(Part 1: Himalayan Mountain Area)		
1.	Northern Himalaya Snow Clad Region	116,000	3.56
2.	Northern Himalaya Alpine Grass and Meadow Region	98,250	3.02
3.	Northern Himalaya Forest Region	131,750	4.08
4.	North Eastern Himalaya Alpine Grass + Meadow Region	16,000	0.05
5.	North Eastern Forest Region	161,000	4.97
	Total	523,000	16.13
	(Part 2: Himalayan Sediment Area) The areas directly fed by the Himalayan rivers indicated		
1.	Punjab–Haryana Alluvial Plains Region	101,250	3.10
2.	Upper Gangetic Alluvial Plain Region	200,000	6.15
3.	Lower Gangetic Alluvial Plain Region	145,500	2.41
4.	Assam Valley Region	88,500	2.73
5.	Rajasthan Desert Region	191,000	5.87
		726,250	22.26
	Grand Total of Himalayan Region	1,249,250	38.39

Present Efforts to Save Himalayas

Eco-development Camps

The Eco-development Camps Programme was undertaken by the Department of Environment during 1981-82. The main objectives of the programme are:

1. To create an awareness in the student and non-student youth and the society about the basic ecological principles and environmental management;
2. To enable student and non-student youth to identify and analyse the root cause of the ecological problems, particular related to human activities and the biosphere;
3. To involve student and non-student volunteers in solving ecological/environmental problems at national and regional levels;
4. To create general awareness about the seriousness of the problems of environment in the community and an atmosphere of acceptance of measures being undertaken to meet the situations and
5. To develop the spirit of national integration among the youth.

Table 24.2: Sediment Data of Major Himalayan Reservoirs

Sl.No.	*Reservoirs*	*Years of Improving*	*Annual Rate of Siltting in ha m/100 sq. km*	
			Assumed	*Observed*
1.	Bhakra	1959	4.29	6.10
2.	Panchet	1956	2.41	10.00
3.	Maithon	1956	1.62	13.10
4.	Mayurakshi	1956	3.61	16.56
5.	Nizamagar	1931	0.29	6.55
6.	Matatila	1958	1.43	4.38
7.	Lower Bhawani	1953	NA	4.19
8.	Shivakisagar	1961	NA	15.24
9.	Tungabhadra	1953	4.29	6.54
10.	Hirakud	1956	2.52	3.57
11.	Gandhisagar	1960	3.61	3.71
12.	Nachkund	1956	3.90	2.57
13.	Ramganga	1974	4.29	18.20
14.	Kangsabati	1965	3.27	3.76
15.	Ghod	1996	3.61	15.40
16.	Dantiwada	1965	3.61	5.14
17.	Ukai	1971	1.47	10.90
18.	Tawa	1974	3.61	6.38
19.	Beas Unit II	1974	4.29	14.30
20.	Narmada	Under Construction	1.55	5.62
21.	Mahi Stage II	–do–	1.29	8.17

Table 24.3: The Damage Caused by Floods During the Period 1953–1981

	Unit	*Total 1953–81*	*Maximum in One Year*	*Average/ Year*
Total area affected	M ha	235.6	19(1978)	8.1
Total population involved	M	795.0	70(1978)	27.4
Damage to Crops	M ha	104.0	10(1978)	3.7
US $ M	7,200.0	1,030(1978)	248.3	
Cattle lost	Thousand	2,826.0	618(1979)	97.4
Human lives lost	Thousand	40.0	11(1997)	1.4
Damage to public utilities	US $ M	3,200.0	582(1981)	110.3
Total damage to crops, houses and public	US $ M	11,800.0	1,700(1978)	406.9*

The activities initiated by the volunteers in these camps included digging of pits and tree plantation, undertaking pre-plantation operations like nursery bed preparation, trenching and fencing and so on. Programmes aimed at promotion of alternative sources of energy like biogas, solar energy are also undertaken. Major emphasis is laid on creation of awareness of environmental problems through field demonstrations, seminars, symposia and so on. Given below are DOE's financial eco-development activities in the country as a whole including the Himalayan areas.

Setting up of Ecological Task Forces

The idea of utilizing ex-servicemen for the restoration of the ecological balance in the Himalayas and other hilly regions has been put forward in the Himalayan Tasks Force and even earlier. The concept of Eco-Developmental Task-Forces of Ex-Servicemen is on the general pattern of Territorial Army. Units were given operational shape soon after the formation of the National Eco-development Board in the Department of Environment. The Task–Forces are raised to undertake ecological restoration work in the environmentally fragile/degraded areas especially in difficult and unapproachable terrains. The Director General, Territorial Army (Ministry of Defence) is actually responsible for the Eco-Task Force.

Eco-Task Force in Uttar Pradesh/Uttarakhand–127 Infantry Battalion (TA)

The creation of the first Eco-development Task Force was approved by the National Eco-Development Board in its Second Meeting held in September, 1982 for deployment in the Shahjahanpur Block (Shivaliks) U.P. The Task Force comprises 200 ex-servicemen with the strength of 42 at headquarters. The Task Force became operational in January, 1983. The Task Force has done a commendable job with the following main features:

1. Plantation of trees,
2. Construction of check dams,
3. Treatment/construction of gully-plugs,
4. Construction of jeepable tracks,
5. Eradication of weeds, and
6. Fodder collection.

The Task Force has completed the job assigned to it and is ready to be deployed for reclamation of mined areas on Mussoorie slopes for which a detailed project report has been prepared calling for increased working hands immediately. The strength of this Task Force, is therefore, being increased by adding one more company of 200 jawans, without augmenting the Headquarter strength.

Eco-Task Force in Himachal Pradesh

Considering the magnitude and intensity of denudation in Himachal Pradesh Himalayas, especially in Lahul-Spiti and Kinnaur districts, the Department of Environment had proposed the creation of a Task Force for deployment in the high altitude, cold desert region of Himachal Pradesh, Availability of suitable species of planting material for that region was, however, a constraint which has been overcome by setting up four nurseries to supply an adequate number of suitable varieties of sapling to the Eco-Task Force when it is made operational in Himachal Pradesh.

Accordingly, an allocation of Rs. 10,203 lakhs had been made in March, 1984 to the State government for the establishment of nurseries. These nurseries are being raised. Meanwhile on request of the state government the Task Force may be raised immediately and planting material diverted in the short run from their existing high altitude nurseries. Accordingly, a Task Force has become operational in Himachal Pradesh with effect from January 1985.

Task Forces During the Seventh and Subsequent 8, 9, 10 Plan Period:

- ✰ Work by Soil Conservation Department (Ministry of Agriculture) in Priority Catchments
- ✰ Integrated Watershed Management in the Catchments of Flood Prone Rivers and Control of Shifting Cultivation.
- ✰ Work by the Forestry Departments (Ministry of Agriculture and now Ministry of Environment and Forests).
- ✰ Work and Recommendations of the National Land Resources Conservation and Development Commission.

The important initiative taken comprises:

- ✰ Envigouring Conservations Society, its chapters and special work on Bioindustrial Watershed (Prof J.S. Bali Has promoted this program since last over 20 years and now time has come for promoting it intensively)
- ✰ Several other state chapters and state Departments like in H.P. and J&K and some North East States have been formed.
- ✰ A Task Force for the 11th 5 year plan has been formed which is working under the leadership of Dr. R.S. Tolia, former Chief Secretary, Uttrakhand.
- ✰ Several NGO have started program of Tree Planting such as HIMCON and emphasis is given on Fruit Planting so as to increase the nutrition and income generation of the farmers and orchardists.
- ✰ A brief video was, presented during the talk.

MIPF–Mountain India Peoples Forum

Mountain are many things to may people; home of the Gods, sacred place of pilgrimage, natural resources repositories, biodiversity treasure groves, aesthetic landscape and much more; but the most important aspect is that they are home to many people of diverse cultures ¼people of the mountains are at the center of concerns in all endeavors). The work of the Mountain India People's Forum will be directed towards securing a better quality of life for people living in the mountain areas who are often marginalized and ignored. The Forum will work for enhancement of material, environmental, social, cultural and spiritual well-being and active engagement of mountain people in civil societies/institutions.

It is estimated that mountains make up a quarter of the world's landscape and are home to at least 1/10th of the worlds people. An additional 2 billion people depend on mountains for much of their food, hydro-electricity, timber and mineral resources. About half of the world's people, as well as surprisingly large share offers biological diversity, rely on mountain watersheds for fresh water. Upton 80 per cent of the planets fresh surface water comes from mountains. Yet in the deliberations of the governments and organizations worldwide, the fate of the mountains has been largely ignored.

Findings from all over the world show that expanding economic pressures are degrading mountain ecosystems, while confronting mountain people with increasing cultural assimilation, debilitating poverty, and political disempowerment. The fact that many people are ethnic minorities or outside the dominant cultures of the plains, and smaller in number, leave their regions poorly represented in the centers of political or commercial power where much of their fate is determined.

India has many mountain and hill tracts. The longest and the highest are the Himalayas which sweep across the north and north east part of the country. Towards the west and flowing into the desert state of Rajasthan are the Aravallis. The Vindhya ranges mark off the northern part of the country from its southern part. The Western Ghats and the Sahyadris follow the west coast and in the south of the country lay the Nilgiris and the Annamalai hills. The East coast of India is flanked by the Eastern Ghats. High or low, the hills and mountains are inhabited by people who have remained to a large extent outside the mainstream of development.

MOTI–Millions of Trees International

Concept–"Millions of Trees" International is working for a healthy world and healthy humans. It is a non-profit organization working for creating awareness, facilitating tree plantation with the vision to synergize human life with nature for a healthy and happy future. Its motto is "Save the Earth for Healthy Life."

We know the whole life of trees is to serve the humanity with their leaves, flowers, fruits, branches, roots, shade fragrance, sap, bark, wood and also their ashes and coal. A tree that lives for 50 years recycles Rs. 4 lakhs worth of fertility and soil erosion control, serves Rs. 10 lakhs worth of air pollution control, creates Rs. 5.3 lakhs worth of food and shelter for birds and animals, besides providing flowers, fruits and timber.

So when even one tree falls or is felled, the community loses something worth "more than Rs. 32.00 lakhs".

Nature has always inspired and challenged human life. The trees are an essential component of life and play a critical role in sustaining the natural and human environment. But the human interventions in nature have created unprecedented environmental hazards. Felling of trees has created tremendous eco-imbalance and if it is not taken care of now, humanity and other creatures are condemned to a disastrous future. The very survival of planet Earth depends on saving trees from destruction.

Mission

1. Restore our natural heritage.
2. Involve the common man, specially the young minds to be a part of making the natural environment healthy.

Methodology

1. Involvement of women led voluntary Agencies for leading such a movement.
2. Identifying spots and organization having sufficient and suitable land for nurseries.
3. Identifying schools, colleges and other interested institutions as partners.
4. Involvement of school and college students and to promote a culture for planting and protecting trees.
5. Develop a linkage with organizations such as "Trees for Life" founded by Balbir Mathur who is leading a global "Awareness and Action Program" for "Tree Planting and Protection"
6. Develop a link with Corporate Bodies and their Associations such as CII, Assochem, PHD, FICCI and other national and international associations for promoting greening of their establishments and helping voluntary agencies in such mission.

Activities

1. Organizing "Tree Plantation" Program mainly in different schools and through the schools and their camps in all areas.
2. Organizing 'on the spot' painting competition, quiz competition and different games which will create more awareness and interest among the school children.
3. Awarding prizes to the best schools that will plant maximum trees and protect them also.
4. Making tree planting a Movement with Women's Group, NGOs including rural and urban area that need such tree planting.
5. Organizing tree plantation in corporate lands.
6. Rejuvenation and re-greening of wastelands.
7. Participate in "OWN/GIFT/DEDICATE A TREE" Project as a part of this Movement.

We intend to accomplish the above objectives through various programs and projects in different institutions and hence increase the awareness among all, specially the young hearts and minds towards such pressing needs of the country, the nation and the world at large.

CARBON SEQUESTRATION
A Life-building, Life-sustaining and Life-enhancing Phenomenon on Earth

Ecosystem Diversity and Carbon Sequestration: Climate Change Challenges and a Way Out for Ushering in a Sustainable Future *Pages 233–239*
Editors: P.L. Gautam, Vir Singh and Uma Melkania
Published by: DAYA PUBLISHING HOUSE, 2009

Chapter 25

Carbon Sequestration: Global Warming Mitigation through Improved Carbon Economy Linked with Photosynthesis

Munna Singh

The coal is one of the world's cheapest fossil fuel energy resources, but the gases produced by the coal are causing global climate change, resulting in global warming. It will affect our health, natural balance, agricultural yields, and the future of coming generations as well. It has already been statistically estimated that one ton of carbon dioxide gets produced when we burn ca. half a ton of coal. The Kyoto Protocol made CO_2 reduction mandatory among 124 countries; most have already accepted the accord. Burning of fossil fuels adds ca. 5.4 billion metric tons (bmt) of carbon each year to the atmosphere, and deforestation further adds another 1.6 bmt annually, increasing the concentration of CO_2. Human activities also emit CFCs, ozone, methane and nitrous oxides. However, CO_2 has received a lot of attention with respect to global warmimg, because 50-60 per cent of anthropogenic greenhouse effect is attributed to this gas. It has grown-up exponentially, with an annual current rate of increase ca. 0.5 per cent (approaching 370 ppm) with the prediction that it will reach ca. 450 ppm by the year 2050. The atmospheric carbon concentration has been rising steadily since the industrial revolution due to anthropogenic production of CO_2 through burning of the fossil fuels. The level of CO_2 is approaching ca. 370 ppm by now from ca. 285 ppm, with an enhancement of CO_2 in the last 150 years.

The biological CO_2 mitigation occurs through all green plants. However, forest trees contribute up to 70 per cent of terrestrial carbon sequestration, and are a major sink for CO_2 globally to maintain and sustain their growth dynamics. Particularly, C_3 plants have shown a positive sign through improved scrubbing of CO_2 enzymatically using photsynthetic enzyme ribulose 1, 5 bis phosphate carboxylase (rubisco) encoded by chloroplast genome (cpDNA). Its active sites have strong differential competition between CO_2 and O_2. The higher availability of CO_2 induces better carbon sequestration/mitigation through photosynthesis, associated with the loss in photorespiration. Thus, photosynthesis may be either up or down-regulated depending on the environmental conditions. However, if carbon emission continues, our most likely adjustment would be to live with it, otherwise a planning is to be framed in time to reduce emission changes for land management and expected energy use as well. Thus, biological CO_2 mitigation for improved carbon economy and learning to adapt with its higher atmospheric availability in relation to climate change must go parallel.

Introduction

Carbon sequestration is of vital importance as the world's economy continues to expand. Coal is one of the world's cheapest and most plentiful energy resource, but the gases produced by the coal causing global climate change, resulting in global warming. It will affect our health, natural balance, agricultural yields, and the future of coming generation as well. It has already been statistically estimated that one ton of carbon dioxide gets produced when we burn down about half a ton of coal (Gavrilov, ENS 2004). Inview of this, Kyoto Protocol made CO_2 reduction mandatory among 124 countries; most have already accepted the accord. Some of the countries are leaping even ahead. Britain has vowed to slash emissions by 60 per cent by the year 2050. Climate change is a greater threat to the world than terrorism as told by Sir David King, Chief Science Advisor to Prime Minister Tony Blair in recent past. DuPont has cut its greenhouse-gases emissions by 65 per cent since 1990, saving hundreds of millions of dollars in the process. General Motors Corporation is spending millions to develop hydrogen-powered cars that don't emit CO_2. The efforts are also giving a boost to set up advance nuclear reactors to produce electricity without any greenhouse-gas emissions. Consequently, all industrial nations ratifying the Kyoto Protocol, which has been implemented w.e.f. 16th Feb., 2005 will curb emissions of six so-called greenhouse/suicidal gases by 2008-2012. These gases are being produced by burning fossil fuels. They stay in the atmosphere, trap heat from the sun irradiance, and thereby enhance the Earth's surface temperature.

It is assumed that internationally, world's energy and the level of CO_2 emission will be 60 per cent higher in 2030 than they are now (ENS, 2004). Thus, climate change can only be tackled through concerted international action. Accordingly, Jack Straw (UK, Foreign Secretary) has given top priority to the climate change in Britain's G8 Presidency in 2005, run by the European Union (bound by the Protocol) with individual Country's Action Plan on energy and climate change. It also includes focus on the large new consumers of energy in developing countries *viz.*, China (ca. 10 per cent global consumption) and India, and producers too such as Russia. This will affect production of major crops and the trend of future agriculture in anticipated frequent droughts and floods. Japanese firms have planned to earn greenhouse gas emissions rights by collecting methane gas getting generated at Chinese coal mines and using it to produce electricity. They are expected to become the first Japanese companies to launch a programme in China under Clean Development Mechanism. More Japanese firms are likely to follow–up the similar projects in China, which produces the largest amount of the global warmimg gases after the United States.

Atmospheric CO_2 Enrichment

The burning of fossil fuels adds ca. 5.4 billion metric tons (bmt) of carbon each year to the atmosphere, and deforestation further adds another 1.6 bmt annually, increasing the concentration of CO_2. Human activities also emit CFCs, ozone, methane and nitrous oxides. However, CO_2 has received a lot of attention with respect to global warmimg, because 50-60 per cent of anthropogenic greenhouse effect is attributed to this gas. It has grown-up exponentially, with an annual current rate of increase ca. 0.5 per cent (approaching 370 ppm) with the prediction that it will reach ca. 450 ppm by the year 2050. The atmospheric carbon concentration has been rising steadily since the industrial revolution due to anthropogenic production of CO_2 through burning of the fossil fuels. The level of CO_2 is approaching ca. 370 ppm by now from ca. 285 ppm, with an enhancement of CO_2 (Roger *et al.*, 2004) in the last 150 years.

Biological CO_2 Sequestration

Forest trees contribute up to 70 per cent of terrestrial carbon sequestration, and are a major sink for CO_2 globally to maintain and sustain their growth dynamics. Particularly, C_3 plants have shown a positive sign through improved scrubbing of CO_2 enzymatically using photsynthetic enzyme ribulose 1, 5 bis phosphate carboxylase (rubisco) encoded by chloroplast genome (cpDNA). Its active sites have strong differential competition between CO_2 and O_2. The higher availability of CO_2 induces better carbon sequestration through photosynthesis, associated with the loss in photorespiration. Thus, photosynthesis may be either up or down–regulated depending on the environmental conditions (Curtis and Wang, 1998). However, if carbon emission continues, our most likely adjustment would be to live with it, otherwise a planning is to be framed in time to reduce emission changes for land management and expected energy use as well. Initially, we must use energy more efficiently, followed by contingency plans for greater reduction in emissions. Thus, both CO_2 mitigation and learning to adapt with its higher atmospheric availability in relation to climate change must go parallel.

The increase in primary greenhouse gas (GHG) *i.e.*, CO_2 occurs along with several other GHGs *viz.*, CH_4, N_2O, SO_2, CFCs and O_3. All these GHGs put an impact on the fate of global climate due to their tendency to trap irradiance, which may result in the increase of temperature up to 3-5°C due to doubling of ambient CO_2, though, it's higher concentrations stimulate photosynthesis in a group of C_3 green plants. For, limitation of CO_2 impairs photosynthetic efficiency and acquisition of adequate biomass yield or plant productivity (Siffel and Braunova 1999). Thus, it extends CO_2 fertilization effect (25-75 per cent), which otherwise gets hindered by photorespiration (Singh *et al.*, 1996). Therefore, better CO_2 harvesting can be made possible by planting C_3 plants to aim better carbon economy/profits in near future, and to cope up with the elevated CO_2 levels based on biologically sustainable CO_2 sequestration. This may promote to acquire superior carbon currency ecofriendly.

The dynamics of the terrestrial ecosystems depends on interactions between a number of biogeochemical cycles *viz.*, carbon, nutrient and hydrological cycles. These may be modified by human actions, which may enhance the levels of greenhouse

Carbon sequestration by green plants/PHOTOSYNTHESIS

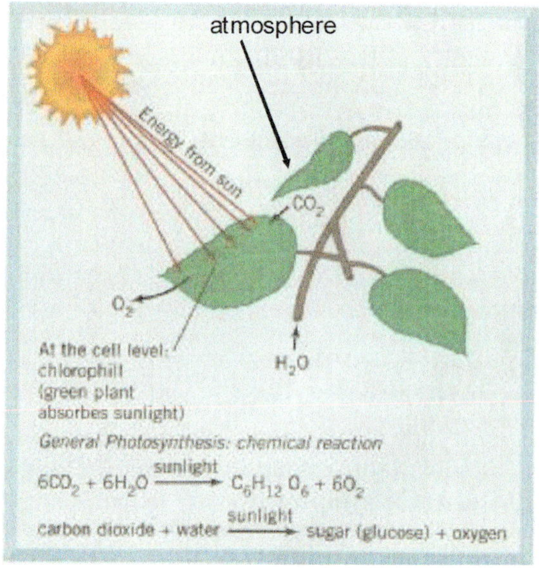

CO_2 harvesting by green plants

gases *viz.*, carbon dioxide (primary greenhouse gas), methane and nitrous oxide. Out of these, especially, carbon is exchanged naturally in the atmosphere through photosynthetic process. Hence, continuous increase of the atmospheric carbon dioxide concentration has attracted attention of several scientists for evaluating its impact globally in relation to plant productivity or biomass yield.

A model green plant shows biological CO_2 sequestration based on photosynthesis. The sun light gets photo-transformed in biochemical energy to promote atmospheric CO_2 sequestration which enters into leaf through stomatal conductance, gets reduced enzymatically in energy rich biomolecules. It promotes plant growth and development, and also extends ability to sustain life on the earth.

CO_2 Sequestration vs Biomass and Future Prospects

Last two decades have fruitfully contributed to the understanding of the consequences of CO_2 enrichment on plants' activities *i.e.* plant response and biomass production. The CO_2 is sequestered or absorbed by trees, their roots, branches, trunks, leaves immature fruits, plankton, soils and water bodies. However, the degree of response is very much variable, depending upon the plant species and place of cultivation/growing site, growth conditions, availability of mineral nutrients, water, temperature and duration of differential levels of CO_2 enrichment. Consequently, it may either up or down regulate the carboxylation efficiency of the crop plants or trees (Urban, 2003). CO_2 is the substrate for photosynthesis, and it has been estimated that due to decrease in photorespiration under elevated CO_2 surroundings C_3 plants have tremendous ability to sequester higher levels of carbon dioxide through improved photosynthesis (Pleijel *et al.*, 2000) due to better affinity of CO_2 with the active sites of rubisco (ribulose 1, 5 bis–phosphate carboxylase) enzyme encoded by *rbcL- rbcS* genes encoded by chloroplast and nuclear genomes. The doubling of atmospheric CO_2 has been shown to stimulate photosynthesis upto ca. 50 per cent. However, lower and elevated temperature may down-regulate chlorophyll fluorescence variable yield (Fv/Fmax) and also the expression of *psbA* gene product, *i.e.* D1 reaction center core protein associated with PSII (Singh 2000; 2004; Singh and Satoh 2003). The up-regulation of specific proteases under elevated temperature seems to be biologically important to remove/digest the damaged proteins (Singh *et al.*, 2005a and b) to maintain the photosynthetic system functional for CO_2 sequestration.

Overall biological CO_2 sequestration regulates net primary productivity, which in turn depends on the balance between carbon uptake by photosynthesis and loss to respiration (Singh *et al.*, 1996; Ziska and Bunce 1997; Gesch *et al.*, 2003; Urban 2003; Jones and Donnelly 2004; Rogers *et al.*, 2004). Thus, carbon economy maintains the balance between carbon acquisition through photosynthesis, and its utilization through respiration. The process of photosynthesis occupies most significant position for carbon sequestration to yield with biomass production and crop/plant productivity. Photosynthesis and productivity both are influenced by a variety of genetic and environmental factors *viz.*, light, available CO_2, temperature, soil water, nutrients and canopy structure. The present CO_2 concentration (ca. 0.035 per cent) is well below the CO_2 saturation level for most of the C_3 plants. It gives a clue for better CO_2 sequestration with a possibility to provide higher CO_2 levels under normal temperature by these plants to enable them for optimal carbon assimilation with the potential expression of photosynthetic genes *viz.*, *rbcL*, *psbA* encoded by chloroplast genome (cpDNA), yields Rubisco protein/enzyme and D1 reaction center core protein associated with PSII (Gesch *et al.*, 2003), primarily required for carbon concentrating mechanism. The higher levels of CO_2 directly enhance substrate availability to rubisco for improved carboxylation with reduced photorespiration in C_3 plants. In contrast to this, most C_4 plants appear to get saturated at CO_2 levels at or just above the normal atmospheric concentrations regardless of their nature of high light adaptability. Basically, intracellular CO_2 maintains rubisco enzyme affinity for carboxylation, regulated by stomatal conductance through step-wise diffusion gradient maintained between the intra-intercellular CO_2 and the atmospheric CO_2 concentrations. Photosynthetic capacity is also determined by the balance between carboxylation capacity and electron transport activity. Photosynthetic efficiency/carboxylation becomes limited at low CO_2 concentrations in C_3 plants (Siffel and Braunova 1999). The carboxylation capacity may also get impaired upon inadequate availability of the CO_2 acceptor *i.e.*, ribulose 1,5- bis-phosphate (RuBP) under low light, associated with impaired rate of photosynthetic electron transport. Thus, potential regeneration of RuBP and dynamic expression of Rubisco (Adam *et al.*, 2004) are essential to sequester the higher CO_2 levels efficiently through natural and unique process *i.e.*, Photosynthesis. However, combination of low light and high CO_2 would favour in-efficient carboxylation or CO_2 assimilation capacity of the plants due to limitation of photosynthetic electron transport.

Eventually, it is conceivable that CO_2 enrichment upto a certain extent at high or saturating PFDs would contribute positively for carbon fixation in C_3 plants. Consequently, CO_2 enrichment has become a practice for commercial growers of vegetable crops such as lettuce, tomato, and several cucumbers cultivation under greenhouse technology. The CO_2 enrichment may also improve growth and biomass yield/harvest index of fast growing trees (poplar) in an open environment with the utmost care to the extent not to put a risk of stomatal closure due to elevated surroundings of carbon dioxide (ca. 700 ppm), and deleterious effect of temperature on photosynthesis. Preference must also be given for the cultivation of fast growing trees (*e.g.* poplar, salix etc.) plants for short-term intensive rotation cultivation under agroforestry to harness their improved ability of CO_2 fixation/assimilation which may bring better economic returns to the farmers keeping the area clean and green. Since, major carbon sinks are forests therefore; afforestation and reforestation activities must be paid adequate attention to retain eco-friendly ecosystems globally to sustain long-term biological sequestration appropriately under changing environmental variables.

Acknowledgements

Thanks to Prof. B.N. Johri, Dept. of Microbiology, CBSH, G.B. Pant University of Agric. and Technolgy, Pantnagar for providing an opportunity to present a concept proposal before DBT-NTPC, New Delhi which has inculcated a keen interest in me towards carbon sequestration.

References

Adam, N.R., Wall, G.W., Kimball, B.A., Idso, S.B. and Webber, A.N., 2004. Photosynthetic down-regulation over long-term CO_2 enrichment in leaves of sour orange (*Citrus aurantium*) trees. *New Phytologist*, 163: 341–347.

Curtis, P.S. and Wang, X., 1998. A meta analysis of elevated CO_2 effects on woody plant mass, form and physiology. *Oecologia*, 113: 299–313.

Gesch, R.W., Kang, I.H., Gallo-Meagher, J.C.V., Boote, K.J., Allen, J.R. and Bowes, G., 2003. Rubisco expression in rice leaves in related to genotypic variation of photosynthesis under elevated growth CO_2 and temperature. *Plant Cell and Environment*, 26: 1941–1950.

Jones, M.B. and Donnelly, A., 2004. Carbon sequestration in temperate grassland ecosystems and the influence of management, climate and elevated CO_2. *New Phytologist*, 164: 423– 440.

Pleijel, H., Gelang, J., Sild, E., Danielsson, H., Younis, S., Karlsson, P.E., Wallin, G., Skarby, L. and Sellden, G., 2000. Effects of elevated CO_2, ozone and water availability on spring wheat growth and yield. *Physiologia Plant*, 108: 61–70.

Rogers, D.J., Allen, P.A., Davey, P.B., Morgan, E.A., Ainsworth,C.J., Bernacchi, G., Cornic, O., Dermody, F.G., Dohlemman, E.A., Heaton, J., Mahoney, X.G., Zhu, E.H., Delucia, Ort, D.R. and Long, S.P., 2004. Leaf photosynthesis and carbohydrate dynamics of soybeans grown throughout their life-cycle under free air carbon dioxide enrichment. *Plant Cell and Environment*, 27: 449–458.

Siffel, P. and Braunova, Z., 1999. Release and aggregates of the light-harvesting complex in intact leaves subjected to strong CO_2 deficit. *Photosynth. Res.*, 61: 217–226.

Singh, M., Yamamoto, Y, Satoh, K., Eira, K. and Aro, E.M., 2005a. *In vivo* light-induced inhibition of PSII and D1 degradation in Synechocystis PCC 6803: role of protease, temperature and darkness. *J. Plant Physiol.*, 162: 1246–1253.

Singh, M., Pooja, S., Nidhi, S., Kalpana, C. and Singh, D., 2005b. Differential expression of photosynthetic efficiency (Fv/Fm) in four genotypes of *T. aestivum* under warmer growth variables. *Physiol. Mol. Biol. Plants*, 11: 327–331.

Singh, M. and Satoh, K., 2003.Site-specific mutations localized in the D–E loop of D1 protein of PSII affect phototolerance in Synechocystis sp.PCC 6803 containing psbAII gene. *Indian J. Biochem. and Biophysics*, 40: 108–113.

Singh, M., 2004. *In vivo* turn-over of D1 reaction center core proteon of PSII extends photosynthetic efficiency under light stress. *Phyiol. Mol. Biol. Plants*, 10: 147–151.

Singh, M., 2000. Turn-over of D1 protein encoded by psbA gene in higher plants and cynobacteria sustains photosynthetic efficiency to maintain plant productivity under photoinhibitory and non-photoinhibitory irradiance. *Photosynthetica*, 38: 161–169.

Singh, M., Chaturvedi, R. and Sane, P.V., 1996. Diurnal and seasonal photosynthetic characteristics of poplar leaves. *Photosynthetica*, 32: 11–21.

Urban, O., 2003. Physiological impacts of elevated CO_2 concentration ranging from molecular to whole plant responses. *Photosynthetica*, 41: 9–20.

Ziska, L.H. and Bunce, J.A., 1997. Influence of increasing CO_2 concentration on the photosynthetic and growth stimulation of selected C_4 crops and weeds. *Photosynth. Res.*, 54: 199–208.

Ecosystem Diversity and Carbon Sequestration: Climate Change Challenges and a Way Out for Ushering in a Sustainable Future *Pages 240–252*
Editors: **P.L. Gautam, Vir Singh and Uma Melkania**
Published by: **DAYA PUBLISHING HOUSE, 2009**

Chapter 26

Carbon Sequestration on Agricultural Lands: Ameliorating Sustainability and Environmental Security

B. Mishra and K.P. Raverkar

In the terrestrial system carbon is mainly held in vegetation and soil. Aquatic bodies like oceans also stocks large volumes of carbon, as does the atmosphere. Soils contain about three times more C (1500 Pg of C to 1 m and 2500 Pg of C to 2 m depth) than vegetation (650 Pg of C) and twice as much as that present in the atmosphere (750 Pg of C; USEPA 1995). Additionally, fossil fuels, *e.g.*, coal, petroleum, and natural gas contain large amounts of carbon which are released upon combustion.

All biogeochemical cycles revolve and dictated by the element carbon. Carbon supports the existence and flourishes the life on the earth. All ecosystems both store and emit carbon back to the atmosphere at lesser or greater extent as they continuously recycle carbon by photosynthesis and respiration.

Carbon (in the form of carbon dioxide) is accumulating in the atmosphere at a rate of about 3.5 billion metric tons per annum as a result of fossil fuel combustion and tropical deforestation. Approximately 22 per cent of the annual global carbon dioxide emissions are due to deforestation, largely in the tropics. The carbon dioxide concentration and other greenhouse gases in the atmosphere are changing speedily due to anthropogenic activities consisting industrialization, fossil fuel combustion, deforestation, wood and residue burning, non-friendly soil cultivation practices etc. During pre-industrial time the CO_2 concentration was around 270 ppm, however the current concentration has reached to 372 ppm which is increasing at the rate of 1.5 ppm per year or 3.2 Pg C per year (Sage, 1995). Such a dramatic rate of increase in the CO_2 content has not been experienced for the past 4, 30,000 years. According to the latest calculations by scientists at the Max Planck Institute for Meteorology in Hamburg, Germany an increase in the level of CO_2 in the atmosphere at the existing rate may shoot up the global temperature by 1.5 to 4.5°C and may result in its related consequences. Unless effective measures are taken to combat and reduce the net emission of CO_2, its concentration in the atmosphere would reach to 800 to 1000 ppm by the end of 21st century (Whipps 1990).

Soil is the biggest reservoir of carbon. Soils can be a source or sink for atmospheric CO_2 depending upon the land use and management practices. The conversion of natural habitats to cropland and pasture, and unsustainable land practices such as excessive tillage frees carbon from organic matter, releasing it to the atmosphere as CO_2 and thus soil acting as a source whereas the photosynthetic activity, through which the plants convert CO_2 into organic forms of carbon such as sugars, starch and cellulose; as a sink. The soil organic carbon (SOC) pool to 1m depth ranges from 30 tons/ha in arid climates to 800 tons/ha in organic soils of temperate regions. The predominant range of SOC is 50-150 tons/ha. The conversion of natural habitats to agricultural ecosystems has resulted in depletion of SOC pool by 60 and 75 per cent in soils of temperate and tropics, respectively. The depletion of SOC pool generally takes place speedily when the decomposition rates are higher than the rates of carbon added to soil. Severe loss of SOC leads to deterioration of soil quality, biomass productivity and also affects the water quality. An overall impact of depletion of SOC levels may be exacerbated by projected global warming (Lal 2004).

The carbon from the atmospheric CO_2 can be transferred to the long lived stable form in the soil and plants through the process of sequestration. The process of carbon sequestration results in the increased SOC and soil inorganic carbon stocks through the judicious implementation of the land use and eco-friendly/good management practices. Presently, photosynthesis is the only viable way to trap CO_2 from atmosphere. Carbon is referred to be sequestered if it is converted into a stable form *i.e.* wood or soil organic carbon.

Agricultural ecosystems represent an estimated 11 per cent of the earth's land surface and include some of the most productive and carbon-rich soils. As a result, they play a significant role in the storage and release of C within the terrestrial carbon cycle (Lal *et al.*, 1995). With a large land area and diverse eco-regions, there is a considerable potential of terrestrial/soil carbon sequestration in India. Of the total geographical area of 329 million hectares (Mha), 297 Mha is the land area comprising 162 Mha of arable land, 69 Mha of forest and woodland, 11 Mha of permanent pasture, 8 Mha of permanent crops and 58 Mha is under other land uses (Table 26.1).

Soil Carbon Stock

India is bestowed with diverse soil wealth having the significant amount of carbon stock (Table 26.2). The carbon stock in different soil orders varies due to climatic conditions, land use pattern, management practices, vegetation cover etc. The carbon stock follows the sequence Aridisols > Inceptisols > Alfisols > Vertisols > Entisols in the top surface 30cm depth (Bhattacharyya *et al.*, 2000). The SOC stock in Mollisols, Oxisols and Ultisols is less in Indian soils. In general, the SOC concentration in soils of arable lands is low and varies between 2-5 g kg^{-1}. Under natural ecosystems the SOC concentration is relatively higher by a factor of 1.4 to 2.5. In Indian soils the SOC pool is estimated to the tune of 108 Pg (Velayutham *et al.*, 2000), however Bhattacharyya *et al* (2000) has reported as 53.4 Pg in first 150 cm depth of soils.

Table 26.1: Pattern of Land Use in India

Land Use	*Area (Mha)*
Total area	328.7
Land area	297.3
Agricultural area	180.8
Arable land	161.8
Forests and wood lands	69.0
Permanent crops	8.2
Permanent pastures	10.9
Irrigated land	57.0

Source: FAO (2004).

Table 26.2: Acreage Under Major Soil Orders of India and Soil Organic Pool

Soil Type/Order	*Area (Mha)*	*Soil Organic Pool (Pg)*
Alfisols	44.42	23.76 (22.0)*
Aridisols	13.16	34.99 (32.4)
Entisols	92.12	7.56 (7.0)
Inceptisols	131.60	25.92 (24.0)
Mollisols	1.65	(< 1)
Ultisols	8.23	0.65 (0.6)
Vertisols	27.92	15.12 (14.0)
Total	319.10	108.00 (100)

*: Figures in parentheses indicate percentage.

Source: Velayutham *et al.* (2000).

Forms of SOC

Soil organic matter comprises of mainly unhumified substances and mummified remains of plant and animal tissues. Soil being a graveyard for all types of organisms, it may contain all the biochemical compounds synthesized by the flora, fauna, animals and plants. The soil organic matter mainly comprises living organisms, fresh residues, and well decomposed residues. These three parts of soil organic matter can be described as the living, the dead and the very dead. The various constituents under the categories of organic matter are given in Table 26.3.

Table 26.3: Constituents of Soil Organic Matter

The Living	*The Dead (Fresh Residue)*	*The Very Dead (Decomposed Residue)*
Micro-organisms (Bacteria, fungi, actinomycetes, viruses, protozoa, algae) Plant roots Insects Earthworms Larger animals (moles, woodchucks, rodents)	Recently deceased micro-organisms, insects, earthworms, aged plant roots, crop residue, recently incorporated manure etc.	Humus: It consists series of highly acidic, yellow colored

The dead constituent of the soil organic matter is also referred as the labile pool/ carbon fraction as it is easily decomposable, susceptible to microbial breakdown and easily oxidizable while the very dead is resistant to further degradation, microbial breakdown and oxidization. The turnover time of different fractions of the soil organic matter ranges between 0.1 to 1000 years (Table 26.4; Carter 2002). Therefore, the carbon sequestration in soil depends upon the amount of the carbon put away in different fractions. More the amount stabilized in the component having the longer turnover time the more carbon sequestration for longer period. Measurement of soil organic matter at any one time is simple; however the changes in the SOC concentrations due to various management practices are difficult to gauze against the relatively large background mass. The indicators which are sensitive to the changes in management practices and organic matter input (Table 26.5) are handy which indicates the direction of change in total mass of organic matter (Carter 2002).

Table 26.4: Estimates of Turnover Time for Different Fractions in Soil Organic Matter

Organic Matter Fraction	*Estimated Turnover Time (in years)*
Organic matter fraction	0.5-2.0
Microbial biomass	0.1-0.4
Macroscopic matter	1-8
Light fraction	1-15
Organic matter in non-aggregated soil	1-7
Organic matter in macro aggregates (>250 μm diameter)	1-23
Organic matter in micro aggregates (20-250 μm diameter)	3-80
Organic matter in silt+clay (< 20 μm diameter)	5-1000

Source: Carter (2002).

Loss of SOC

Among Asian countries, India has the lowest average concentration of SOC in agricultural soils (Kyuma 1988). The alluvial soils of Indo-Gangetic Plains are inherently fertile and high in K, however intensive cultivation of rice-wheat; rice-maize and taking away of almost all crop residues lead to continuous decline in SOC levels (Nambiar 1994). The Long Term Fertilizer Experiments in India has indicated that the intensive cropping and imbalance fertilizer use especially fertilizer N alone resulted in declined SOC concentration irrespective of cropping system and type of the soil (Swarup 1998). Accelerated erosion due to wind and water is also a major cause of terrestrial carbon loss. In India the extent of soil degradation is highest due to desertification (68.1 Mha) followed by water erosion (32.8 Mha), wind erosion (10.8 Mha), water logging (8.5 Mha) and salinization (7.0 Mha) (FAO 1994; Dregne and Chou 1994). The amount of sediments displaced by soil erosion is to the tune of 21 Pg year^{-1} (Lal 2005). Assuming SOC concentration of 8-12 g kg^{-1} total carbon displaced every year is 168-252 Tg. Out of this if 20 per cent of carbon is mineralized and emitted into the atmosphere as CO_2, the estimated emission of C through erosion during 1998 in India is 34-50 Tg per year (Lal 2005), whereas the emission of C through fossil fuel from India in 2000 was 292 Tg. The emission of C from fossil fuels has increased dramatically since 1970 onwards due to industrialization, intensive agricultural practices etc.

Table 26.5: Selected Attributes of Soil Organic Matter Sensitive to Soil Management

Organic Matter Attribute	*% Soil Organic C*	*Methodology*
Macro organic matter C and N (particulate C and N)	15-45	Soil dispersion or sieving
Light fraction C and N	3-45	Soil densimetric fractionation
Microbial biomass C and N	1-3	Soil fumigation or extraction
Mineralizable C and N	3-30	Soil incubation

Source: Carter, (2002).

Importance of Carbon Sequestration in Soil

Soil carbon sequestration is an important and immediate sink for removing atmospheric carbon dioxide and slowing global warming. Carbon in the form of organic matter is a key element to healthy soil. It is estimated that each ton of soil organic matter releases 3.667 ton of CO_2, which is lost into the atmosphere. Similarly, the build-up of each ton of soil organic matter removes 3.667 ton of CO_2 from the atmosphere (Bowen and Rovira 1999). In natural habitats, carbon from plants is deposited in the soil through roots and plant residues. Carbon sequestration in soils is a climate-change mitigating strategy mainly based on the assumption of absorbance or transfer of atmospheric CO_2 in the soil can be increased while the release of carbon from soil back to the atmosphere is reduced. This has the potential to combat and reduce the CO_2 level in atmosphere and to slow down the global warming and mitigate climate change. Soils can sequester around 20 Pg C in 25 years, more than 10 per cent of the anthropogenic emissions. The carbon sequestration not only takes care of the global warming, climate change but it also provides various benefits to the soil, crop and environment quality, prevention of erosion and desertification and for the enhancement of above as well as below ground biodiversity. It holds the promise of win-win options and has the potential to provide the benefits at individual/local, national and global scale. The benefits of SOC sequestration at various levels are interlinked and mainly dictated from bottom to the top *i.e.* local to global level. The potential benefits at various scales are interlinked/overlapped (Table 26.6).

Table 26.6: Potential Benefits of Soil Carbon Sequestration

Local/Farmer Scale	*National Scale*	*Global Scale*
Improved food security	Improved food security	Improvements in global climate change
Improved resource base for future generation	Better agricultural sustainability	Reduced CO_2 concentration in atmosphere
Reduction in soil degradation and improved soil fertility	Improvement of environmental issues (air and water quality)	Biodiversity build up
Improved crop, timber and livestock yield	Economic returns	Economic returns
Improved soil and food quality	Social security	Social security
Economic returns		
Social security		

Strategies for Terrestrial Carbon Sequestration

To mitigate the climate change and global warming the CO_2 concentrations in the atmosphere needs to be combated and efforts be made to reduce the same to the level which can sustain the life on the earth and also keep the environment congenial. Various strategies comprising good management practices can be adopted to achieve the goal. Several processes dictate the storage of carbon in soils. The amount of carbon stocked in soil system depends on the rate and magnitude of the process. These processes can be influenced by land use and agriculture management systems and practices. The potential of carbon sequestration in Indian soils is 77.9 to 106.4 Tg yr^{-1} (Table 26.7)

Table 26.7: Potential of C Sequestration in Indian Soils

	Process	*Potential(Tg yr^{-1})*
A	Soil organic carbon	
	☆ Restoration of degraded lands	9.8-13.9
	☆ Erosion prevention	33.6-50.4
	☆ Agricultural intensification	12.7-16.5
B.	Secondary carbonates	21.8-25.6
	Total	77.9-106.4

Source: Lal (2005).

In agricultural soils more carbon can be sequestered by adopting the various recommended/good management practices. The practices which enhance biomass production, add increased amount of below ground biomass to the soil, slows decomposition rates, enhances activity and species diversity of soil flora and fauna, improves soil fertility, strengthens the mechanisms of elemental cycling, and takes care of the soil and water are the good management practices. The good management practices comprising efficient land use, diversified cropping, conservation tillage, efficient nutrient management, afforestation/agroforestry, administered grazing, erosion control, use of cover crops and rehabilitation of degraded soils lead to the more carbon sequestration. The important strategies which have an impact on the process of carbon sequestration in agricultural soils are dealt hereunder:

Land Use

Natural habitats hold more carbon as compared to the agricultural ecosystems. More amount of the carbon can be sequestered through the adoption of the alternate land use practices *viz.*, agroforestry, horticulture, silviculture. Gupta and Rao (1994) have suggested that more organic carbon can be sequestered in soil/degraded soils through appropriate land use planning after adopting the suitable rehabilitation measures. Alternative land use systems *viz.* agroforestry (0.71 and 0.73 per cent), agro-horticulture (0.73 and 0.74 per cent) and agro- silviculture (0.48 and 0.56 per cent) had more effective soil organic carbon sequestration than sole cropping systems at 0-15 and 15-30 cm soil depth, respectively. Grasslands can also act as a significant carbon sink with the implementation of improved management strategies. Sustaining production in grassland ecosystems is the maintenance of soil organic matter (SOM), which can be strongly influenced by management practices. Many management techniques intended to increase forage production may potentially increase SOM, thus sequestering atmospheric carbon. Further, conversion from either cultivation or native vegetation into grassland could also sequester atmospheric carbon (Conant 2001). An another approach to enhance the carbon sequestration in soil is through the bio-strategies comprising afforestation, ecosystem restoration, energy plantation, raising of plantation/crops having the deep and prolific root system, growing species having more content of cellulose and other recalcitrant material. For harnessing the more CO_2 from the atmosphere and keep away in soil the alternate land use practices need to be employed. The farmers shall be educated for bringing more acreage of agricultural lands under the agroforestry, agro-horticulture and silviculture system. The adoption of practice of growing grasses for a period of 3-4 years after 5-6 years of cropping leads in increase/maintenance of SOC levels.

Tillage, Organic Residue and Crop Rotation

Tillage and soil carbon are negatively related indicating greater the tillage more oxidation of carbon resulting in enhanced CO_2 concentrations in the atmosphere and depletion of organic carbon in soil. About 58 per cent carbon get sequestered in soil organic matter under zero till system. Different no-till systems result in varying soil disturbance, but any system that reduces tillage substantially can increase soil carbon and followed by its consequential benefits. In a study carried out to quantify the potential of carbon sequestration for different crops in combination with reduced tillage and/or enhanced crop rotation complexity it was found that zero tillage can sequester 57±14 g C m^{-2} yr^{-1} (West and Post 2002). Enhancing crop rotation complexity sequestered an average 20±12 C m^{-2} yr^{-1}. They further reported carbon sequestration rates are expected to reach peak due to conversion of conventional tillage to zero till or no till in 5-10 years with SOC reaching a new equilibrium in 15-20 years while the enhanced complexity in crop rotation may reach a peak in 40-60 years.

An efficient management of the crop residue either through mulch or incorporation is must for mitigating the challenge of increasing SOC levels. These practices are crucial to improve infiltration capacity of soil, reducing run-off rate, and soil loss through detachment and transportation. Large quantity of plant residue to the magnitude of 440 Tg is produced in India. Out of this 90 per cent is a cereal residue. This resource can be utilized as mulch at a threshold level of 4 Mg mulch ha^{-1} (Lal 2005) and incorporations depending upon the soil types, weather conditions, crop grown, physiographic conditions etc. An incorporation of crop residue builds up the SOC levels in comparison to either removal or burning. Under Indian condition the utilization of crop residue for improving the carbon sequestration does not seem to be practical as the residues are alternatively used as fodder, fuel, construction material and other purposes. However, for betterment of the soil quality and carbon sequestration needs to be enhanced through sparing the residue for incorporation in soil. Lal (2005) has reported that agricultural intensification in India can lead to 12.7–16.5 Tg yr^{-1} of SOC sequestration.

Continuous removal of crop residue for meeting the various requirements along with the mining of nutrients coupled with low or imbalanced application of fertilizers leads to degradation and desertification of soils and ecosystems.

Impact of Fertilizers

Fertilization affects soil carbon mainly through increased crop biomass. However, the carbon: nitrogen ratio in stable organic matter remains constant within a range of about 8-10:1. It means that for every 8-10 kg of carbon sequestered in soil organic matter, 1 kg of nitrogen must accompany it. This tie-up of nitrogen reduces nitrogen for crops. In the absence of sufficient nitrogen stable formation of soil organic matter via soil microbial degradation of crop residues will be affected and thus little carbon may be sequestered. Legume crops such as alfalfa, peas, lentils and chickpea could serve as alternative sources for nitrogen. Addition of leaves from out side in the form of residue also mirror as incorporation of green manure in the soil. Green leaf manuring with *Glyricidia*, Karanj (*Pongamia glabra*) and aak (*Calotropis* spp.) are popular in south India which mirrors an impact of crop residue incorporation and improves soil physico-chemical and biological properties. Potato foliage containing 2.59 to 2.68 per cent N, contribute 30-32 Kg N/ha when used as green manure with autumn sugarcane. Yadav and Prasad (1986) reported that sugarcane trash (5 t/ha) in combination with 75 kg N ha^{-1} produced yield equivalent to 150 kg N ha^{-1}. The beneficial value of trash lies in its capacity to minimize N losses from soil to increase soil organic carbon and conserve soil moisture.

Nitrogen additions affect the dynamics of SOC in soil. Application of nitrogen results in increasing the vegetative production and stabilizing the soil organic carbon.

Long term studies in Oregon indicated that 18 per cent of the N applied to a wheat-fallow system was incorporated in to the organic fraction (Rasmussen and Collins 1991).

Organic Manures

Use of organic manures such as FYM, composts, biofertilizers in combination with inorganic fertilizers in agriculture supports the carbon sequestration in soils through addition of increased amount of biomass and stimulating the growth and activities of soil biota. Swarup *et al.* (1999) reported that integrated use of fertilizer, organic and green manures and biofertilizers are better management strategies for future rational land use towards productivity without any environmental hazard. Goswami *et al.* (1998) suggested that the organic carbon in soil could be sequestered through green manuring and application of farmyard manure. However, the rate of organic matter application to the soils would vary according to the crops, climate, availability and quality of organic matter. It is observed that by continuous application of FYM and green manure there has been substantial improvement of OC under different soils and cropping systems (Table 26.8).

Table 26.8: Impact of Different Organic Amendments on Soil Organic Carbon Under Different Cropping Sequences

Cropping System	*Organic Amendments*	*Organic Carbon (per cent)*
Maize-wheat (25 yrs)	Control	0.51
	FYM	2.49
Cotton-sorghum (45 yrs)	Control	0.56
	FYM	1.14
Ragi-cowpea-maize (3 yrs)	Control	0.30
	FYM	0.64
Rice-rice (10 yrs)	Control	0.43
	50 per cent from inorganic + 50 per cent through green manuring (*Sesbania aculeate*)	0.90
Rice–wheat (3 yrs)	Control	0.44
	FYM	0.54
Rice–wheat (7 yrs)	Fallow	0.23
	Green Manuring (*Sesbania aculeate*)	0.37

Source: Swarup *et al.* (1999).

In rice-rice wheat system the SOC levels built up gradually when *S. rostrata* included and incorporated at flowering stage (Ramesh and Chandrasekharan 2004). In long term experiments an application of NPK along with FYM resulted in increase of soil organic carbon levels in Alfisols, Inceptisols, Inceptisols and Vertisol (Table 26.9).

Table 26.9: Organic Carbon (g kg^{-1} Soil) in Different Soils as Affected by Continuous Application of Manure and Fertilizer

Treatment	*Inceptisol (Udaipur)*	*Inceptisol (Ludhiana)*	*Vertisol (Jabalpur)*	*Vertisol (Bangalore)*	*Alfisol (Ranchi)*	*Mollisol (Pantnagar)*
Control	6.83	2.6	5.3	4.06	2.9	5.1
N	7.23	3.2	5.9	4.39	3.6	8.1
NP	7.13	3.2	6.9	4.83	4.0	7.6
NPKS	8.10	3.7	8.0	5.01	4.3	8.5
NPK+FYM	9.92	5.2	9.4	5.97	4.6	15.2
Initial	8.0	3.1	5.4	4.6	4.5	14.8
Years after	9.5	34	34	17	34	33

Source: Singh (2007); Singh *et al.* (2004).

Benefits of recommended management practices are visible in the soils capable of supporting the crop production more pronouncedly as compared to that of in severely degraded soils depleted in SOC stock. An optimum level of organic carbon stock is needed to maintain the physical, chemical and biological health of the soil. SOC plays innumerable roles for maintaining the soil healthy in addition to the sink for CO_2 and other greenhouse gases. There is a positive relationship between the amount of SOC and crop yield even in high input commercial agriculture. Aggarwal *et al.* (1997) reported that increased levels of SOC have the similar effect on crop yields as that of increased inputs of fertilizers. Increased levels of SOC generally have positive impact on crop yields in soils having clay content < 10 per cent. In soils of tropics and northern latitudes the critical level of SOC is 1.1 and 2.0 per cent, respectively below which the crop yields may reduce by 20 per cent (Aune and Lal 1997; Greenland *et al.*, 1975). In Kenya, an 18 years study showed the quantum jump in the maize and bean yield when the stover was retained and fertilizers and manure were applied over the no external input. It also increased the SOC stocks in 15 cm depth by approximately 5 t ha^{-1} (Kapkiyal *et al.*, 1999).

Carbon Trading

Carbon sequestration in soils also has great potential in regard to the international trading of carbon credits. Carbon trading has open up the realm of balancing the carbon emissions of one nation by supporting carbon sequestration in some other part of the world. It implies the trading of permits of CO_2 or other GHG emissions. The carbon trading in EU countries may reach to 1 billion dollars by 2010 (Johnson and Heinen 2004). Clean Development Mechanism (CDM) of Kyoto Protocol provides the platform for trading C credits between developed country and a developing country. Trading C credits between two developed nations can be done as per the Joint Implementation clause of Kyoto Protocol. Carbon trading not only provides an opportunity to enhance the farm incomes but it is also a good incentive for improving the soil and environment quality, rehabilitate wastelands.

Epilogue

Soil organic carbon is central to the global carbon cycle and its dynamics are mainly dictated by the soil management practices which influence the physical, chemical and biological properties. Due to enhanced and intensive cultivation practices the organic matter level in soils is declining over the period and also contributing to the increased concentration of CO_2 in the atmosphere. The soil unfriendly or intensive management practices expose the soil to accelerated erosions leading to declined sustainability and subsequently the degradation of soils. During pre-industrial era air contained about 270 ppm CO_2 concentration while the current concentration has reached to 372 ppm increasing @ 1.5 ppm every year. Burning of coal and oil has a major influence on atmospheric CO_2. The increased concentration of CO_2 leads to global warming and climate change influencing the overall environment. This effect can be reversed up to certain extent by transforming the atmospheric CO_2 to SOC employing the good management practices in agriculture. This would result in increased SOC stock *vis-a-vis* sustainability. Carbon trading concept will provide an opportunity of additional income to the farmers in addition to the rejuvenation of the wastelands/degraded lands and improve the environment. Therefore, to combat the increasing concentration of CO_2 in atmosphere a holistic system approach is warranted for the sustainability of the agricultural soils and environmental security.

References

Aggarwal, A.K., Kumar, P. and Powell, J.F., 1997. Use of crop residue and manure to conserve water and enhance nutrient availability and pearl millet yield in an arid tropical region. *Soil and Tillage Research*, 41: 43–51.

Aune, J.B. and Lal, R., 1997. Agricultural productivity in the tropics and critical limits of properties of Oxisols, Ultisols and Alfisols. *Tropical Agriculture*, 74: 96–103.

Bhattacharyya, T., Pal, D.K., Velayutham, M., Chandran and Mandal, C., 2000. Total carbon stock in Indian Soils. Issues, priorities and management. Paper abstracted in *International Conference on Land Resource Management for Food, Employment and Environment Security*, Nov. 9–13, New Delhi, India.

Bhattacharya, T., Pal, D.K., Mandal, C. and Velayutham, M., 2000. Organic carbon stock in Indian soils and their geographical distribution. *Current Science*, 79: 655–660.

Bowen, G.D. and Rovira, A.D., 1999. The rhizosphere and its management to improve plant growth. *Adv. Agron.*, 66: 1–102.

Carter, M.R., 2002. Soil quality for sustainable land management: Organic matter and aggregation interactions that maintain soil functions. *Agronomy Journal*, 94: 38–47.

Conant, R., Paustian, K. and Elliot, E., 2001. Grassland management and conversion into grassland: Effects on soil carbon. *Ecological Applications*, 11(2): 343–355.

Dregne, H.E. and Chou, N.T., 1994. Global desertification dimensions and costs. In: *Degradation and Restoration of Arid Lands*, (Ed.) H.E. Dregne. Texas Tech. University, Lubbock, TX., pp. 249–281.

FAO, 2004. *Production Yearbook in Agriculture*. FAO, Rome Italy.

Goswami, N.N., 1998. Soil organic matter and organic residue management for sustainable productivity: Key note address. *Bull. Indian Soc. Soil Sci.*, 19: 1–13.

Greenland, D.J., Rimmer, D. and Payne, D., 1975. Determination of structural stability class of English and Welsh soils, using a water coherence test. *Journal of Soil Science*, 26: 294–303.

Gupta, R.K. and Rao, D.L.N., 1994. Potential wasteland for sequestering carbon by reforestation. *Current Science*, 66: 378–380.

IPCC, 2001. The Scientific Basis. In: *Contribution of Working Group I to the Third Assessment Report of the Inter-governmental Panel on Climate Change*, (Eds.) J.T. Houghton, Y. Ding, D.J. Griggs, M. Noguer, P.J. vander Linden, X. Dai, K. Maskell and C.A. Johnson. Cambridge University Press, Cambridge, United Kingdom and NY, USA, pp. 881.

Kapkiyal, J.J. *et al.*, 1999. *Soil Biol. Biochem.*, 31: 1773.

Kyuma, K., 1988. Paddy soils of Japan in comparison with those in tropical Asia. In: *Proceedings of First International Symposium on Paddy Soil Fertility*, held at Chiagmai, Thailand, December 6–13, p. 5–19.

Lal, R., 2004. Soil carbon sequestration impacts on global climate change and food security. *Science*, 304: 1623–1627.

Lal, R., 2005. Carbon sequestration and climate change with specific reference to India. In: *Proceedings of International Conf. on Soil, Water and Environmental Quality: Issues and Strategies*, held at New Delhi, January 28–February 1, pp. 295–302.

Lal, R., Kimble, E., Lewin, E. and Stewart, B.A., 1995. *Advances in Soil Management and Greenhouse Effect*. Lewis Publisher, Bocan Raton, p. 93.

Nambiar, K.K.M., 1994. *Soil Fertility and Crop Productivity under Long-Term Fertilizer Use in India*. Indian Council of Agricultural Research, New Delhi.

Ramesh, K. and Chandrasekharan, B., 2004. Soil organic carbon building and dynamics in rice–rice cropping system. *Journal of Agronomy and Crop Science*, 190: 21–29.

Rasmussen, P.E. and Collins, H.P., 1991. Long-term impacts of tillage, fertilizers and crop residue on soil organic matter in temperate semiarid regions. *Advances in Agronomy*, 45: 93–134.

Sage, R.F., 1995. Was low atmospheric CO_2 during the Pleistocene a limiting factor for the origin of agriculture. *Global Change in Biology*, 1: 93–106.

Singh, M., 2007. Important issues and lessons learnt from LTFE experiments for organic matter management in tropical soils. In: *Lecture Notes of Winter School on*

Soil Organic Carbon Stocks and Soil Organic Matter Management in Relation to Soil Quality and Climate Change, (Eds.) Misra, A.K. and A.K. Biswas, held at Indian Institute of Soil Science (ICAR), Bhopal, pp. L17 (1–8).

Singh, M.V., Wanjari, R.H. and Adhikari, T., 2004. Nutrient dynamics, crop productivity and sustainability under long-term fertilizer experiments in India. In: *AICRP–LTFE*, Indian Institute of Soil Science (ICAR), Nabibagh, Bhopal, pp. 1–120.

Swarup A., 1998. Emerging soil fertility management issues for sustainable crop productivity in irrigated systems. In: *Proceedings of a National Workshop on Long-Term Soil Fertility management Integrated Plant Nutrient Supply*, (Eds.) Swarup, A., Damodar Reddy, D. and R.N. Prasad, pp. 54–68.

Swarup, A., Manna, M.C. and Singh, G.B., 1999. Impact of land use and management practices on organic carbon dynamics in soils of India. In: *Global Climate Change and Tropical Ecosystems*, (Eds.) R. Lal, J.M. Kimble, H. Eswaran and B.A. Stweart. Lewis Publishers, Boca Raton, Fl, pp. 261–281.

USEPA, 1995. *Inventory of US Greenhouse Gas Emissions and Sinks, 1990–1994*. Washington DC.

Velayutham, M., Pal, D.K. and Bhattacharyya, T., 2000. Organic carbon stocks in soils of India. In: *Advances in Soil Sciences*, (Eds.) R. Lal, J.M. Kimble and B.A. Stewart. CRC/Lewis Publishers, Boca Raton, Florida, USA, pp. 71–95.

Whipps, J. M., 1990. Carbon economy. In: *The Rhizosphere*, (Ed.) J.M. Lynch. Wiley, New York, pp. 59–97.

Yadav, R.L. and Prasad, S.R., 1986. *Indian Journal of Sugarcane Technology*, 3: 24–28.

Ecosystem Diversity and Carbon Sequestration: Climate Change Challenges and a Way Out for Ushering in a Sustainable Future *Pages 253–263*
Editors: P.L. Gautam, Vir Singh and Uma Melkania
Published by: DAYA PUBLISHING HOUSE, 2009

Chapter 27

Soil Carbon Sequestration: A Potential Approach to Climate Change Mitigation

J.S. Chauhan, Bineet Singh and J.P.N. Rai

With the advent of the industrial revolution, there has been a drastic increase in the atmospheric concentration of carbon dioxide (CO_2) and other greenhouse gases (GHGs) which is resulted due to change in the land use pattern and increasing combustion of fossil fuels. The atmospheric concentration of CO_2 has increased from 280 ppmv in 1750 to 367 ppmv in 1999 and is currently increasing at the rate of 1.5 ppmv/year or 3.3 Pg C/year (1 Pg = petagram = billion ton) (IPCC 2001). Emissions due to land use change include those by deforestation, biomass burning, conversion of natural to agricultural ecosystems, drainage of wetlands and soil cultivation. These main activities have resulted in decreased soil organic carbon (SOC), which influences soil fertility adversely. Thus, it is necessary to restore land use practices with effective management strategies like conversion of marginal lands into restorative land uses, adoption of conservation tillage with cover crops and crop residue mulch, nutrient cycling including the use of compost and manure, and other systems of sustainable management of soil and water resources. These management approaches will lower down the atmospheric CO_2, thereby slowing global warming and mitigating climate change, which in turn may result in positive impacts on food security, agro-industries, water quality and the environment.

Keywords: *Carbon sink, Soil organic carbon, Climate change.*

Introduction

Carbon enters the soil as roots, litter, harvest residues, and animal manure. It is stored primarily as soil organic matter (SOM). The density (weight/volume) of carbon is highest near the soil surface. But much of the most recently deposited SOM decomposes rapidly, releasing CO_2 to the atmosphere. Some carbon becomes stabilized, especially in the lower part of the soil profile. Balanced rates of input and decomposition determine steady state carbon fluxes. However, in many parts of the world, agriculture and other land-use activities have upset the natural balance in the soil carbon cycle, contributing to an alarming increase in carbon release from soils to the atmosphere in the form of CO_2.

Carbon sequestration in soils is a climate change-mitigating strategy based on the assumption that movement, or flux, of carbon from the air to the soil can be increased while the release of carbon from the soil back to the atmosphere is decreased. In other words, certain activities can transform soil from a carbon source (emits carbon) into a carbon sink (absorbs carbon). This transformation has the potential to reduce atmospheric CO_2, thereby slowing global warming and mitigating climate change.

At the same time, carbon sequestration can confer substantial benefits on the people in a region, such as farmers and small landholders, who directly manage the soil carbon pool. Improved land and soil management practices that help sequester carbon in soils can lead to higher soil fertility, increased yields, and other outcomes that aid local populations economically, environmentally, and socially.

Why are Atmospheric Levels Increasing?

Levels of several important greenhouse gases have increased by about 25 per cent since large-scale industrialization began around 150 years ago (Figure 27.1). The global fossil emission of carbon has reached more than 6500MmT C/year emission as a total *i.e.* petroleum, coal, natural gas and cement production in 2000 (Figure 27.1).

Concentrations of carbon dioxide in the atmosphere are naturally regulated by numerous processes collectively known as the "carbon cycle". The movement ("flux") of carbon between the atmosphere and the land and oceans is dominated by natural processes, such as plant photosynthesis. While these natural processes can absorb some of the net 6.1 billion metric tons of anthropogenic carbon dioxide emissions produced each year (measured in carbon equivalent terms), an estimated 3.2 billion metric tons is added to the atmosphere annually. The Earth's positive imbalance between emissions and absorption results in the continuing growth in greenhouse gases in the atmosphere.

Carbon Dioxide and Climate Change

Carbon dioxide (CO_2) is one of the main culprits of greenhouse gas emissions linked to global warming. CO_2 contributes to 60 per cent of total global warming caused by greenhouse gas emissions. Industrialization has brought with it the bane of fossil-fuel burning, deforestation, automobile exhaust, etc. This has added to an increase in the concentration of atmospheric CO_2. It is vital to reduce greenhouse gas emissions such as CO_2, CH_4 and N_2O for arresting global warming and climate change. The combustion of fossils fuel during the 20th century had increased the emission of CO_2 in the atmosphere drastically. Currently its concentration had released to 370

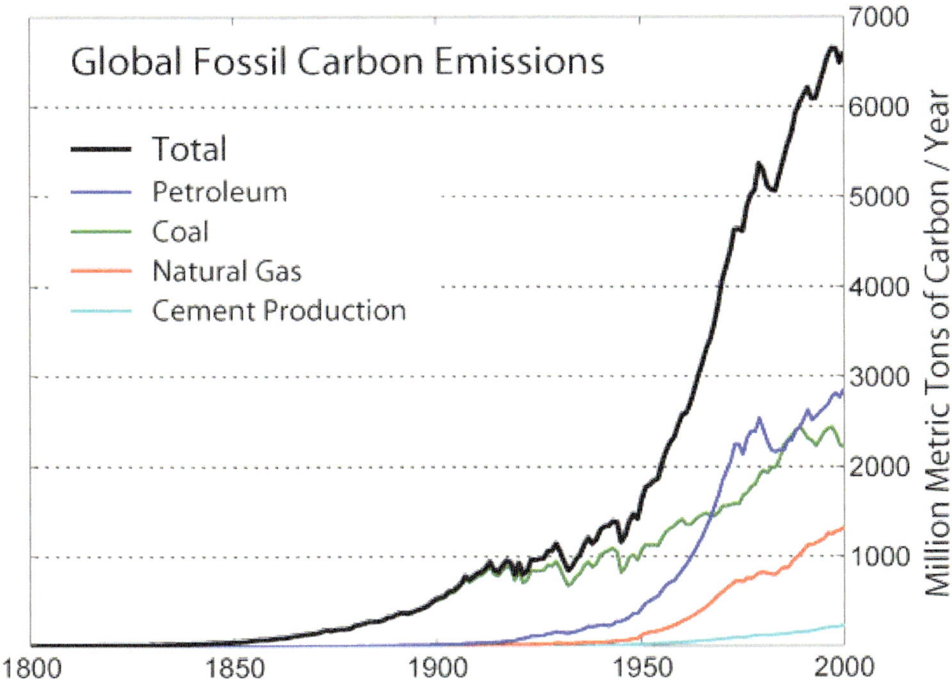

Figure 27.1: Global Carbon Dioxide Emissions 1751–2000 (IPCC-1994)

ppm that is 31 per cent since 1750 from fossils fuel burning and land use changes, which created to a threat to global warming. Its present rate of increase per year is 0.4 per cent which is more than other greenhouse gases. The anthropogenic increase of GHGs in the atmosphere has led to increase the global surface temperature to 0.6 degree since the late 19th century, with the current warming rate of 0.17 degree C per decade (IPCC 2001).This change in temperature has affected the life system on earth. The emission of greenhouse gases due to human activities disturbed the balance between the rate of greenhouse gases entering the atmosphere and the rate of gases lost.

The concentration of carbon dioxide in the atmosphere depends on the rates of exchange of carbon dioxide between the atmosphere and the terrestrial biomass, the oceans, and the carbon in deposits of fossil fuels and other geological formations. Economic growth and population pressure, especially since the industrial revolution, have greatly increased the input to the atmosphere of carbon from fossil fuel deposits and from terrestrial biomass, by deforestation and other land use changes. The processes which remove carbon from the atmosphere have not increased proportionately. Consequently, atmospheric concentrations of carbon dioxide have risen to levels which are unprecedented in the recent geological history of the earth.

Table 27.1: Change in the Main Greenhouse Gas Concentration Since Pre-industrial Period (1750)

Gas	Current (1998) Amount by Volume	Increase Over Pre-industrial (1750)	Percentage Increase	Radiative Forcing (W/m^2)
Carbon dioxide	365 ppm	87 ppm	31 per cent	1.46 (~1.532)
Methane	1,745 ppb	1,045 ppb	150 per cent	0.48
Nitrous oxide	314 ppb	44 ppb	16 per cent	0.15

Source: IPCC (1994).

The amount of carbon dioxide in the atmosphere in 2005 (379 ppm) exceeds by far the natural range of the last 650,000 years (180 to 300 ppm(IPCC 2007). This has given rise to concern about climatic change. This concern has produced policy responses in the form of the United Nations Framework Convention on Climate Change (UNFCCC) and its Kyoto Protocol. These agreements aim to limit the accumulation of carbon dioxide and other greenhouse gases in the atmosphere and to stabilize concentrations at levels which would prevent dangerous anthropogenic interference with the climate system. In principle concentrations may be stabilised by any combination of policies which decrease emissions of carbon dioxide to the atmosphere, or increase rates of removal and storage, also called *sequestration.* The initial policy debate concentrated mainly on emissions reduction, but since the Kyoto Protocol there has been increased emphasis on the contribution that sequestration could make.

Carbon and Sink

There are five important global carbon pools among which oceanic pool (38,000 Pg) is the largest followed by geological pool (5000 Pg; 4000Pg of coal pool and 500 Pg of each oil and gas) pedological pool (soil carbon pool, 2500 Pg) biotic pool (560 Pg) and the atmospheric pool(760 Pg). The average atom of C spends about 5 yrs in the atmosphere, 10 yrs in vegetation (including trees), 35 yrs in soil, and 100 yrs in the sea (Lal 2004). Increase density of C in the soil and depth of C in the profile, decrease decomposition of C and losses due to erosion are important measures to increase the soil organic carbon.

About 20 per cent of the anthropogenic emissions of CO_2 is contributed by agriculture and land-use change (Dumanski and Lal 2004). Average soil organic carbon (SOC) in the top 30 cm of native soil worldwide is approximately 15 Mg/ha. However, when cultivated, about 20–30 per cent of this C is released to the atmosphere within the first 20 years in temperate regions, and 50–75 per cent in the tropics (Dumanski and Lal, 2004).

Soil Organic Carbon as Carbon Sinks

SOC play an important role in soil profile. It is the dead or decay parts of living biota accumulated in soil. It holds about 1500 Pg C to a depth of 1 meter (Eswaran *et al.*, 2000). The ultimate source of organic matter in soils is CO_2 fixed by plants, including leaf litter, roots, and root exudates. The activity of soil fauna (especially fungi and microbial communities) metabolizes some of these substrates and transforms others into more resistant organic compounds (collectively referred to as humus). Soil carbon pool consists of soil organic carbon (SOC) and soil inorganic carbon (SIC). The stabilization and fate of organic matter residues is affected by the quality of the original plant substrate, and the physical environment in the soil. It is a better source for principal plant nutrients (Macro and micro), a good promoter of soil aggregation and increases water holding capacity of soil. The rate of SOC storage is highly dependent on weather conditions because of it influence on net primary production (carbon input) and rate carbon decomposition (Campbell *et al.*, 2001a).The reservoir of soil carbon has been proposed as both a significant source and sink of atmospheric CO_2. A soil source results when net decomposition exceeds C inputs to the soil, either as a result of human activities such as clearing forests for agriculture or because of increased decomposition rates due to global warming.

Soil organic matter promotes good soil structure which creates soil spaces for roots to travel through the soil for air and water. The space allow microorganisms to turn the nitrogen in the air into nitrate and ammonia and soil carbon dioxide contained in these air spaces increases plant growth. SOC helps plant and microbial growth through growth stimulating compounds and root growth by making it easy. Native levels of soil organic matter for any particular site are determined largely by the latitude location on the earth and by the annual precipitation received. Native soil organic matter levels will generally increase as you move either north or south of the equator.

Carbon Sequestration

Carbon sequestration in terrestrial ecosystems can be defined as the net removal of CO_2 from the atmosphere into long-lived pools of carbon. The pools can be living, aboveground biomass (*e.g.*, trees), products with a long, useful life created from biomass (*e.g.*, lumber), living biomass in soils (*e.g.*, roots and micro-organisms), or recalcitrant organic and inorganic carbon in soils and deeper subsurface environments. It is important to emphasize that increasing photosynthetic carbon fixation alone is not enough. This carbon must be fixed into long-lived pools. Otherwise, one may be simply altering the size of fluxes in the carbon cycle, not increasing carbon sequestration.

The biosphere, geological formations and the oceans are the three main reservoirs which sequester carbon from the atmosphere. Each of these reservoirs is in principle accessible to policy intervention. Biospheric carbon may be increased by policies which promote afforestation or increase soil carbon content. Carbon dioxide may be captured and sequestered by injection into geological formations including depleted oil and gas fields, saline aquifers or unmineable coal seams. The ocean reservoir can be increased by fertilising plankton to increase the rate of transfer across the sea surface, or by direct injection of carbon dioxide at depth. Carbon sequestration from afforestation occurs while trees are growing and ceases when they mature, although sequestration of carbon in the forest soil may continue at a much lower rate. In productive forests, trees are often harvested before they mature so that sequestration is not halted. The sequestration effect is potentially reversible due to the vulnerability of forests to fire, pests etc. Additional benefits can arise, *e.g.* from improved biodiversity and new recreational opportunities, but increased water uptake by trees can in some situations (*e.g.* in dry tropics) have impacts on supplies of groundwater and river flow.

The level below which atmospheric concentration of carbon dioxide should be stabilised to prevent dangerous anthropogenic interference with the climate system is not known precisely, but the European Union has suggested 550 parts per million (ppm), or about twice the pre-industrial level. Achieving this by emissions reduction alone would require a cut of about two thirds in total anthropogenic emissions. The cut by developed countries would need to be even greater, when projected economic growth by developing countries is taken into account. About 200 years would be available for these changes, depending on the emissions path followed.

Role of SOC Sequestration in Climate Change Mitigation

Within the scope of global change related to human-induced release of greenhouse gases to the atmosphere, global carbon transfers from and to the soil have been receiving increasing attention for more than a decade (Schlesinger *et al.*, 2000). The carbon pool in the world's soils is three times that in the atmosphere (IPCC 2001). Any change in the belowground pools resulting from changes in land-use (conversion to crop or pasture, afforestation), can thus have major impacts on carbon concentrations in the atmosphere. Most of these carbon flows are mediated by SOM, and have led to the numerous studies of the carbon-oxidation to carbon-fixation balance in relation to land use and climate change. Whilst these studies are generally undertaken at the plot scale, they have been conceived and applied to the issues of global climate change (Schlesinger *et al.*, 2000). It is therefore now accepted that among the assessments made of the impacts of changes in cropping pattern or other aspects of land use change carbon sequestration capacity is an important factor. The models developed to explain SOM dynamics have thus become essential tools for studies of carbon sequestration in soil. Nonetheless figures published for below ground carbon sequestration are seldom accurate, and this underlines among other things, how limited our knowledge about SOM storage determinants and location remains (Schlesinger *et al.*, 2000; Lal 2001). To help mitigate global warming, a variety of means of artificially capturing and storing carbon as well as of enhancing natural sequestration processes are being explored. Atmospheric carbon dioxide and other greenhouse gases act to trap heat that is reflected from the earth's surface. This buildup of heat could lead to global warming. Through carbon sequestration, atmospheric carbon dioxide levels are reduced as soil organic carbon levels increase. If the soil organic carbon is undisturbed, it can remain in the soil for many years as stable organic matter. This carbon is then sequestered or removed from the pool available to be recycled to the atmosphere. This process reduces CO_2 levels in the atmosphere, reducing the chances of global warming.

Beyond its role as a key compartment in nutrients cycles and as a regulator of climate change, SOM has also come to be valued for its influence on a wide range of so-called "ecosystem services" including those of water availability and quality, the erodibility of soil, and as a source of energy for soil biota acting as biological control agents of pest and diseases of plants, livestock and even humans (Swift *et al.*, 2004).

There is an extensive amount of evidence indicating that the earth's climate has warmed during the past century. Foremost among this evidence are compilations of the variation in global mean sea surface temperature and in surface air temperature over land and sea. Supplementing these indicators of surface temperature change is a global network of balloon-based of atmospheric temperature since 1958. Apart from temperature, changes in the extent of alpine glaciers, sea ice, seasonal snow cover, and the length of the growing season have been documented that are consistent with the evidence that the climate is warming. Less certain, but also consistent, changes appear to have occurred in precipitation, cloudiness, and interannual temperature and rainfall variability.

Why the Concentration of Carbon is Decreasing in Soil?

It is estimated that man-made changes in land-use have, until now, produced a cumulative global loss of about 200 Pg carbons. Land use changes include deforestation, conversion of natural ecosystem to agricultural land and biomass burning. Widespread deforestation has been the main source of this loss, estimated to be responsible for nearly 90 percent of losses since the mid-nineteenth century. Losses primarily occur due to the relatively long-term carbon sinks of forests being replaced by agricultural land. The conversion of land from forested to agricultural land can have a wide range of negative effects as far as greenhouse gases emission is concerned. Soil disturbance and increased rates of decomposition in converted soils can both lead to emission of carbon to the atmosphere, with increased soil erosion and leaching of soil nutrients further reducing the potential for the area to act as a sink for carbon. Deforestation and biomass burning, cultivation of soil, by plowing and other tillage methods, also enhances mineralization of SOC and releases CO_2 into the atmosphere (Reicosky *et al.*, 1999). Tillage increases SOC mineralization by bringing crop residue closer to microbes where soil moisture conditions favor mineralization (Gregorich *et al.*, 1998). Tillage not only disturbs the soil profile but also emit fuel gases because of machinery and tractor use.

Conant *et al.* (2001) in a review reported rates of soil C sequestration as high as 1.01 Mg C/ha/year for conversion of cultivated land to grassland. Since the industrial revolution, global emission of carbon are estimated at 270 ± 30Pg (1Pg = petagram = 1billion ton) due to fossil fuel burning and 136 ± 55 Pg due to land use changes (Lal, 2004). Traditional method used in agriculture like biomass burning and residue removal, conventional tillage or deep ploughing, bare fallow, practice of monoculture, use of chemicals etc. are main reason for the increase of carbon in the atmosphere. In addition to its impact on decomposition of SOM (Trumbore *et al.*, 1996), macroclimate has a large impact on a fraction of the SOC pool which is active (Franzluebbers *et al.*, 2001a). Conversion of natural to agricultural ecosystems increases maximum soil temperature and decreases soil moisture storage in the root zone, especially in drained agricultural soils (Lal 1996).

CO_2 Mitigation Potential of Different Land Uses

To mitigate the climate change now, it is important to follow the recommended management practice (RMP) suggested by the scientists. The traditional method of agriculture should be replaced by the following RMP.

- ☆ Residue of crop should be returned to the soil as mulch.
- ☆ Conservation or zero tillage should be opted against conventional tillage
- ☆ Crop should be grown during off season.(fallow period)
- ☆ Integrated nutrient management and integrated pest management should be practiced.
- ☆ Organic farming should be followed.
- ☆ Crop rotation over continuous monoculture should be adopted.
- ☆ Deforestation should be replaced by afforestation, reforestation and agroforestry
- ☆ Reduce overgrazing
- ☆ Improve forage quality.

Table 27.2 shows the mitigation potential of different land use pattern with respect to barren land. The barren land can store 20.0t/ha SOC and its mitigation potential is considered to be 1.00, compared with it the forestry have highest potential to reduce the climate change followed by agroforestry, plantations, agriculture and pasture. Hence, agroforestry can be a better option for farmer to cope with the climate change.

Table 27.2: CO_2 Mitigation Potential of Different Land Uses with Barren Land as a Base

Land Use	*SOC Store t/ha (up to 30 cm depth)*	*Mitigation Potential*
Barren land	20.0	1.00
Pasture	40.0	2.00
Agriculture	66.0	3.30
Plantations	80.5	4.02
Agroforestry	83.6	4.18
Forests	120.0	6.00

Source: Jha (2000).

Lulu and Insam (2000) observed positive effect of agroforestry with *Sesbania* on the SOC pool. Carbon accumulation rates in tropical agroforestry systems ranges from 4-9 MgC/ha/year more above ground than in the soil (Dumanski and Lal 2004) Deforestation and reforestation can be better tools for increasing the carbon sequestration. World cropland has soil C sequestration potential of 0.45-0.57PgC/ year (Lal and Bruce 1999).

Impacts of Management Practices

The recommended management practices (RMP) is important to conserve environment to have sustainable agriculture and development. The traditional method of farming should be eliminated completely. The RMP will have the following positive impacts:

- ☆ Reduce the soil erosion due to proper crop cover.
- ☆ Enhance biodiversity of soil micro fauna and flora
- ☆ Improve soil structure and fertility by increasing the nutrient availability as a result of residue return to soil.
- ☆ Increase production
- ☆ Enhance water storage because of better water holding capacity of humus.
- ☆ Enhance food security
- ☆ Rehabilitation degraded land by afforestation and reforestation.
- ☆ Reduce animal pressure
- ☆ Reduce the rate of enrichment of atmospheric CO_2 by more soil carbon sequestration and hence decrease global warming.

The adoption of agricultural management practices capable of sequestering C is hampered both by environmental (weather, etc.) and socio-political factors. The latter constraints, including the supply and demand for agricultural products, production costs, subsidies, incentives to reduce environmental impacts and social, aesthetic and political acceptance for changes, may well be the most important factors in deciding whether or not suggestions are applied by producers. It must be understood though, that in the end, producers will only adopt new management practices if it is found to be economically feasible. Analyses of these factors are highly complex, and studies on this are in their infancy (Cole *et al.*, 1996).

Conclusion

The concentration of greenhouse gases in the atmosphere is increasing drastically which is resulting in the global climate change and global warming. Hence there is an urgent need to find some mitigatory measure for protection of environment from this change. Soil carbon sequestration can be one of such measure to lower down the concentration of CO_2 which is an important GHGs. Globally, under the terms of the Kyoto Agreement, several countries committed to reduce its greenhouse gas emissions in the first commitment period (2008–2012) below the 1990 baseline year levels. The agricultural sector is expected to help meet this target, by reducing their own GHGs emissions, and increasing carbon sequestration. To enhance carbon sequestration it is necessary to apply proper recommended practices globally in accordance with the climatic condition of a particular area. It is important that the carbon sequestrated should remain in any ecosystem (aquatic, forests, soil, grassland etc.) for a longer period of time.

References

Campbell, C.A., Selles, F., Lafond, G.P. and Zentner, R.P., 2001a. Adopting zero tillage management: Impact on soil C and N under long-term crop rotations in a thin Black Chernozem. *Can. J. Soil Sci.*, 81: 139–148.

Cole, V., Cerri, C., Minami, K., Mosier, A., Rosenberg, N. and Sauerbeck, D., 1996. Agricultural options for mitigation of greenhouse gas emissions. In: *IPCC Working Group II*, (Eds.) Watson, R.T., Zinyowera, M.C., Moss, R.H. and D.J. Dokken. *Climate Change 1995. Impacts, Adaptations and Mitigation of Climate Change: Scientific Technical Analysis.* Cambridge University Press, UK, pp. 75–771.

Conant, R.T., Paustian, K. and Elliott, E.T., 2001. Grassland management and conversion into grassland: Effects on soil carbon. *Ecol. Appl.*, 11: 343–355.

Dumanski, J. and Lal, R., 2004. *Soil Conservation and the Kyoto Protocol Facts and Figures.* Agriculture and the Environment, Environment Bureau, Agriculture and Agri-Food Canada, Ottawa, ONTARIO.

Dumanski, J., Desjardins, R.L., Tarnocai, C., Montreal, C., Gregorich, E.G., Kirkwood, V. and Campbell, C.A., 1998. Possibilities for future carbon sequestration in Canadian agriculture in relation to land use changes. *Climatic Change*, 40: 81–103.

Eswaran, H., Reich, F.P., Kimble, J.M., Beinroth, F.H., Padamnabhan, E. and Moncharoen, P., 2000. Global carbon stocks. In: *Global Climate Change and Pedogenic Carbonates*, (Eds.) Lal, R., Kimble, J.M., Eswaran, H. and B.A. Stewart. CRC/Lewis, Boca Raton, FL.

Franzluebbers, A.J., Haney, R.L., Honeycutt, C.W., Arshad, M.A., Schomberg, H.H. and Hons, F.M., 2001a. Climatic influences active fractions of soil organic matter. *Soil Biology and Biochemistry*, 33: 1103–1111.

Gregorich, E.G., Greer, K.J., Anderson, D.W. and Liang, B.C., 1998. Carbon distribution and losses: erosion and deposition effects. *Soil and Tillage Research*, 47: 291–302.

Inter-governmental Panel on Climate Change, 2001. *Climate Change: The Scientific Basis.* Cambridge Univ. Press, Cambridge, UK.

Inter-governmental Panel on Climate Change. Retrieved on 2007 April 9.

Jha, M.N., 2002. *Forestry and Climate Change: Assessing Mitigation Potential and Costs.* India International Centre, New Delhi.

Lal, R., 2001a. Fate of eroded soil carbon: emission or sequestration. In: *Soil Carbon Sequestration and the Greenhouse Effect*, (Ed.) R. Lal. Soil Science Society of America Special Publication, Vol. 57. Madison, WI, pp. 173–181.

Lal, R., 1996. Deforestation and land use effects on soil degradation and rehabilitation in western Nigeria. II: Soil chemical properties. *Land Degradation and Development*, 7: 87–98.

Lal, R., 2004a. Soil carbon sequestration to mitigate climate change. *Geoderma*, 123: 1–22.

Lal, R., 2004b. Soil carbon sequestration impacts on global climate change and food security. *Science*, 304: 1623–1627.

Lal, R. and Bruce, J.P., 1999. The potential of world cropland soils to sequester C and mitigate the greenhouse effect. *Environmental Science and Policy*, 2: 177–185.

Lulu, B. and Insam, H., 2000. Effects of cropping systems and manuring on soil microbial and C status of a Vertisol in Ethiopia. *Bodenkultur*, 51: 107–114.

Reicosky, D.C., Reeves, D.W., Prior, S.A., Runion, G.B., Rogers, H.H. and Raper, R.L., 1999. Effects of residue management and controlled traffic on carbon dioxide and water loss. *Soil and Tillage Research*, 52: 153–165.

Schlesinger, W.H. and Andrews, J.A., 2000. Soil respiration and the global carbon cycle. *Biogeochemistry*, 48: 7–20.

Swift, R.S., 2001. Sequestration of carbon by soil. *Soil Science*, 166: 858–871.

Ecosystem Diversity and Carbon Sequestration: Climate Change
Challenges and a Way Out for Ushering in a Sustainable Future *Pages 264–266*
Editors: **P.L. Gautam, Vir Singh and Uma Melkania**
Published by: **DAYA PUBLISHING HOUSE, 2009**

Chapter 28

Enhancing Carbon Sequestration: Pondering over Some Strategies

Shiwani Bhatnagar and AK Karnatak

Global warming could remake the world's climate zones with some polar and mountain climates disappearing altogether and previously unknown ones emerging in the tropics. Climates will get disappeared, not just regionally but from the global set of climates and the species which live in these climates would really have nowhere to go as the system changes. Climates zones are likely to shift away from the equator towards the poles as the earth warms are predicted to happen by up to 8°C at some latitudes by the end of this century. Polar bears and ring seals which depend on Artic ice could be among those species threatened by the shifting of climate zones (Times of India, 2007). When snow and ice melts their ability to reflect sunlight goes away too accelerating the warming effect. Therefore, the change in temperature in the artic and Antarctic region will be greater.

The accelerated greenhouse effect is caused by increases in the tropospheric concentrations of radioactive gases such as CO_2, CH_4, N_2O, etc. resulting from the anthropogenic activities such as fossil fuel combustion, deforestation and agricultural practices.

Global warming poses a serious threat to not only human beings, flora, fauna but also on the environment. So, we need to make all possible efforts to save the planet by maintaining and enhancing its greenery and helping the excess of free atmospheric CO_2 go into sink. It's not easy because CO_2 is no longer the gas we just exhale. Now, we inhale it all the time. And it is not just CO_2 that is destroying the climate; other greenhouse gases have a share too. But the scientists think differently. They say that the earth is dying and even if we turn off all lights, ground all jets, stop all traffic, halt all trains, turn off all turbines, and shut down all workshops, the damage cannot be controlled.

Inter-governmental Panel on Climate change (IPCC), in its first report, projected as a "best estimate" that temperatures, stoked by human emissions led burning fossil fuels, would raise the temperature up to 1.8-4.0 Celsius this century. Scientists have suggested various techniques through which global warming can be avoided but there is no foolproof solution for the problem. The increase in atmospheric concentration of CO_2 by 31 per cent since 1750 from fossil-fuel combustion and land use change necessitates identification of strategies for mitigating the threat of the attendant global warming (Lal 2004). Deforestation, biomass burning, conversion of natural to agricultural ecosystems, drainage of wetlands and soil cultivation are some causes of emissions due to land use change. Carbon sequestration as an efficient mean for offsetting greenhouse gases (GHGs) emissions has generated a high level of interest over a decade. Carbon sequestration has been suggested as a potentially low cost means to rectify atmospheric concentrations of GHGs. Agriculture is both a sink for carbon as well a major emitter of CO_2 and other potent greenhouse gases such as nitrous oxide and methane. Agricultural soil C sequestration can be exaggerated through changes in land use or changes in production practices. Changes in agriculture land use and management practices alone could potentially sequester between 75 and 208 million metric tons of C per year in agricultural soils (Lal 1999, 2004). In the dry land grain production systems, changes in the land use such as converting crop land to permanent grasses and changing the management practices, such as reduced fallow, may be economically practicable where afforestation–the conversion of non-forest land to forest–is not.

Crop rotations play an important role in the maintenance of soil quality and productivity, and affect soil carbon accumulation. Adoption of recommended management practices on agricultural soil and restorative land use can reduce the rate of enrichment of atmospheric CO_2 while having positive impacts on food security, agro-industries, water quality and the environment.

Removal of atmospheric CO_2 by plants and storage of fixed C as soil organic matter is referred to as soil carbon sequestration and the strategies to increase SOC density in soil improves depth distribution of SOC and stabilizes SOC by encapsulating it within stable micro-aggregates, to protect C from microbial processes. Some recommended management practices in relation to soil organic carbon sequestration are: (*i*) adopting conservation tillage, no-till and mulch farming; (*ii*) residues used as surface mulch; (*iii*) growing cover crop during the off-season; (*iv*) crop rotations; (*v*) judicious use of off-farm input; (*vi*) integrated nutrient management and nutrient cycling; (*vii*) precision farming, integrating trees and livestock with crop production; (*viii*) drip, furrow or sub irrigation; (*ix*) integrated pest management; (*x*) conservation reserve program; (*xi*) restoration of degraded soils through land use change.

Agricultural wastes and bio-energy crops can be used to produce energy by charring them in specialized power plant and bio-char (biomass derived black carbon) as a by-product. In this process the majority of energy is converted in to hydrogen, electricity or bio-oil. This is a revolutionary process which makes energy production a net sink of atmospheric carbon dioxide and provides organic matter to enrich soil.

Biosphere greenhouse gas (GHG) management consists of preserving and enhancing terrestrial carbon pools and producing biomass as a fossil fuel substitute (Steven 2002). Soil texture and structure, rainfall, temperature, farming system, and soil management decides the rate of soil carbon sequestration. Agricultural soils prospectively provide a way of mitigating the increasing atmospheric concentration of CO_2. Some other strategies to increase the soil carbon pool includes restoration of soil and regenerating woodland, manuring and sludge application, improved grazing, water conservation and harvesting, efficient irrigation, agroforestry practices, and growing energy crops on spare lands. C storage and the potential for C sequestration are influenced by both the quantity and the quality of soil C inputs. C sequestration can also be affected by changes in tillage intensity and crop rotations which change the soil physical and biological conditions and also the amounts and types of organic inputs to the soil. Therefore, the task of soil management should be to restore soil organic carbon in order to enhance soil structure and fertility and to help counter the atmospheric greenhouse effect.

Carbon sequestration and biomass production together can make an effective contribution to the stabilization of atmospheric GHGs concentrations. As the concern increases about the adverse impact of anthropogenic emissions of greenhouse gases on the global climate it becomes necessary to reduce these emissions significantly. From the perspective of fossil fuels which is the source of substantial portion of these anthropogenic greenhouse gases emissions, the industry has three main options

- ✰ Improving the efficiency of energy production and use;
- ✰ Reducing carbon content of fuels through increased use of non-carbon fuels (*i.e.* renewable and nuclear) and natural gas; and
- ✰ Sequestering the emission of carbon dioxide.

The third option, carbon sequestration, is increasingly seen as a cost effective strategy for achieving deep reduction in carbon dioxide (and carbon) emissions. Carbon sequestration is a critical third option for addressing green gashouse emissions, along with increased energy efficiency and expanded use of low carbon fuels (Beecy and Kuuskraa 2000).

References

Beecy, D. and Kuuskraa, V., *et al.*, 2000. *Perspective on the Potential Role of Geologic Option in a National Carbon Management Strategy.*

Lal, R., 1999. Soil management and restoration for C sequestration to mitigate the accelerated greenhouse effect. *Progress in Environmental Science*, 1(41): 307–326.

Lal, R., 2004. Review on soil carbon sequestration to mitigate climate change, *Geoderma*, 123.

Steven, A., 2002. National policies for biosphere greenhouse gas management: Issues and opportunities. *Environment*, 30(5).

Times of India, 2007. New unknown climate zones by 2100. *The Times of India*, 28 March.

ENHANCING CARBON SEQUESTRATION IN NATURE
Potential Technological and Institutional Mechanisms, Carbon Trading and Policies

Ecosystem Diversity and Carbon Sequestration: Climate Change Challenges and a Way Out for Ushering in a Sustainable Future *Pages 269–281*
Editors: P.L. Gautam, Vir Singh and Uma Melkania
Published by: DAYA PUBLISHING HOUSE, 2009

Chapter 29

Coastal Wetland Ecosystem in Sequestering Carbon Directly by Geological Repositories and Phytoplankton Fertilization: Workable Strategies for Maintaining Ecological Integrity

Alok Mukherjee

The higher rates of Carbon sequestration in salt marshes, soils of tidal salt and lower methane emissions make coastal wetlands more valuable C sinks than other ecosystems in a warmer world. With rise in sea level owing to global warming, landward expansion of coastal wetlands in gentle topography shall provide potential sink of atmospheric carbon dioxide. This terrestrial wetland carbon pool globally stores 44.6 $TgCyr^{-1}$, with an average Carbon sequestration of 210 g $CO_2\,m^{-2}\,yr^{-1}$.

Urbanization and industrialization of coastal areas with 40 per cent offshore oil production from Indian Ocean coastal states has called for capturing gases after combustion and reinjecting them into subsurface settings where they will be retained for geological period of time. CO_2 is removed directly from natural gas, point source (Sleipner process), by an active amine process and injected by pipeline system into permeable in use and abandoned oil reservoirs. It is added by sorption in residual shale and sludge. The use of CO_2 for enhanced oil recovery is the only known large scale use where CO_2 has an economic value that may be significant compared to the deposition costs. Elevated CO_2 enhances plant biomass in mangroves.

Sustainable fisheries release micro and macro-nutrients in coastal water which are helpful in fertilization of phytoplanktons, the primary member of foodchain.

Pure CO_2 stream is also released in coastal water depths. Bioshield of mangrove, casuarinas, salicornia, laucoena, atriplex, palms, bamboo, halophytes are carbon sinks. Workable strategies to maintain integrity of coastal ecosystem necessary to secure preservation of life support system and avoid drainage of wetland by conversion to agricultural and forest patches have been dealt with.

Introduction

Wetland ecosystem in a coastal region comprises of floodplains, mangroves, sea grass beds, salt marshes, lagoons, intertidal flats etc. For land-water interface they are the intricate characteristic. Carbon sequestration is the capture and long-term storage of carbon dioxide and other greenhouse gases that would otherwise be emitted into the atmosphere. Practices of carbon sequestration in coastal ecosystem have been dealt with in this paper. Four steps are generally taken for sequestering carbon in this ecosystem. Firstly, drainage of wetlands is avoided along lower topographical regions, used for agricultural purposes. Secondly, the mangrove and coral reef reclamation along coasts and land use is monitored. Thirdly geological sequestration is observed in deep saline aquifers, depleted oil and gas fields, unminable coal, peat measures and for enhancing oil recovery. Lastly for stimulating the growth of phytoplanktons which consume great amounts of CO_2, nutrients are added to the coastal surface waters.

Techniques

Before techniques are recommended, the basic situation and utility of wetlands have to be emphasized. Wetlands comprising of swamps, marshes, fens, bogs, estuaries and tidal flats have been hardest hit today. Globally, the world has lost half of its wetlands with most of the destruction having taken place over the past half century. The loss of these productive ecosystems is doubly harmful to the environment. Their importance as a highly productive ecosystem with a unique and diverse flora and fauna has been recognised. Wetlands not only store water and transport nutrients they act as natural pollution filters, soaking up and diluting harmful substances such as nitrogen and phosphorus from fertilizers, residues from pesticides such as DDT, heavy metals such as lead, cadmium and raw sewage from human settlements.

Wetlands have been drained for agricultural purposes and setting industries in coastal region as a result of urbanization. Many coastal wetlands have been converted into a salt-making flat near Bombay, India. Not only is there less water available for ecosystems, pollution from agriculture, industries and municipalities continues to increase, reducing the ability of freshwater systems to process and dilute it. The science of coastal habitat restoration has been advanced by setting navigation channels amongst eroding wetlands in coastal region and introducing pesciculture there. Land use change monitoring frequently is needed in the coastal strip to monitor the changes which have occurred within the wetland area. Housing, commercialization and density of vegetation or green patches can well be detected by CIR photographs or IRS 1C LISS-111, say at a spatial resolution of 23.5m. The morphometric drainage analysis has to be carried out with areal survey for orientation of channels.

Mangroves and coral reefs have a very special relationship and the location of mangroves in the tropics often run parallel to coral reefs. While mangroves thrive in brackish, nutrient-rich waters, coral reefs need clear, nutrient poor waters. Mangroves purify the water of silt, human waste and nutrients, providing the clean water that corals need to thrive. Coral reefs protect mangroves and seagrass beds in certain areas, which are the breeding and nursing grounds of various economically important fauna. Along Gulf of Mannar, from Tuticorin to Rameswaram Island are endowed with a combination of ecosystems including mangroves, seagrass, seaweeds and corals reefs. Different types of reef formations have also been observed along the periphery of the islands *viz.* fringing reef, patch reef and coral pinnacles. Carbon accumulation in soils can be greatly improved by reclamation of mangroves, conservation management, such as no-till and replanting with grasses. This carbon sequestration occurs because there is less soil disturbance and more carbon is added to the soil. Coral reefs avoid high intensity coastal erosion by subduing the tidal waves. Corollary benefits of carbon sequestration are increased soil fertility, reductions in erosion and increases in soil quality. Grading or plowing removes native plants like salt reed, smooth cord grass, glasswort, salt grass, sea lavender, cat tail, salt meadow grass and exposes subsurface clay. It not only destroys this habitat but also allows non-native or exotic plant populations to take hold. Plants, such as purple loosestrife and giant reed, that provide little food and habitat value for wildlife, are quick to colonize disturbed soils and are difficult, to remove.

Figure 29.1: Mangrove Forests of Sunderbans, India

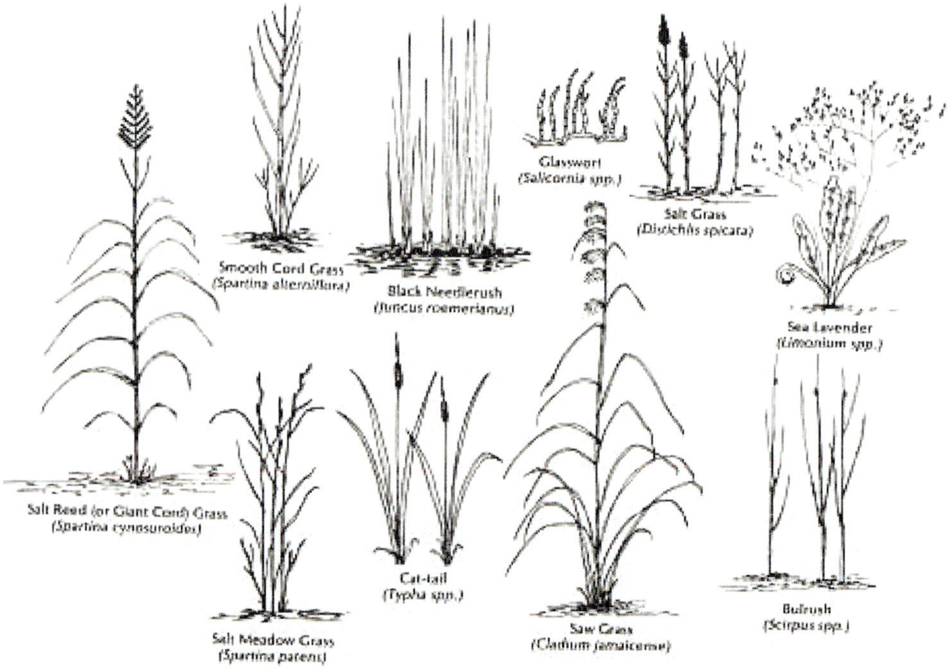

Figure 29.2: Coastal Wetland Plant Species

Figure 29.3: Coral Reefs Protecting Mangroves and Sea Grass Beds

Before I say about Geologic repositories and their potential for sequestration in coastal landforms within continental shelves, I must mention about the various ways of carbon storage and mode of geological facies. The process of Geologic Sequestration is carried out in (i) depleted oil fields, (ii) depleted gas fields, (iii) deep saline aquifers, (iv) unmineable coal measures and EOR (Enhanced Oil Recovery) mode.

The methodology used in Depleted Oil and Gas Reservoirs is the injection of compressed carbon dioxide into a depleted oilfield creating a CO_2 "flood" that forces the remaining oil into a well where it can be captured; the CO_2 remains behind, safely and permanently stored beneath the earth's surface. In Unmineable Coal Seams, coal beds typically contain large amounts of methane-rich gas. The current way to recover coal bed methane is to depressurize the bed, usually by pumping water out of the reservoir. An alternative approach is to inject pressurized carbon dioxide into the bed. The porous coal surfaces absorb CO_2 more easily than methane, so the CO_2 displaces the methane and remains sequestered in the bed. In Underground Saline Formations, carbon dioxide can also be pumped into underground saline, or brine, formations. Underground brine formations are so common that geologists believe they could provide enough space to store all the CO_2 captured from fossil fuels burned in the 21st century. CO_2 injection technology, works like this: from point source carbon dioxide is captured by amine process (Slipners), liquified, transported through pipelines to reservoirs. It is required that the reservoirs be of a depth of greater than

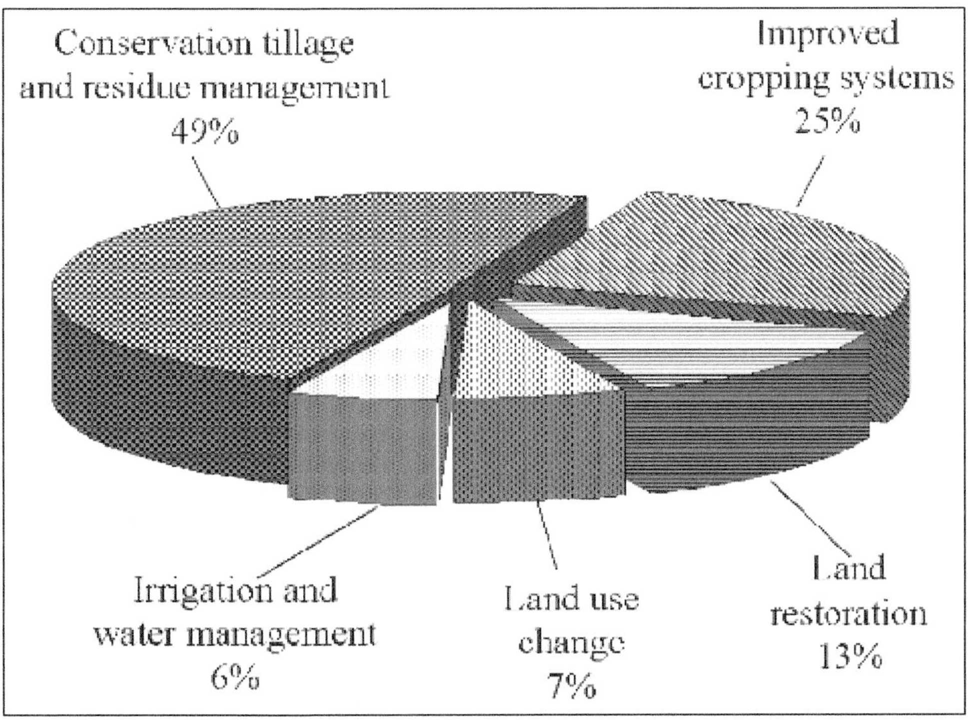

Figure 29.4: Various Forms of Conservation Management in a Coastal Ecosystem

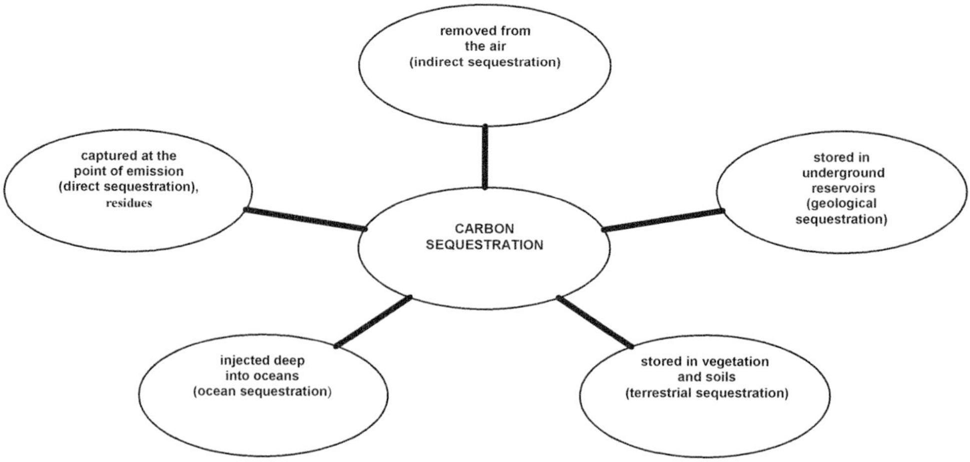

Figure 29.5: Various Techniques of Capturing Greenhouse Gases

800m, to allow the CO_2 to be in supercritical state. By this method the capacity of the reservoir can be increased. However, although the pressure needs to be high, the temperature must be low; those areas which have a low geothermal gradient would be desirable. The petroleum industry has long used CO_2 injection to get more oil from depleted reservoirs. If the pressure is high enough in these formations, the CO_2 and oil become completely miscible, leading to highly efficient oil recovery. At lower pressures CO_2 displaces oil without mixing together to form a single fluid phase. This too enhances recovery by reducing the oil's viscosity and by swelling, as some fraction of the CO_2 dissolves in the oil. While some of the carbon dioxide comes back up with the oil, much remains underground. Operations can be modified so that more of the CO_2 remains underground after the enhanced-recovery project is complete.

Salt cavern placement could acquire a priority sequestration position in Coastal region. Pure salt is essentially impermeable (<10-19 m^2). Also, bounding strata usually have pores plugged by crystalline salt for some distance beyond the massive salt bed. This reduces risk of CO_2 escape through porous media, leaving only the well bore and the possibility of roof rock fracturing as possible leakage paths from the cavern. Salt is naturally self-sealing by virtue of creep behavior. Fissures or fractures created by drilling and dissolution processes will heal and a sound permeability seal will be reestablished in a relatively short time frame. In some environments (*e.g.* western and eastern ghats, India), salt strata can provide storage sites where other suitable unfractured geological sites are not easily found. CO_2 can be stored in relatively shallow, cool salt caverns in supercritical form, allowing a far more concentrated storage capacity than storing CO_2 in solution, as a free gas, or through adsorption onto a substrate with micro-porosity, such as coal or oil shale. The rate of filling or emptying of salt caverns is not limited by porous media flow capabilities, only by the flow capacity of the tubular delivery system. The CO_2 can be quickly accessed and is available in pure form at any future time. Even if slow leakage occurs, there are several security mechanisms through which the CO_2 will be retained at depth in the geological media and not enter the atmosphere suddenly. Saline formations in coastal wetlands are layers of porous rock that are saturated with brine. They are much more commonplace than coal seams or oil and gas bearing rock, and represent an enormous potential for CO_2 storage capacity. Saline formations tend to have a lower permeability than do hydrocarbon-bearing formations, and work is directed at hydraulic fracturing and other field practices to increase injectivity. They contain minerals that could react with injected CO_2 to form solid carbonates. The carbonate reactions have the potential to be both a positive and a negative. These formations can increase permanence but they also may plug up the formation in the immediate vicinity of an injection well. Injection techniques are sought that promote advantageous mineralization reactions.

The Geological parameters below are studied before going for salt cavern or reservoir selection.

1. Depth
2. Permeability/Hydraulic Conductivity
3. Formation Thickness
4. Net Sand Thickness
5. Percent Shale
6. Continuity
7. Top Seal Thickness
8. Continuity of Top Seal
9. Hydrocarbon Production
10. Fluid Residence Time
11. Flow Direction
12. CO_2 Solubility Brine
    12a. Temperature
    12b. Pressure
    12c. Salinity
13. Rock/Water Reaction
14. Porosity
15. Water Chemistry
16. Rock Mineralogy

The long-term fate of CO_2 in a reservoir is monitored. For volumetric reasons, it is desirable that CO_2 is stored geologically in the supercritical phase. The critical point of carbon dioxide is 73.8 bars and 31°C which imply supercritical behaviour below depths of about 800 m in the geosphere. Supercritical CO_2 is lighter than formation fluids at these depths and will thus rise via buoyancy to the top of the reservoir. A CO_2 'bubble' may spread laterally beneath the surface of the sealing formation above the reservoir. With time, and with increasing contact area with the formation fluid,

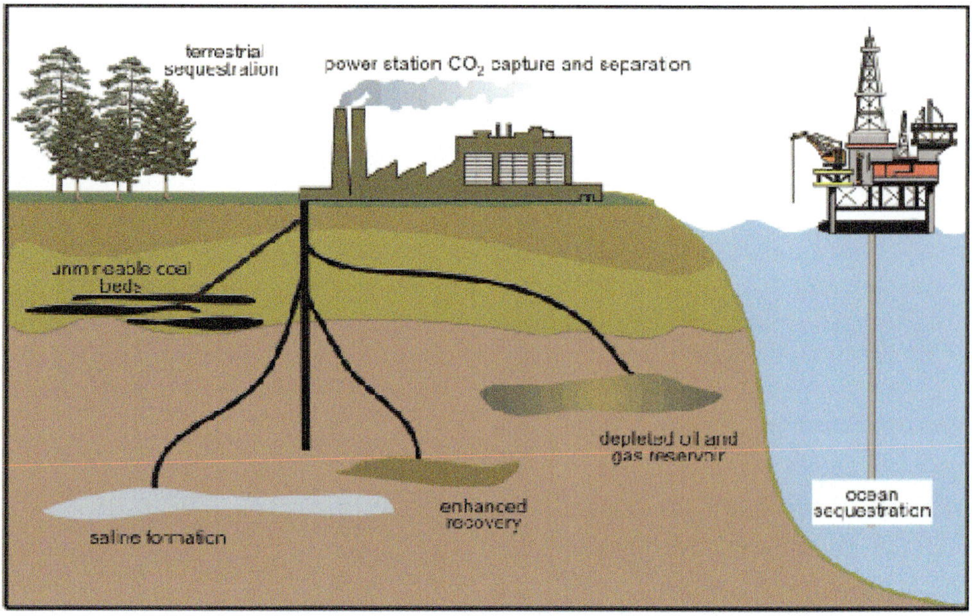

Figure 29.6

CO_2 will dissolve, such that dissolution into the aqueous phase may be complete over a timescale of a few thousand years. Reactions with minerals in the aquifer/reservoir system may also 'fix' CO_2 in the solid phase subsequently. Migration of the CO_2 may occur, either laterally and up-dip within the reservoir formation itself, or vertically through confining sealing formations. Driving forces include diffusion, buoyancy, and regional hydraulic gradients. The degradation of cement seals and the corrosion of metal liners of any boreholes penetrating the disposal reservoir are important issues.

One thing that has to be kept in mind and given prior thought is the The Potential and Hazards of CO_2 leakage. A 0.01 per cent annual leakage rate from 5000 Gt CO_2 stored worldwide in geological reservoirs would be 0.5 Gt CO_2 per year, which is considerably greater than the total annual volcanic emissions of CO_2. Leakage rates of 1 per cent imply that most of any stored carbon dioxide would return to the atmosphere after only 400 years.Carbon dioxide is a non-toxic gas, with ambient concentrations in the atmosphere of about 370 ppm. Their study has demonstrated that humans may tolerate concentrations of CO_2 up to 10000 ppm (1 per cent CO_2) with no apparent physiological effects. At concentrations between 3 and 5 per cent, there are effects upon the rate of respiration, and above 5 per cent, physical and mental capabilities may be impaired, with potential loss of consciousness. Exposure to atmospheric concentrations above 10 per cent results in loss of consciousness and possible coma or death. Sequestration is not a cheap deal. Power plants are not cheap to run: about 25 per cent of the energy they make is used just to keep them operating. Once the CO_2 arrives at its destination it must be compressed into a liquid so it can be injected into the ground—a step that typically consumes 20 per cent of the energy yielded by burning coal in the first place. Then a kilometre-deep hole must be drilled and the CO_2 injected. From that day on, the geological formation must be closely monitored; should the gas ever escape, it has the potential to kill. The largest disaster caused by CO_2 occurred in 1986, in Cameroon, central Africa. A volcanic crater-lake known as Nyos belched bubbles of CO_2 into the still night air and the gas settled around the lake's shore, where it killed 1800 people and countless thousands of animals.

Phytoplanktons approach is one of the most effective sequestering processes. Phytoplanktons recycle the other 50 per cent or more of the carbon dioxide and oxygen of Earth, and could grow rapidly near the surface of the oceans. Phytoplankton, like tree leaves and pine needles, take in carbon dioxide from the air and use the carbon to make plant structure, which is the very beginning of the food chain. If fertilized properly, they could multiply rapidly and may reverse the process of climate change. By enriching the coastal zones and oceans with iron on a large scale, enough plankton could grow to reduce atmospheric carbon dioxide and mitigate global warming. Later research on trace metals in polar waters showed coincidence between areas with elevated iron concentrations and high productivity, large plankton biomass, and increased nutrient drawdown. HNLC waters phytoplanktons display iron deficiency, preventing plankton blooms despite abundant nutrients. HNLC waters phytoplanktons display iron deficiency, preventing plankton blooms despite abundant nutrients. During ice ages, much of the world's water was locked into glaciers and the world becomes a drier, dustier place. During these periods, greater

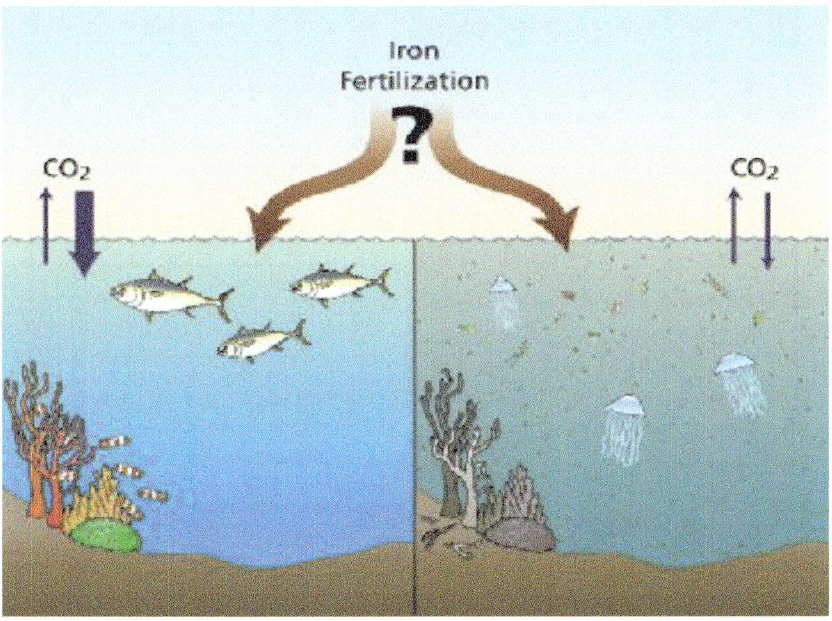

Figure 29.7

amounts of iron are swept into the oceans by wind, resulting in larger plankton blooms that assimilate large quantities of carbon dioxide, ultimately decreasing the atmosphere's ability to retain heat and prolonging ice ages.

Iron fertilization stimulates phytoplankton growth. The water mass fertilized with iron is found to turn into absorption area of atmospheric CO_2. The quickness of response, increase of stock and responding species of phytoplanktons are quite different among various experiments. Factors for controlling the response of phytoplankton to iron fertilization are:

As a result of the examination of the difference in phytoplankton response among the experiments, the following three factors were cited.

(1) Biological factor: the phytoplankton species responding to iron fertilization

(2) Physical factor: the depth of surface mixed layer

(3) Chemical factor: the retention time of fertilized iron

As future development the amount of carbon fixation by in-situ iron fertilization should be estimated more accurately by identifying the factors controlling phytoplankton growth and carbon fixation efficiency. Additionally, the amount of carbon fixation and the change in ocean environment by repeated iron fertilization should also be estimated with consideration of global ocean circulation. This can result in an increase in fish catch by a factor of 400 or more. A 53,000-square-mile ocean area might see the fish catch go to 50 million tons per year.

However the side effects need also to be expressed, the stimulation of the phytoplankton that we saw was a result in part of the uncoupling of the food web: the natural predators of these tiny plant cells did not have time to respond to their increased food supply and keep it in check. Yes, we know there were more phytoplanktons when iron was added, but they were different phytoplankton than were there before. The entire community shifted to a newer group of species, and we do not know what the consequences of this would be to the ecosystem in the long term. Different species at the base of the food web implies different species at the top. So if you fertilize for fish, you will probably get different fish than you intended. Will they be species that we want? Blooms occur on coastal areas because that's where nutrient enrichments occur. If we enrich the nutrients in the open ocean, we will have blooms there, too. And even if the blooms are not due to toxic species, not all species of algae favor copepod reproduction. Iron fertilization results in diatom bloom. Copepods feeding on diatoms have lower counts of eggs. And diatoms can clog the gills of marine animals. The release of dimethyl sulfide from plankton growth into the atmosphere might take place. This chemical oxidizes to form sulphate aerosols. These aerosols are highly hygroscopic and hence could increase cloud cover if introduced in sufficiently large quantities. The increased cloud cover could lend toward shifting the global radiation budget and lead to a cooling trend for Earths mean temperature. Iron fertilization is one of the cheapest means of sequestering carbon from the atmosphere that is currently known, running about $5.00 per ton. Many other methods considered, such as seaweed farming would cost upwards of $200.00 per ton. Fertilization could also generate greenhouse gases more potent than carbon dioxide.

Methane, for example, is released when phytoplankton decompose. Phytoplankton blooms also produce dimethyl sulfide, which alters cloud cover, and methyl halides that erode the ozone layer: "The possibility of unintended climate and atmospheric change is a crucial argument against ocean fertilization."

Although phytoplankton is one of the smallest organisms in our ecosystem, it is also one of the most important. Not only is it an indicator of environmental change, but it also serves as the base of the marine food chain. It is important to evaluate the effects of changes on the environment on phytoplankton because it supports so many creatures. A change in the smallest of organisms can foreshadow larger changes to come. Some of these changes have already been seen.

Lastly I should comment on the issues of distributive justice to be incurred in the sequestration process in a global scenario. Carbon credits need to be verified, monitored and certified according to certain requirements such as The Kyoto Protocol to be purchased, traded or sold. The current price of a metric ton of carbon sequestered is about \$5 to \$10. The world has changed and there is clear understanding that the rich and the emerging rich world needs to make the transition to a low carbon economy. After the Kyoto targets expire in 2012. Say each country on a 'per capita' basis would be allocated an emission cap–an 'assigned amount' of emission permits–totalling some target per capita amount (a fraction of today's global average emissions per person) multiplied by the country's (present day) population. For example, in India–whose current emissions per inhabitant are much lower than the world average–this would entail a considerable surplus of emission permits. And as long as there are surplus permits, India would hence not be forced to introduce emission mitigation measures to stay within its assigned amount. Indeed, under an international trade in

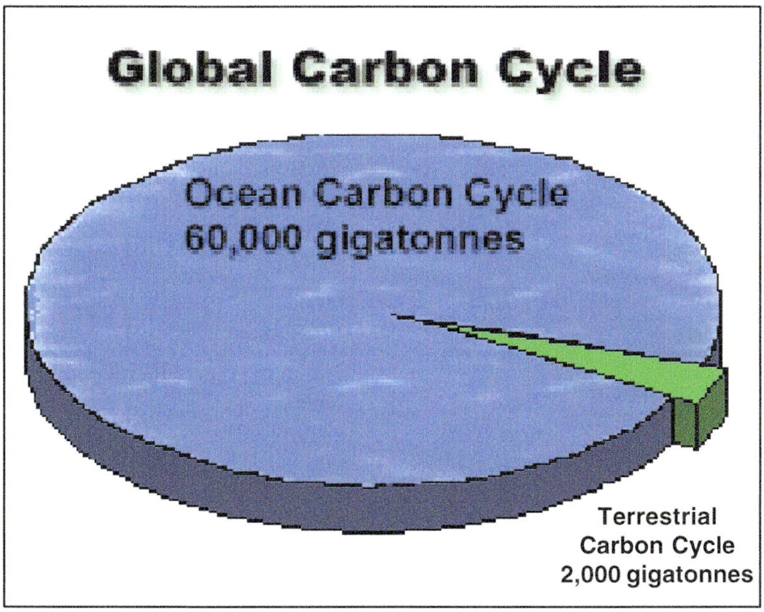

Figure 29.8

such permits, India could legitimately earn significant export revenues from the sale of these surplus permits. Anyone emitting more than the agreed average target would occupy part of the Indian ecological space of someone in India who is emitting less. Carbon-rich either reduce their carbon footprints to give the carbon-poor their fair share of the domestic ecological space, or pay an appropriate compensation.

Conclusion

After a vivid description of sequestering carbon dioxide in Coastal Wetland ecosystem, it may be concluded that since a vast population of the world lives in coastal region, and Indian Ocean adjoining regions have 40 per cent of offshore drilling, in-situ CO_2 sequestration can be practiced. The environment has to be made healthier with more fresh oxygen. Today time has come for developing countries to either look for carbon credits and get more foreign exchange, by practicing afforestation or ask for cheaper technology from developed countries. We can either reduce emission or sequester carbon or develop alternate sources of energy. It appears as if all are spoken at random in the same platform.

References

Bachu, S., 2000. Sequestration of carbon dioxide in geological media: Criteria and approach for site selection. *Energy Conversion and Management*, 41(9): 953–970.

Bachu, S., 2001. Geological sequestration of anthropogenic carbon dioxide: Applicability and current issues. In: *Geological Perspectives of Global Climate Change*, (Eds.) L. Gerhard, W.E. Harrison, and B.M. Hanson. AAPG Studies in Geology, 47: 285–304.

Beecy, D., Kuuskraa, V. and Schmidt, C., 2001. A perspective on the potential role of geologic options in a national carbon management strategy. In: *First National Conference on Carbon Sequestration*, Conference Proceedings, Available at http://www.netl.doe.gov.

Behrenfeld, M.J., Bale, A.J., Kolber, A.S., Aiken, J. and Falkowski, P.G., 1996. Confirmation of iron limitation of phytoplankton photosynthesis in the equatorial Pacific Ocean. *Nature*, 383: 508–511.

Benson, S., Apps, J., Hepple, R., Lippman, M. and Tsang, C.F., 2002. Health, safety and environmental risk assessment for geologic storage of carbon dioxide: Lessons learned from industrial and natural analogues. Lawrence Berkeley National Labs, 2002. Available at http://www.co2captureproject.org/reports/reports.html.

Benson, S.M., Gasperikova, E. and Hoversten, M., 2004. Overview of Monitoring Techniques and Protocols for Geologic Storage Projects. *IEA Greenhouse Gas R&D Programme Report*, PH4/29. 99 pages.

Boyd, P.W. *et al.*, 2004. The decline and fate of an iron-induced subarctic phytoplanktons bloom. *Nature*, 428: 549–553.

Environmental Challenges and Greenhouse Gas Control for Fossil Fuel Utilization in the 21st Century. (Eds.) M. Mercedes Maroto-Valer *et al.*, Kluwer Academic/

Plenum Publishers, New York, 2002: *Sequestration of Carbon Dioxide by Ocean Fertilization*, pp 122. by M. Markels, Jr. and R.T. Barber.

Holt, T., Jensen, J.I., Lindeberg, E, 1995: Underground storage of CO_2 in aquifers and oil reservoirs. *Energy Conv. Mgmt.*, 36: 335–338.

Intergovernmental Panel on Climate Change: An Introduction to Simple Climate Models Used in the IPCC Second Assessment Report, Feb. 1997.

IPCC, 2005. *IPCC Special Report on Carbon Dioxide Capture and Storage*. Prepared by Working Group III of the Intergovernmental Panel on Climate Change, (Eds.) Metz, B., O. Davidson, H.C. de Coninck, M. Loos, and L.A. Meyer. Cambridge University Press, Cambridge, United Kingdom and New York, NY, USA, 442 pp.

Lindeberg, E., 1997. Escape of CO_2 from aquifers. *Energy Conv. Mgmt*, 38(Supplement): 235–240.

Lindeberg, E. and Holloway, S., 1999. The next step in geo-storage of carbon dioxide. In: *Greenhouse Gas Control Technologies*, (Eds.) Riemer, P., Eliasson, B. and A. Wokaun. Elsevier Science Ltd., pp 145–150.

Martin, J.H. and Fitzwater, S.E., 1988. Iron deficiency limits phytoplankton growth in north-east Pacific subarctic. *Nature*, 331: 342–243

Solomon, Semere, 2006. Carbon dioxide storage: Geological security and environmental issues case study on the Sleipner Gas Field in Norway. The Bellona Foundation. Retrieved November 7th, 2006, from http://bellona.no/filearchive/fil_Paper_Solomon_-_CO2_Storage.pdf

Tsuda, A., *et al.*, 2003. A meso-scale iron enrichment in the western sub-arctic Pacific induces large centric diatom bloom. *Science*, 300: 958–961.

Ecosystem Diversity and Carbon Sequestration: Climate Change Challenges and a Way Out for Ushering in a Sustainable Future *Pages 282–293*
Editors: **P.L. Gautam, Vir Singh and Uma Melkania**
Published by: **DAYA PUBLISHING HOUSE, 2009**

Chapter 30

Carbon Sequestration: Mitigating Environmental and Socio-economic Impacts of Global Warming and Climate Change

Vikram S. Rathore

Global warming, *i.e.* slow but steady rise in the mean global temperature in the post-industrial years (1850 onwards), is the consequence of accumulation of "greenhouse gases" (carbon dioxide, methane, nitrous oxide, and CFCs) in the atmosphere emitted by indiscriminate use of fossil fuels, modern agricultural practices and other human developmental activities. "Greenhouse gases", by virtue of their capacity to absorb outgoing infrared radiations, create a blanket over the atmosphere, thus causing rise in the temperature. This has caused an increase of about 0.74°C during past century. The United Nation's Intergovernmental Panel on Climate Change (IPCC-IV 2007)), in its latest report, has predicted a further rise in the global mean temperature up to 3-5°C by the end of 21st century. The IPCC report has predicted drastic changes in climatic factors which would vary from region to region. The earliest effects of this, some of which are already being registered, are melting and shrinking of mountainous glaciers and permafrost causing rise in ocean and sea levels. Rivers which draw water from the glaciers would initially overflow, followed by their drying out water scarcity as the glaciers disappear. Major river deltas around the world may turn into infertile areas, with the consequent decline in agricultural productivity. Desertification would intensify in many areas. Rise in sea levels would engulf coastal mega cities, forcing human population there to migrate (eco-refugees). Natural forests, grasslands and other vegetation may change drastically, with appearance of new species assemblages. An estimated 34 to 50 per cent of present biodiversity may become threatened or go extinct. Rise in ocean temperatures and U.V. radiations would cause bleaching and extinction of many coral species. Island and coastal nations would be hit hard by flooding, and some may disappear.

Increased microbial and chemical pollution would cause outbreaks of vector-borne diseases and environmental stresses would lead to an increase in psychological disorders. Human infrastructure and industries, like energy production and distribution, telecommunication, transport and navigation, fisheries and food processing, would suffer gravely along with loss of cultural heritage, natural wonders of the world and socio-economic fabric of affected countries. The IPCC predicts an apocalyptic scenario in the next few decades, if remedial measures to reduce the emissions of greenhouse gases, or to maintain it at the present level, are not initiated without any loss of time.

A number of options have been proposed to meet the challenges posed by the global warming and associated climatic changes and their impacts. One group of these measures includes technology development for more efficient production (coal gasification, oxy-fuels, etc.) and transmission of energy, modifying manufacturing processes, use of alternate renewable energy sources, more efficient refrigeration, heating and domestic appliances, vehicles and transport systems. On the other hand, carbon sequestration, which refers to long-term storage of carbon in natural ecosystems (forests, soil, and oceanic biomass) or by artificial means (injection of carbon dioxide in deep sea layers, mineral sequestration, geological sequestration) offer vast potential for reducing further build up of carbon in the atmosphere. Recent advances in genetic engineering and tissue culture techniques have opened door for development of novel forest trees and phytoplankton with higher photosynthesis and carbon sequestration potential.

Introduction

The term "Greenhouse Effect" was coined by J Fourier (1827) because it mimics the "heat trapping" effect of glass walls of the greenhouse which block the outward passage of infrared radiations, causing rise in the temperature of its inside air. In the same manner, in case of open atmosphere, higher levels of certain gases, which absorb infrared radiations, trap the heat close to the earth surface, which would otherwise escape to the outer atmosphere. The net result of this is a progressive rise in the earth's mean temperature (Global Warming). The gases which are responsible of this warming process are collectively called as "Greenhouse Gases". The main greenhouse gases causing global warming are carbon dioxide, methane, nitrous oxide, chlorofluorocarbons and ozone. Amongst all these gases, carbon dioxide is the most important because it contributes maximally to the overall global warming (up to 59 per cent) and has a long life (residence time). In the pre-industrial years (1750-1800 A.D.) the concentrations of the carbon dioxide in the atmosphere was only 280 ppm, and other greenhouse gases were also present at sufficiently low levels to cause any significant global warming. However, the advent of industrial era (post 1800 A.D.) resulted in steady rise in the atmospheric concentration of greenhouse gases, particularly CO_2, which has reached an all time high level of 379 ppm. Level of methane has also doubled since pre-industrial level (from 0.79 ppm to 1.7 ppm). CFCs, which were non-existent in the atmosphere in pre-industrial years, have reached average level of above 300 ppm. This unprecedented rise in the atmospheric levels of "Greenhouse Gases" is mainly due to various types of industrial, agricultural and other developmental activities in the post-industrial years. Rise in CO_2 levels has been due to indiscriminate use of fossil fuels (coal, petroleum etc) for energy generation, automobiles, railroads, and biomass burning. There is a close parallel between global man-made CO_2 emissions and atmospheric CO_2 during the past two centuries particularly during the past century. Rise in methane levels is attributable to surge in rice-paddy culture, wet-lands and land fills to store industrial wastes, and increased biomass burning. Major sources of nitrogen oxides in the atmosphere in the past century are use of nitrogenous fertilizers, super sonic air crafts, biomass and fossil fuel burning. CFCs are the by-products of refrigeration and other industrial process (Anid and Sombrock, 1994). Thus, there is strong evidence that past human activities in these spheres are directly responsible for accumulation of "greenhouse gases" in the atmosphere and their combined effect on global warming. The latest report of IPCC in February 2007 clearly states that it is almost 90 per cent certain that mankind was to blame for most of warming in the last half-century.

Global Warming and Climate Change

A slow but steady rise in global mean temperature of even 1-3°C would have serious effects on the regional climates. The effects would not be uniform but vary from region to region, depending on its geographical location, latitude, longitude and altitude. Some regions would be more affected than the others. However, precise information on this aspect is presently not available, and most predictions are based on model studies under various scenarios of rise levels of greenhouse gases, particularly CO_2, temperature increase and other contributing factors. However, taking into account results of various studies and, the IPCC group II in its report projected an increase in global mean surface temperature of about 1-3.5°C by 2100 A.D. A more recent report of IPCC group IV has predicted even a higher increase in global mean temperature of about 1.4-5.8°C by the end of this century. Whatever the actual rise, it is certain that global mean temperature are on the rise and that would impact many climatic-associated factors regionally, including, intensity, time and duration of precipitation, There would be changes in frequency, intensity, timing, rate of change, distribution and other characteristics of extreme climate,events like drought, floods, storms, heat/cold waves (Obasi 1996). These climatic changes are predicted to be associated with major shifts in world's climatic zones including arid, semi-arid, sub-humid, humid, etc. (IPCC II 2001). The overall impact of global warming and associated climate change could be apocalyptic and may bring about many disastrous changes in earth's environment, natural ecosystems, human populations, socio-economic fiber, human health, depending upon the magnitude and rate of change in global warming and climate.

Impacts of Climate Change on Environment

Climatic changes caused by rise in global mean temperature would induce many changes in various components of earth's environment, as most of them are sensitive to climate change Important among these are described below briefly- based on the predictions made by U.N. sponsored International Panels on Climate Change–Group I–IV during the last two decades (1988-2007), and other studies undertaken on global warming and its possible impacts by scientists all over the world:

Hydrosphere

Global rise in temperature would lead to most serious impacts on the distribution of water in the hydrosphere. Rising temperatures would cause melting/shrinking of world's mountainous glaciers. The IPCC Report (2007) has predicted that if the present rate of rise in temperature continues, Himalayan glaciers would shrink to almost one-fifth of their present volume, and could disappear completely by the end of this century. Significant shriveling of many Himalayan glaciers including Pindari, Bara, Shigri, Dorian, Meola, Sonapani, Milam, and Zemu glaciers, ranging from 10-36 meters, has been recorded. Other major glaciers of the world, *e.g.* Andean Glacier in South America, would meet the same fate. Permafrost (permanent sheet of ice on land) is also predicted to melt in higher latitude Canada and Siberia. A recent NASA study has revealed a decline of 7-10 per cent decline per decade in the arctic permafrost.

Glacial and permafrost melting would lead to increased seasonal flow of rivers fed by the glaciers, leading to flooding in delta areas. Most vulnerable deltas around the world include: Ganges-Brahmaputra, Godavari-Krisna, Mahanadi deltas in India; Shatt-el Arab delta in Saudi Arabia; Changjiang, Zhujiang Chaophrya in China and S.E. Asia; Senegal, Niger and Nile in Africa; Rhine in Europe; Mississippi and Grijalva in North America and Amazon and Sao Francisco in North America continents.

It is predicted that for the next 3 decades, there would be a sharp rise in the water levels of the glacial rivers, but as the glaciers disappear, the rivers would go dry or carry only small volumes of water. This would cause severe water scarcity and drought in river deltas and valleys. The soil fertility would deteriorate due to lack of silt deposits by the rivers. As the water tables go down, a severe scarcity of water would ensue.

Initial melting and increased flow of water through the feeding rivers is certain to witness significant rise in sea levels The latest IPCC report (IPCC 2007) has predicted a rise of 15-59 cm in sea level by the turn of century. However a recent study by U.S. Geological Survey, using a new mapping technique, states that sudden surge of up to 30 meters in sea level due to universal melting of glaciers can not be ruled out.

Rise in ocean and sea levels would inundate many coastal areas and island country and force large scale migration of human settlements there (discussed in more detail later). Warming of ocean waters due to increasing temperatures would expand the sea water thereby further raising sea levels and changing oceans circulation.

Natural Ecosystems and Biodiversity

Natural ecosystems are repositories of all living organism and natural resources on the planet earth. However, most ecosystems are extremely sensitive to climate changes. Rise in temperature, precipitation patterns and other climatic factors would severely influence both terrestrial and aquatic ecosystems. And thereby biome distribution on the earth could be greatly y altered in the next century. Even a sustained rise of 1° C would be sufficient to significantly change the existing forest ecosystems. Growth and rejuvenation of many forest species, would affect the functions and composition of many forests. Doubling of CO_2 levels with consequent changes in temperature and water availability, has been predicted to change at least one-third of broad leaf vegetation, with greatest change occurring at high latitudes and minimum in the tropical regions (Obasi 1996) Disappearance of some existing forests and establishment of new ones with altered species composition and assemblages could take place. Rangeland and grassland ecosystems would change, depending on their geographical location. For tropical rangelands, a change in temperature is not predicted to cause major changes in species composition and productivity, except lowering of nutritional value of forages due to increase C:N ratio. However, temperate grasslands would be subjected to changes in growing season and boundary shifts. Disappearance or shrinking of mountainous glaciers, permafrost and snow cover, could lead the vegetation to shift to higher elevations, while some species could become extinct due to lack of suitable habitat according to IPCC II (2001).

Aquatic ecosystems around the world would experience major influence of global warming and associated climate changes. Some lakes in mountainous regions may burst due to excessive amount of water from glaciers, as is being seen in some lakes in Bhutan. However, biological productivity of high altitude, in general, would increase due to increased temperatures, and the same could happen for temperate lakes. However, in low altitude lakes, rise in temperature could lead to extinction of non-tolerant species.

Change in global temperature, precipitation and rise in sea levels, would result in marked changes in geographical distribution of existing wetlands of the world. Coastal ecosystems like mangroves and salt marshes would be severely affected by increase in sea levels, erosion of shores, increase salinity, coastal flooding, microbial contamination, and nutrient availability. Coastal flooding may completely wash out some mangroves, salt mashes and reeds. Loss of habitat may make it difficult for the costal wet lands and mangroves to re-establish.

Climate variations and changing rainfall patterns would result in extension of deserts and many new areas converting to desert, due to increased temperatures and lack of precipitation. Deserts are expected to become hotter and drier, affecting many desert species. In some areas, irreversible desertification may occur, due to unfavorable climatic and soil conditions (*e.g.* soil compaction and erosion).

Agro-ecosystems

Agro- ecosystems are the man-managed eco systems for the production of cereals, pulses, oil-seeds, vegetables, fibers, medicinal plants and host of other economically useful species. Rise in temperature, precipitation (periodicity, intensity, duration) changes would affect agricultural production, depending on the region (Chen 1990; Perri and Swaminathan 1992). Sub-tropical and tropical areas would witness most serious impacts on crop productivity. There is evidence that even slight changes in temperatures at specific growth stage of the crop such as wheat, sunflower etc. could significantly reduce the yield as well as the nutritional quality of the produce. (Terannum and Sullivan 1989; Subrahmaniyam and Rathore 1994). Available research data indicate that even a 0.5°C rise in winter temperature would reduce wheat yield in India by 15-20 per cent. Overall agricultural productivity of the world may register a decline of 25 to 30 per cent in the next three decades. The first report of

IPCC Group VI (2007) has predicted that in some parts of Sub-Saharan region, agricultural yields may be reduced to one half, while failure of rains could devastate cotton production in Pakistan. India may suffer also in agricultural productivity in the Gangetic plains due to rising temperature and scarcity of water resources by 2030. Higher temperatures and less severe winters may, however, improve agricultural productivity in high latitude regions (Canada, Russia and Scandinavian countries) Rise in sea temperature and sea levels would have adverse impacts on reproduction, migration and productivity of fish and other sea foods (Anid and Sombrock 1994) However, high latitude fisheries could register positive impact due to faster growth rates and decrease in winter temperatures. Cattle and dairy productivity would be adversely affected due to scarcity of forage, water and lack of proper sanitary conditions in flooded areas.

Biodiversity Loss

Global warming and climate change are going to have significant adverse impact on world's flora and fauna. Most vulnerable species are those habituating coastal areas, island countries and desert species. Some of the predicted effects include (*i*) 30 per cent of world's coral reefs are at risk due to rise in temperature of water, increased U.V. radiations, and acidification, according to *Wikipaedia–The Free Encyclopedia*; (*ii*) mangrove vegetation species such as in Sunderbans of Eastern India may be completely lost by rising seas; (*iii*) increasing sea levels from melting glaciers in high latitude regions would threaten many fish species; (*iv*) sea level rise would threaten Caribbean Hawksbill turtle populations; (*v*) bleaching and death of organisms which keep the coral reef intact and alive is predicted in Indian and South African coasts; (*vi*) largest tiger population in Sunderbans in east Indian coast faces a great risk of extinction; (*vii*) many species of East African Marine Region are threatened by rise in temperature and U.V. radiation; (*viii*) desert species living at the limit of heat tolerance may become threatened or extinct due to rise in temperature and drier conditions; (*ix*) Adele penguins in arctic are threatened due to loss of their major food (krill and shrimp-like invertebrates) by melting glaciers; and (*x*) top biodiversity hot spots in the world would lose many rare and indigenous species.

Impacts of Climate Change on Human Population and Socio-economic Fabric

Human Population

Human beings will suffer in innumerable ways consequent to rise in temperature and climate changes (Chen 1990). Most vulnerable regions are human settlements in coastal areas and island countries. It is feared that rising sea levels would prove extremely disastrous for about 1 billion human population, which lives in low lying areas. A sea level rise of 16 feet has been predicted to affect 669 million people and would flood 2 million square km of land. A 100 feet rise in sea level would prove extremely disastrous. Millions of human being living in mega coastal cities and low lying areas would be forced to migrate to safer areas as "eco-refugees", creating scarcity of residences, food and other basic necessities. IPCC-IV report has predicted that if the current trend of greenhouse gases continues unabated, about one-third of human population (about 3.3 billion) around the world would be short of water, and about 600 million people would face shortage of, food.

Climate changes would result in many adverse impacts on human health. Cardio-vascular and cardio-respiratory mortality and illness are feared to rise. Rise in temperature would induce higher psychological disorders, Increase in vector-borne infectious disease like dengue, malaria, yellow fever and viral encephalitis are predicted to increase due to increased geographical range and more favorable weather conditions for vectors. Increase in air pollution, pollen grains and moulds would lead to greater incidence of allergic disorders, morbidity and mortality. Shortage of drinking water and food, would add to human misery (IPPC-II 2001).

Infrastructure

Global warming, rise in sea levels, flooding, and other climate changes, combined with problems of human migration, would have many negative impacts on many industries and infrastructure built over the decades. Most serious negative impacts would occur on energy production and transmission, construction, road transport, sea navigation, food processing, dairy, fishery, tourism and insurance industries. Cultural assets and heritage, particularly those situated in coastal and low lying regions, could suffer irreparable loss. The IPCC-IV Report has estimated that global GDP may decline by 3 per cent in the next 25 years, even if GHGs are stabilized at the present levels. As succinctly stated by Dr. R.K. Pachauri, Chairman, IPCC group IV and Director of the New Delhi-based Tata Energy and Resource Institute, "It is the poorest of the poor in the world who are going to be worst hit by the impacts of global warming and climate change".

Indian Scenario

India, along with other developing Asian countries, faces very high risk of flooding, water shortages, hunger, and human health in the next few decades, as the impacts of global warming unfold themselves. Mega cities and towns along the eastern and western coasts of the countries may be devastated by rising sea levels (up to 40 cm rise by the end of this century), wind storms and hurricanes. Some impacts of this nature are already happening, *e.g.*, increased rise of sea levels in Bay of Bengal, high speed cyclones in Andhra Pradesh and Orissa coasts, and rapid melting of many Himalayan glaciers. Food grain productivity of northern India, particularly of wheat, could decline by 30 per cent in the next three decades. Per capita water availability in the country may register 30 per cent drop by middle of this century. Vector-borne diseases like dengue, malaria and viral encephalitis, may assume epidemic proportion. Hydroelectric power generation from Himalayan Rivers in Northern India would suffer as the flow of water is considerably reduced by rapidly melting of glaciers. Temperature rise, coastal flooding and loss of habitat in many areas of the country could result in extinction of almost one-quarter of country's flora and fauna. Increase in ocean temperature and acidification would lead to bleaching and loss of coral reefs and shells along the country's coastal seas. Fish and other marine life are also endangered by these changes. Coastal mangroves and salt marshes could be completely destroyed by oceanic flooding. Transportation and sea navigation may be hit hard. Mumbai is predicted to be one of the most seriously affected cities in the world as revealed in a study by a team of scientists at University of Columbia, U.S.A. The high vulnerability was attributed to poor drainage and location of Mumbai in flood prone zone, along with poor stability of landfills on which the city has been built. Being home to many industries and considered as financial capital of India, the projected impacts on Mumbai would have a serious effect on national industrial production, GDP, and FDI.

Mitigation Measures for Global Warming and Climate Change

The most important question that arises, considering the world-wide devastating impacts of global warming and climate change in the next two to three decades and continuing till the end of this century and even further, is how can we proceed to control the global warming, if not completely check, and mitigate the catastrophic consequences. The report of IPCC-II is pertinent in this matter which concluded that "Significant reductions in net greenhouse gas emissions are technically possible and can be economically feasible. These reductions can be achieved by utilizing an extensive array of technologies and policy measures that accelerate technology development, diffusion and transfer in all sectors including energy, industry, transportation, residential, commercial, and agriculture/forestry sectors". The IPCC-IV report (2007) has also emphasized that we have only less than a decade to stabilize greenhouse gases in the atmosphere and arrest further rise in temperatures beyond 2°C from that in pre-industrial era. The report has estimated that it would require not more than 0.1 per cent of the world's GDP per annum to develop necessary technology to arrest the impacts of global warming and climate change. The next important question is: What are the technologies and processes which can bring significant reduction in global warming and what is their present state of development? Further, what policy measures are being taken at the national and international level to achieve these goals? These aspects are briefly outlined in the following account, based on the IPCC reports and other available literature on the subject as also mentioned in the US Department of Energy Report on Global Warming and Climate Change (Schuman 2002):

Two broad groups of technologies and processes have been proposed to achieve reduction in further build up of greenhouse gases or stabilize it in the near future, particularly CO_2, in the atmosphere and arrest rise in temperatures. One of these groups of measures advocates development and use of technologies and processes which would reduce emission of CO_2 and other greenhouse gases in the atmosphere. The other group of technologies is based on development and utilization of processes based on carbon sequestration, *i.e.*, removal of carbon from the atmosphere by a variety of artificial and natural means and sequestering (long-term storage) of carbon into various sinks.

Reduction in CO_2 Emissions in the Atmosphere

A number of technologies have been advocated to reduce emission of greenhouse gases in the atmosphere, particularly CO_2, the most important GHG applicable mostly to energy generation, transport, and energy utilization in industrial and home energy consumption sectors (Schuman 2002). Important among these include: (*i*) Alteration in power generation technologies including reduction in use of fossil fuels, switching from coal to natural gas which has lowest CO_2 emission per unit energy produced, decarburization of fuels and flue gases, switching to coal liquefaction and oxy-fuel technologies, development of alternate/renewable energy resources like solar, wind, hydro, nuclear, geothermal, biomass, tidal etc. These are already contributing about 20 per cent of world's total primary energy production; (*ii*) improvement in energy use efficiency particularly in industries by means of new technologies *viz.*, cogeneration, cascading, more energy saving motors and electronics, material recycling, etc.; (*iii*) modifying manufacturing processes through use of alternative feedstock, solvents, halocarbon refrigerants, etc.; (*iv*) development of energy efficient and non-fossil fuel based transport systems, including hybrid automobiles, use of lighter and smaller and air-resistant vehicles and transportation systems; (*v*) development of alternative agricultural practices to reduce methane emission from flooded rice-paddy cultivation, and ruminants, nitrogen oxide emissions from nitrogenous fertilizers and soils; (*vi*) change in life styles, particularly in advanced countries; (*vii*) development of more efficient electronic gadgets, insulation devices; and (*viii*) better construction technology for more efficient heating and refrigeration of buildings, water supply lines, lighting and home appliances etc. These are some of the many technologies and processes by which modern civilization could bring reduction in greenhouse emissions significantly and at the same time enjoy the amenities of life without endangering its own future. The US Department of Energy Report on Global Warming and Climate Change suggests a paradigm shift in our thinking and psychology about the ecological inter-relationships among different components of the environment, and the need to maintain the fine balance of the nature.

Carbon Sequestration Technologies

Carbon sequestration technologies depend on long-term storage of carbon emitted by various human activities into various environmental components (sinks, or the carbon reservoirs in which the carbon concentration is continuously increasing) to check further accumulation of GHGs in the atmosphere and consequent rise in global temperatures (Lal and Kimble 1997; Schuman 2002). This includes both artificial capture and storage of CO_2 (artificial sequestration), as also improving carbon storage in natural sinks (natural sequestration). A brief description of these two types of sequestration and their potential for mitigation of global warming follows:

Natural Sequestration

There are three important sinks for natural carbon sequestration, (Lal and Kimble 1997; Schuman 2002) namely (i) Forests, (ii) Oceans, and (iii) Soil. The terrestrial ecosystems (forests and soils) contain 75 per cent of the total carbon on earth, with 53.6 per cent C in soils and 21.4 per cent in forests and other vegetation and, therefore, are most important sinks for carbon sequestration. Oceans account for the remaining 25 per cent C on earth.

Forests

Total forests and woodland area of the world in the year 2000 was estimated at 4.17 billion hectare and permanent pastures at 3.14 billion hectare. Significant carbon sequestration can be achieved by improving the management practices of forests and rangelands (Lal and Kimble 1997) including, (*i*) conservation of existing forest covers and slowing down deforestation, (*ii*) tree plantation and regeneration of natural forest areas, (*iii*) promotion of agroforestry and social forestry in rural areas, *iv*) management of forest and pasture soils and better fertilizer use efficiency, and (*v*) degraded land use for forestation and pastures (more efficient management of nurseries and sapling distribution practices.

Carbon sequestration through growing trees is particularly attractive because it is relatively inexpensive and cab be practiced on a large scale. Therefore, this approach has received international support in formulating climate policies. Besides, more extensive use of lumber in building construction, greater use of timber and forest industry by-products as fuels to replace fossil fuels could also contribute significantly in carbon sequestration and reducing GHG emissions. Furthermore, use of new forest species having higher photosynthesis and lower photorespiration rates by using transgenic and tissue culture technology could result in greater carbon sequestration by forest ecosystems. India, with about 20 per cent of its land area under forest cover, could utilize this resource for achieving its carbon sequestration goals.

Soils

Soils of the earth contain the largest pool of organic carbon in the form of humus. In grassland, however, most of the organic carbon is retained by roots. Soils also contribute to CO_2 pool of atmosphere through practices such as ploughing, burning, grazing etc. Management practices have been found to promote carbon conservation. As the soils contain more than one-half of the total carbon on earth, they also offer greatest potential for C-sequestration. Many soils and crop management practices have been found to result in enhanced carbon conservation in soil. Enhancement of carbon sequestration requires increment in the passive or non-labile soil organic carbon (SOC) pool. This can be achieved by soil surface management, soil fertility and water conservation. Conservation tillage, *i.e.* tillage by means other than by ploughing, has been found to be an effective practice for soil carbon sequestration through increasing micro-aggregations and pushing soil organic carbon to deeper sub-soil layers (Henderson 2000). Grazing has also been found to increase SOC in upper layers of the soil in several rangelands (Le Cain *et al.*, 2000 and Rice 2000). Increase in SOC has been reported by annual burning and grazing (Reader, Schuman and Bowman 1998). Application of N and other nutrients to rangeland also increase SOC by increasing litter production (Sampson and Scholes 2000).

Microbial population of the soil also plays an important role in determining SOC through interaction between microbial community and soil properties (Anid and Sombrock 1994). Other agricultural practices found to improve C-sequestration in soil include residue mulching, cover cropping and crop rotation.

Oceans

Oceans are the largest active CO_2 sink on earth. The CO_2 in the oceans is driven by different CO_2 solubility in sea water and its circulation (solubility pump), on one hand, and biological processes (biological pump) that transport CO_2 from outer euphotic zone to interior of ocean, where it is finally converted to oil and natural gas at sea floor (Lal and Kimble 1997). Since oceans contain about one-fourth of total carbon on earth, it is a potential sink for CO_2-sequestration. Enhancement in sequestration efficiency of oceans is, therefore, a desired goal. This could be archived by increasing stimulating photosynthetic carbon fixation and growing the plankton in the euphotic zone (photosynthetically active zone). Since sea waters are generally deficient in iron, addition of hematite (micron size iron ore or ferric sulphate) has been shown to enhance plankton growth in certain seas. It has been found that addition of each iron atom results in enhancing sink for ten thousand to a hundred thousand carbon atoms in oceans (Lal and Kimble 1997). Addition of other nutrients particularly nitrogen and their effects on carbon sequestration in coastal waters needs to be explored. Research is also needed for selection of efficient CO_2 sequestration micro-algae and biofixation.

Radiotracer technique using carbon-14 labeled bicarbonate as substrate offers efficient method for evaluation of C-sequestration efficiency of oceanic phytoplankton and micro algae. Transgenic technology needs to be extended for obtaining more effective sequestration strains and species for enhancement of oceanic C-sequestration.

Artificial Sequestration

Sequestration which does not use natural processes for C-sequestration is referred to as artificial sequestration (Lal and Kimble 1997). It involves, firstly, capture of carbon dioxide, and secondly, preventing re-release of captured CO_2 in the atmosphere. The net result is prevention of CO_2 accumulation in the atmosphere. Several types of carbon dioxide capture technologies are presently under development to make artificial sequestration an effective tool for C-sequestration, including: (*i*) gas separation membranes to remove and capture CO_2 from industrial flue gases, (*ii*) absorption of CO_2 in amine-based solvents, (*iii*) separation and capture of CO_2 from natural gas, (*iv*) cryogenics, (*v*) coal gasification and oxy-fuel based power plants, and (*vi*) use of hydroxides for scrubbing CO_2 from exhaust gases. Technology is also under development for disposal of captured CO_2 in deep sea depths (below 1 km) in solid form for long-term deposition. However, impacts of this on benthic organisms need to be studied (Lal and Kimble 1997 and Schuman 2002).

Geological or Geo-Sequestration

It is another type of artificial sequestration, which could be used for direct injection of captured CO_2 in underground geological formations such as used oil fields saline aquifers, and unutilizable mines(Lal and Kimble 1997).

Mineral Sequestration

It is yet another option for deposition of CO_2 for long-term storage in a non reactive state. It is based on the principle that mineral silicates, when reacted with acids, form mineral carbonates, a non- reactive form of carbon like natural limestone.

Serpentine, Mg-olivine, and Fe-olivine are some of the proposed minerals which could be used for converting captured CO_2 in the non-reactive state for long-term sequestration.

In conclusion, carbon sequestration offers a wide range of choices which could be used effectively to sequester carbon for mitigation of global warming. Many of the options are easy to operate and do not require advanced technology, like forestation, conservation of existing forests and several agro-ecosystems. Others, such as separation membranes, deep sea deposition, etc. need further developmental research management practices, mentioned above. However, it could be safely stated that these sequestration technologies would play a very important role in the future in overall efforts towards control of global warming and climate change.

Acknowledgements

The author expresses his gratitude to Prof P. L. Gautam, Vice Chancellor, G.B. Pant University of Agriculture and Technology, Pantnagar for providing financial support for attending the conference. He is also highly thankful to Dr. (Mrs) Uma Melkania and Dr. Vir Singh for their encouragement and hospitality extended during the conference. He extends his appreciation to all the faculty members and students of the Department of Environmental Science for their co-operation and help during his stay for the conference.

References

Anid, P.J. and Sombrock, W.G., 1994. In: *Non-CO_2 Greenhouse Gases: Why and How to Control*, (Eds.) Van Ham J. *et al.* Kluwer Acad. Publsh., Dordrecht, pp. 475–86.

Chen, R.S., 1990. *Env. Impact. Asses. Review*, 10(4): 335–355.

Henderson, D.C., 2000. *M.Sc. Thesis*. Univ. of Alberta (Canada).

IPPC-II, 2001. *Summary of the Policy Makers: Scientific, Technical, Analysis of Impacts, Adaptations and Mitigation of Climate Change.*

Lal, R. and Kimble, J.M., 1997. *Nutrient Cycles in Agro-ecosystems*, 49: 243–253.

Le Cain *et al.*, 2000. *J. Range Management*, 53: 199–200.

Obasi, G.O.P., 1996. *Our Planet*, 5: 5–8.

Perry, M.L. and Swaminathan, M.S., 1992. In: *Confronting Climate Change: Risks, Implications and Responses*, (Eds.) I.M. Mintzer. Camb. Univ. Press, pp. 113–26.

Reader, J.D., Schuman G.E. and Bowman, R.R., 1998. *Soil Tillage Res.*, 47: 339–349.

Rice, C.W., 2000. *Agricultural Research Services*. USDA, NRCS, Raleigh N.C USA, p. 83.

Sampson, R.H. and Scholes, R.J., 2000. In: *Land Use, Land Use Changes and Forestry Special Report*, (Eds.) Watson, R.T. *et al.* IPCC, IGPOCC, U.K, pp. 183–283.

Schuman, G.E. *et al.*, 2002. *Env. Pollution*, 116: 391–396.

Subrahmanyam and Rathore, V.S., 1994. *J. Agro. Crop. Sci.*, 172: 188–93.

Terramun, A.H. and Sullivan, J.H., 1989. In: *Proceedings of Second North American Conference on Preparing for Climate Change*. Washington DC, USA, pp. 203–207.

Ecosystem Diversity and Carbon Sequestration: Climate Change Challenges and a Way Out for Ushering in a Sustainable Future *Pages 294–297*
Editors: **P.L. Gautam, Vir Singh and Uma Melkania**
Published by: **DAYA PUBLISHING HOUSE, 2009**

Chapter 31

Forest Management: Carbon Mitigation and Social Issues

Govind Singh Kushwaha

The most challenging problem humanity has ever faced on this planet is that of global warming and consequent climate change. It is clear from the fact that global temperature is increasing due to human activities that releases CO_2 and other greenhouse gases in the atmosphere. The problem of increasing atmospheric CO_2 can be addressed in a number of ways. This paper examines issues related to mitigating the global warming problem through forest management. Forest management, which so far has been on the periphery of our thinking, has the potential to resolve this dilemma and capture synergies between local and global environment. Forest management has an important role to play in slowing, halting and reversing the processes of deforestation and release of CO_2 and in promoting rural development.

Introduction

Most of the forests in India (about 94 per cent) are in the public domain. A very large population of the country, particularly in rural areas, depends on forests in a number of ways. There are yet forest-dependent communities who have no future without these habitats. Forests are not just a source of livelihood, but play much greater role which is vital for the existence of life on Earth. Forest management therefore should be a critical aspect of our resource management policies. The pertinent issue is how to manage forests and who should manage the forests?

In some of the areas, such as the fragile mountains, forests are yet of critical value, for these habitats would assume a total disaster without these natural resources. Hill and mountain communities are especially sensitive towards forest-related issues. The ecofriendly mountain communities have in the past always ensured that the rich and diverse forest stretches are preserved and protected. For instance, in the hill State of Uttarakhand, a forest eco system is insulated against the forces of eco-devastation mainly due to the efforts of the hill communities.

More than 70 per cent of the people from hills are fully depended on the forests. They cover more than 65 per cent of the geographical area of Uttarakhand. The other hilly states apart from Uttarakhand also project almost a similar picture. They also serve as a central part of our village culture and their use has been regulated through number of social institutions.

The hill communities have devised certain ideal practices for the conservation and use of the forests. Over the years, such finest practices have become an inherent part of the traditions and customs of the communities resulting in high forest cover. India's forests, the foundation of the nation's ecological security, are being lost to an excess of commercial enterprises at an alarming rate. The latest statistics released by the Forest Survey of India shows that the country has lost over 26,000 sq. km of its dense forests during 2001-2003. The massive loss of forest species which lovers over 3000 species of flowering plants and 200 species of animals of the country which has already been categorized as threatened have added to the decimation of biodiversity. We, indeed, cannot afford to witness this dismal environmental scenario.

The status of forests and trends in forest management show that the Uttarakhand forest resources are under tremendous pressure, due to degradation of forests. The production of timber from the forests had declined significantly in the last few years. Inadequate management over that period had resulted in poor socio-economic conditions of the rural populations associated with the forests. Shrinkage of forests does not only result in the reduced production of its products but also in its vital functional attributes along with the carbon sequestration which is also one of its significant functional attributes. Reduced rate of the carbon sequestration results in increase CO_2 concentration in the atmosphere, and from here starts the vicious cycle that might prove ghastly for the living planet in distant future, if the current trends were allowed to continue unabated.

Carbon release in the atmosphere does not seem to be a natural phenomenon. A major problem faced by the society is that of the global temperature which is believed to be rising due to human activities. The major culprit is thought to be burning fossil fuel, which releasiesncreasing amount of CO_2 in the atmosphere. If it is an anthropogenic phenomenon, it is well revealed that it can be addressed only by anthropogenic means.

Forests to Sequester Carbon

Carbon sequestration would be best dealt with forest management. Forest biomass, forest floor litter and soils are the natural reservoirs where CO_2 can be dumped. Forest management can significantly contribute to carbon sequestration and the types of management that are best suited to this task. It would also have bearing on a relative increase in stock of carbon held captive in the forest ecosystem over what would have occurred in the absence of such management. Increases in the stock of carbon could be accomplished as a result of an increase in the forest biomass and increase in forest soil carbon directly. Finally, if the stock of long-lived wood products increases, the carbon which is held captive in wood products stock would also increase.

There are a number of activities that could lead to an increase in forest area and subsequent enhancement in carbon sequestration rates. Forest area can be increased by halting deforestation, bringing more land area under new forests, regenerating the degraded forests/uncultivated lands, converting croplands into agroforestry systems, increasing forest density to increase forest biomass per unit area, etc.

Social Aspects of Sustainable Forest Management

As the forests play crucial role thanks to human ways for their management. Human input is crucial for the management and protection of forests, and even more so for the restoration and rehabilitation of degraded forests. People are, however, also the most devastating agents of destruction and over-use of forests. Change of forest land to industrial, areas grazing areas, and infrastructure, mines, etc. is also of great concern. It is now widely accepted that many of the fundamental causes of forest destruction and degradation are of a social nature. Poverty, resulting from an extremely uneven distribution of wealth, from the lack of access to or control over resources, in particular to land, and from the lack of alternatives to a livelihood based on marginal agriculture, is probably the single most important driving force for the destruction of forests.

There is a functional as well as an ethical link to the social component of sustainable development and the equitable sharing of the proceeds of economic growth. Forests need to be socially beneficial in order to contribute to the objective of sustainable development. Benefits derived from the existence and management of forests and accruing to people living in and around them may also be a precondition for the conservation of the forest. While the notion of social benefit is thus critical to the mission of sustainable forest management and indeed to the very survival of many forest areas, it is this aspect that is least well defined and least well covered in the criteria and indicators for sustainable forest management formulated in international forest policy.

Generally hilly people or other communities interact closely with forests. There are forests-dependent communities and many native and tribal peoples who derive their economic livelihood and often their cultural and spiritual identity from the forests (Byron and Arnold, 1997).

Changes in the Management System

The success of forest protection is based on sound institutional mechanism. With regeneration the growing forest required efficient management.

Rules Pertaining to Access and Use

- ☆ Annual cleaning and thinning operation before rainy season to ensure steady supply of fuel wood to the villagers.
- ☆ The people of villages are allowed to collect forest products free of cost and without permission.

- ☆ The forest dependants are allowed to sell fuel wood and other products.
- ☆ The members are free to take wood for cremation without cost or permission.
- ☆ The neighboring villages can get bamboo and timber with permission from the authority and by paying an amount. Special concession is made when the material thus sought is to be used for community festivals or after instances of fire or accidents.
- ☆ Constructing a waterhole in the middle of the forest to provide drinking water to the animals.
- ☆ To Bann hunting in the forests.
- ☆ Renovation work of a decaying pond in the forest. It is helpful to provide irrigation facility to the agricultural land.

Reference

Byron, Neil andA rnold, Michael, 1997. *What Futures for the People of the Tropical Forests?* (Bogor, CIFOR), Working Paper No. 19, November.

Ecosystem Diversity and Carbon Sequestration: Climate Change Challenges and a Way Out for Ushering in a Sustainable Future *Pages 298–308*

Editors: **P.L. Gautam, Vir Singh and Uma Melkania**

Published by: **DAYA PUBLISHING HOUSE, 2009**

Chapter 32

Enhancing Carbon Sequestration in India: Economic Issues and Mechanisms

A.K. Singh and Virendra Singh

Forest ecosystems have potential to capture and retain large volumes of carbon over long periods, as trees absorb carbon through photosynthesis process. A young forests, when growing rapidly, can sequester relatively large volumes of additional carbon roughly proportional to the forest's growth in biomass. A mature forest acts as a reservoir, holding large volumes of carbon even if it is not experiencing net growth. Forest management can thus have an influence on carbon sequestration. Present estimates indicate that with appropriate policies the carbon pool in the terrestrial system could increase by up to 100 Giga Tons of Carbon (GtC) by the year 2050 compared to the level of carbon that would be sequestered without such policies (IPCC, 2001). This is equivalent to about 10 to 20 per cent of projected GHG emissions from fossil fuel consumption during the same period. Reducing deforestation, expanding forest cover, expanding forest biomass per unit area, and expanding the inventory of long-lived wood products are some of the activities that could help global community realizing the carbon sequestration potential of forest to mitigate global climate ecosystem.

Economics of Carbon Sequestration

Carbon sequestration is the process through which agriculture and forestry practices remove carbon dioxide from the atmosphere. The term "sink" is also used to describe agriculture and forestry land that absorb CO_2–the most important global warming gas emitted by human activities. Sequestration activities can help to prevent global climate change by enhancing carbon storage in trees and soils, preserving existing tree and soil carbon, and by reducing emissions of CO_2, methane (CH_4) and nitrous oxide (N_2O). Increasing the quantity of the carbon sequestered or stored in soils and biomass is an alternative to reduce the emission of carbon and other greenhouse gases in an overall strategy change and its negative economic and environmental effects.

Early works on economics of carbon sequestration focused on examining whether expansion of forest sinks could play a major role in the effort to slow the accumulation of atmosphere CO_2. These studies typically used a hypothetical government programs such as subsidies, or government purchases, to promote forestry management practices like a forestation of agricultural land, or conservation of forestland, and attempted to estimate cost of sequestration by assessing the costs of various inputs to production including land, labor and material. But most of the studies do not consider the cost of storing carbon in agricultural soils. The marginal cost of sequestering additional unit of carbon would be expected to raise as the quantity sequestered increases. Consequently the cost will increase of the carbon content in agricultural land.

Producer-level Development Costs

When producers choose to undertake a carbon-sequestering activity, they will need to change land management practices, land use, or both. The costs of such a change is the difference in net revenue and cost streams plus the difference in any long-term, fixed-cost requirements. Major elements of this cost include the differences in crop yields under existing and alternative land management, the changes in input costs, the costs of needed new equipment, and the salvage value of discontinued equipment.

Producer Adoption Inducement Costs

Even if they are economically and agronomically attractive, alternative land uses or methods of land management may not be adopted because of learning time, investment costs, increased risk, and other factors. For example, many cases exist in which the adoption of reduced tillage has been limited, despite calculations that this practice would increase average income. Producer adoption inducement costs represent any inducements needed to overcome barriers to adoption of the project practices that are above and beyond development costs.

Market Transaction Costs

The producer development costs and adoption inducement costs are only part of total commodity costs. In particular, offsets still need to be conveyed to the buyer. There are several additional costs to consider in this conveyance, such as broker commission costs, measurement and monitoring, enforcement, and insurance fees protecting against adverse outcomes.

Government Role

The government may offset some proportion of the costs through land-use subsidies or subsidies for sequestration practices.

Co-benefits

Carbon sequestration practices can have implications for environmental quality and income distribution. For example, adoption of reduced tillage can lead to reductions in soil erosion and consequently improve water quality, while payments to farmers may alleviate needs for income support. Such effects may justify a governmental role in subsidizing sequestration practices.

Discounts

Offsets created by projects designed to reduce net emissions need to fit in the compliance structure of the global emission accounting scheme. Consequently, the quantity of greenhouse gas offsets will be paid subject to discounts due to system rules and compliance requirements. Key concerns in this regard involve additionally, uncertainty, leakage, and permanence, as discussed in the Kyoto Protocol.

Uncertainty

Society and buyers recognize that there will be variability in sequestration quantities as a result of climate and other factors. To create a safety margin that protects against potentially penalized shortfalls, a discount may be desired that falls below the expected average offsets created by the project.

Leakage

Project credits should be reduced by the extent to which actions to enhance sequestration alter production and create market conditions (*e.g.* price effects) that induce emission increases elsewhere.

Permanence

Sequestered carbon is stored in a volatile form. Future changes to alternative practices can cause re-emission of some or all of the sequestered carbon to the atmosphere, and ongoing maintenance costs may be needed for a practice after it has ceased to influence carbon uptake. Furthermore, practices may be contracted under a lease, not a permanent arrangement. A discount may arise reflecting the potential for volatility, the existence of required maintenance costs, or the need to re-contract for offsets after a lease expires.

Economic Components for Determining the Cost of Carbon Sequestration

The components of the incremental greenhouse gas offset include the following:

- Project-induced carbon sequestration in the soil or in standing plants or trees over time.
- Project-induced net reductions in CH_4 and N_2O emissions.
- Project-induced net savings in fossil fuel emissions due to changes in system inputs.
- Project-induced net savings in carbon releases from manufacturing agricultural inputs due to changes in use of agricultural inputs.

The risks of the worst impacts of climate change can be substantially reduced if greenhouse gas level in the atmosphere can be stabilized between 450 and 550 ppm CO_2 equivalent (CO_2e). The current level is 430 ppm CO_2 e today, and it is rising at more than 2 ppm each year. Stabilization in this range would require emissions to be at least 25 per cent below current levels by 2050, and perhaps much more. Ultimately, stabilization at whatever level requires that annual emissions be brought down to more than 80 per cent below current levels. This is a major challenge, but sustained long-term action can achieve it at costs that are low in comparison to the risks of inaction. Central estimates of the annual costs of achieving stabilization between 500 and 550 ppm CO_2e are around 1 per cent of global GDP, if we start to take strong action now. Costs could be even lower than that if there are major gains in efficiency, or if the strong co-benefits, for example from reduced air pollution, are measured. Costs will be higher if innovation in low-carbon technologies is slower than expected, or if policy-makers fail to make the most of economic instruments that allow emissions to be reduced whenever, however it is cheapest to do so. It would already be very difficult and costly to aim to stabilize at 450 ppm CO_2 e. If we delay, the opportunity to stabilize at 500–550 ppm CO_2 e may slip away. The costs of taking action are not evenly distributed across sectors or around the world. Even if the rich world takes on responsibility for absolute cuts in emissions of 60–80 per cent by 2050, developing countries must take significant action too. But developing countries should not be required to bear the full costs of this action alone, and they will not have to. Carbon markets in rich countries are already beginning to deliver flows of finance to support low-carbon development, including through the Clean Development Mechanism. A transformation of these flows is now required to support action on the scale required.

With the help of economic models, the review estimates that if we don't act, the overall costs and risks of climate change will be equivalent to losing at least 5 per cent of global GDP each year, now and forever. If a wider range of risks and impacts is taken into account, the estimates of damage could rise to 20 per cent of GDP or more. In contrast, the costs of action–reducing greenhouse gas emissions to avoid the worst impacts of climate change–can be limited to around 1 per cent of global GDP each year. The investment that takes place in the next 10-20 years will have a profound effect on the climate in the second half of this century and in the next. Our actions now and over the coming decades could create risks of major disruption to economic and social activity, on a scale similar to those associated with the great wars and the economic depression of the first half of the 20th century. And it will be difficult or impossible to reverse these changes. Because climate change is a global problem, the response to it must be international. It must be based on a shared vision of long-term goals and agreement on frameworks that will accelerate action over the next decade, and it must build on mutually reinforcing approaches at national, regional and international level. If no action is taken to reduce emissions, the concentration of greenhouse gases in the atmosphere could reach double its pre-industrial level as early as 2035, virtually committing us to a global average temperature rise of over 2°C. In the longer term, there would be more than a 50 per cent chance that the temperature rise would exceed 5°C. This rise would be very dangerous indeed; it is equivalent to the change in average temperatures from the last ice age to today. Such a radical change in the physical geography of the world must lead to major changes in the human geography–where people live and how they live their lives.

Even at more moderate levels of warming, all the evidence–from detailed studies of regional and sectoral impacts of changing weather patterns through to economic models of the global effects–shows that climate change will have serious impacts on world output, on human life and on the environment. All countries will be affected. The most vulnerable the poorest countries and population will suffer earliest and most, even though they have contributed least to the causes of climate change. The costs of extreme weather, including floods, droughts and storms, are already rising, including for rich countries. Adaptation to climate change–that is, taking steps to build resilience and minimize costs is essential. It is no longer possible to prevent the climate change that will take place over the next two to three decades, but it is still possible to protect our societies and economies from its impacts.

Carbon Sequestration Estimates for Forestry

Changes observed in the agricultural land category have been minimal since 1970, and have almost stabilized in India. Though there has been no marked increase in the land area under the agricultural sector, the yield per hectare for the total food grain in the country has increased from 0.872 t/ha during 1970–71 to 1.6 t/ha during 1996–97. As a result, the food grain production has increased from 82 million tons (mt) during 1960–61 to an all-time high of nearly 212.9 mt during 2001–02. The total food grain requirement of India is expected to increase to about 269 mt by the year 2010. At present, increase in area for food grain production is constrained by bio-physical factors such as soil, climate etc. Over the last three centuries, forests have decreased by 1.2 billion hectares *i.e.* 19 per cent and grasslands by 560 million hectares. This has resulted due to an increase in croplands. Area under the forests in 1950s was approximately 40.48 m ha. It marginally increased and reached up to 63.92 m ha during 1970s, 64 m ha during 1980s and is almost constant since then. India has nearly 19.1 per cent of its land area under forests. This excludes land under tree plantations. About 43.11 per cent of land area is devoted to croplands. The remaining 113.535 m ha accounts for land in forest, fallows, restored forest and degraded land.

Bhadwal and Roma Singh used the model for carbon sequestration taking into account all relevant variables in the seven selected categories, say population and current growth rate of the population; land-use and biomass; forest management and uses; forest harvesting practices; agricultural production; fuel wood requirement and project management for the total carbon content of Indian forests for the selected years. The estimated and model simulated values are then compared for validation. It was observed that the trend of year-to-year variations in carbon is well simulated by the model. Slight deviations in the model-simulated and estimated carbon storage values are perhaps due to some unaccounted factors such as natural calamities, etc. and estimation of its future carbon sequestration potential under different land-use patterns. The sensitivity of carbon sequestered to changes in land-use patterns varies regionally. Shifts in land-use have a major effect on terrestrial carbon storage, and the changes associated with over exploitation of land resources. It is estimated that carbon storage in the vegetation in India was 8.6 bt in the year 1880, 7.7 bt in 1920, 6.39 bt in 1950, 5.23 bt in 1970 and 4.39 bt in 1980 carbon storage in the vegetation in the country from 1880 onwards shows a decreasing trend.

Mechanism for Enhancing the Carbon Sequestration

There are some ways and means to enhance the carbon sequestration in the environment. These means can be grouped in to two parts. First is effective implementation of act and regulation. The second one is identifying cultural practices and technologies. A number of independent initiatives are springing up at national and international levels that use various approaches. Although this diversity can lead to innovation and provide lessons, it may also reflect conflicting interests that need to be reconciled.

Legal and Institutional Mechanism

A global carbon market requires the support of an institutional infrastructure that can increase investors' confidence and reduce transaction costs in international trading. This infrastructure may include national offices, regulatory agencies, and establishment of trust funds, trading platforms such as exchanges, brokers, certifiers and insurers. For example, establishing a national carbon registry can help to prevent double selling of carbon credits and also provide transparency for prices that are critical to fair negotiations. Preventing leakage requires an institutional capacity to enforce laws. This capacity also creates greater incentive to invest in sustainable forestry practices such as Reduced Impact Logging that result in higher carbon retention. However, much of the economic benefit from this is in higher future yields– provided that access to forests can be controlled over the full rotation period.

Efforts are afoot and systems are in place such as the Kyoto Protocol, European Union Emission Trading System (EU ETS), the United Kingdom Emission Trading System (ETS), Chicago Climate Exchange (CCX) in USA, Regional Greenhouse Gas Initiative (RGGI) in North Eastern United States, New South Wales Carbon Market (NSW) in Australia for the sale and purchase of carbon credits in order to put a check on the emission of greenhouse gases. Quite a few corporates, especially the power companies and private players, brokers and agents are operating in the spot and forward markets for trading in carbon credits. Under the Cap-and-Trade regime, European Union Allowances (EUAs) have been allocated in the first phase in order to control CO_2 emissions from various enterprises. In the phase II, the caps are going to be at much lower level of emissions. Market in carbon trading has developed quite fast. Carbon Credit Markets operate like share markets and stock exchanges in respect of spot purchases and future hedging. The value of carbon market was more than US \$10 billion in 2005 and is expected to be between US \$25 to 30 billion in 2006. Average price of the 79 million tonnes of carbon credit traded in the first quarter of 2006 is reported to be US \$11.45 per ton CO_2 (Capoor and Ambrosi 2006).

Identifying Promising Technologies for Soil Carbon Sequestration

Land managers can enhance soil sequestration by (a) increasing the rates of organic matter input, (b) partitioning carbon to longer-lived pools, and (c) increasing the longevity of all or selected carbon pools (most widely employed management practices). Depending on the process involved, variations are expected in the length of time the enhancement method is effective (time to saturation) and in the average residence time and susceptibility to disturbances that would release this sequestered carbon back to the atmosphere (permanence). The amount of soil carbon in various forms is determined by the balance between inputs of organic accumulation and losses of carbon through decomposition, leaching to groundwater, and erosion. At a basic level, the mechanisms controlling these processes and their balance can be tied to soil-forming factors. Disturbance or management practices also exert considerable influence on the amount of soil carbon, both through direct effects on inputs and losses and through indirect effects on the factors controlling these fluctuations. However, the complex interactions and feedbacks among controlling mechanisms and processes lead to uncertainties in predicting changes in soil carbon. It is critically important to identify, develop, and quantitatively evaluate appropriate technologies and their effects on processes that determine soil carbon dynamics.

Understanding the Effects of Technologies on Carbon at the Site Scale

Several widely used land management practices that, when applied to plough tilled cropland, have demonstrated significant carbon sequestration potential. Most of the increases in soil carbon associated with these practices result from reversing processes by which traditional management has depleted the soil carbon stocks that accumulated under native perennial vegetation (Cole *et al.*, 1997). There is sufficient information from intensive studies of these soil management practices to gain insight into the processes that are involved in increasing soil carbon content.

Cropping Intensification

Elimination of fallow periods (dry fallow in seasonally dry environments, winter fallow in cold environments), use of high-yielding crop varieties, and the wide scale application of inexpensive fertilizer and other soil amendments greatly increase the amount of organic matter produced, thereby increasing organic matter into the soil (Parton *et al.*, 1995, Buyanovsky and Wagner 1998 and Lal *et al.*, 1998). Precision agriculture, an intensive approach to optimize crop yields relative to fertilizers and other inputs, results in increased soil carbon inputs. A large component of the increased crop residue input resulting from cropping intensification is readily decomposed and does not add to long-term soil carbon increases. A portion remains to be converted into humus, contributing to increasing long-term soil organic carbon pools. The increases in soil carbon due to cropping intensification alone are often limited in amount and result in saturation within a few decades. Much greater increases in soil carbon are frequently obtained if manures (biologically altered inputs) are applied (Jenkinson 1990). Such amendments not only increase the amount of organic carbon input but appear to contain a larger fraction of organic carbon materials that are more resistant to decomposition than unaltered plant material.

Conservation Tillage

Part of the decrease in soil organic carbon that occurs when native vegetation is ploughed for row crops results from mechanical disruption of soil aggregates. Soil structure plays a dominant role in the physical protection of soil organic matter

(SOM) by controlling microbial access to substrates, microbial turnover processes, and decomposer food web interactions. Relatively labile material may become physically protected from decomposition if it is incorporated into soil aggregates or deposited in micro pores that are inaccessible even to bacteria. Macro aggregates (at least 0.25 mm in diameter) are sensitive to soil disturbance, but micro aggregates (less than 0.25 mm in diameter) are generally more stable, appear to turn over more slowly, and are more resistant to disturbance making use of soil fractionation procedures to isolate micro aggregates from the macro aggregate structure of soil, and also taking advantage of the shifts in the natural abundance of stable carbon isotopes following a change in grassland type, has found that micro aggregates facilitate the creation of chemically resistant organic mineral associations with relatively long residence times. Reductions in tillage intensity allow aggregation processes to be re-established, thereby rebuilding this physical protection. Recent research suggests that management practices resulting in decreased disturbance (*e.g.*, no-till cultivation or establishment of perennial vegetation) generally increase fungal-dominated pathways in organic matter cycling, which may increase the residence time of microbial residues and lead to their buildup in SOM (Bardgett and McAlister 1999, Stahl *et al.*, 1999 and Bailey *et al.*, 2002). In addition, decreases in disturbance and the accompanying changes in plant communities can increase the amount of mycorrhizal fungal biomass, which is also biochemically recalcitrant but is derived directly from plant photosynthates rather than from the rendering of detritus residues. An additional benefit of conservation tillage is the reduction of wind and water erosion. Even if soil mineral particles are not removed, erosion processes cause disruption of soil aggregates and loss of particulate organic matter. These losses reduce soil water holding capacity (WHC) and nutrient regeneration and can result in reduced crop productivity. The direct and indirect effects of erosion result in soil carbon losses that must be counteracted by increased irrigation or fertilizer application if soil carbon is to be maintained or increased (Lal 1995).

Perennial Vegetation

Perennial vegetation establishment on previously plough-tilled cropland results in substantial increases in soil carbon (Gebhart *et al.*, 1994; Post and Kwon 2000). Even without additional management, the rates at which soil carbon increases after the establishment of perennial vegetation.

Carbon sequestration and reductions of methane (CH_4) and nitrous oxide (N_2O) emissions can occur through a variety of agricultural and forestry practices. Likewise, carbon can be lost and CH_4 and N_2O emitted to the atmosphere through a number of land-use changes and practices. The other alternative is increasing the rates of removal of CO_2 from the atmosphere through growth of terrestrial biomass (*e.g.* forests). According to the studies conducted on carbon mitigation, the most promising management practices for CO_2 mitigation are reforestation in the temperate latitudes, and agroforestry and natural reforestation in the tropics as well as the forestry sector. In the forestry sector also, we need to identify the best forestry option, which fulfils the demands of wood and its products along with increasing carbon sequestration.

Table 32.1: Key Agricultural Practices for Enhancing Carbon Sequestration

Key Agricultural Practices	*Typical Definition and Some Examples*	*Effect on Greenhouse Gases*
Conservation and buffers	Grasses or trees planted along streams and croplands to prevent soil erosion and nutrient runoff into waterways.	Increases carbon storage through sequestration.
Conservation tillage on croplands	Typically defined as any tillage and planting system in which 30 per cent or more of the crop residue remains on the soil after planting. This disturbs the soil loss, and therefore allows soil carbon to accumulate. There are different kinds of conservation tillage systems, including no till, ridge till, minimum till and mulch till.	Increases carbon storage through enhanced soil sequestration, may reduce energy-related CO_2 emissions from farm equipment, and could affect N_2O positively or negatively.
Grazing land management	Modification to grazing practices that produce beef and dairy products that leads to net greenhouse gas reductions (*e.g.* rotational grazing).	Increases carbon storage through enhanced soil sequestration and may affect emissions of CH_4 and N_2O.
Bio-fuel substitution	Displacement of fossil fuels with bio-mass (*e.g.* agricultural and forestry wastes, or crops and trees grown for biomass purposes) in energy production, or in the production of energy-intensive products like steel.	Substitute carbon for fossil fuel and energy-intensive products. Burning and growing of biomass can also affect soil N_2O emissions.

Agriculture and forestry practices that preserve and sequester carbon, as well as reduce emissions of other greenhouse gases (GHGs), can also affect the quality of soil, air, and water and wildlife habitat. Likewise, land-use changes and management practices that are intended to conserve and improve the broader environment can influence carbon sequestration rates and emissions of GHGs. In general, carbon sequestration practices in agriculture and forestry can be consistent with other environmental objectives, but there may be tradeoffs in some cases.

Conclusion

The idea of using intentional sequestration of carbon in soil through land management to mitigate rising atmospheric CO_2 concentration is relatively recent. Not much is known at this point about the ease of accomplishing significant mitigation and the amount of CO_2 that might be mitigated. The technological capability for increasing carbon sequestration is at hand, many co-benefits seem likely, the potential magnitude of the results appears promising, and initial cost estimates appear to be low. As a result, there is a rising demand to know precisely how much carbon may be sequestered in soil, how quickly this sequestration can take place, and what other environmental and economic impacts will occur as a result. There are, however, many considerations beyond the technological capability and potential benefits that will determine the rate and cumulative magnitude of soil carbon sequestration. Researchers' understanding of the biological, edaphic, and physical environmental conditions that influence the potential amount and permanence of soil carbon is growing rapidly. This knowledge is being incorporated into mathematical models of soil carbon dynamics that allow the extrapolation of information across many conditions and provide a basis for predictions of future soil carbon sequestration. The net greenhouse gas emissions of different soil carbon sequestration methods, the costs of delivering offsets to buyers, and the ancillary environmental issues must also be evaluated. Finally, the acceptance of sequestration methods by land managers and the public will be a significant factor in determining the rate of soil carbon sequestration. The willingness of public and private buyers to use soil carbon sequestration methods to achieve net greenhouse gas reduction in the atmosphere will depend on the costs and economic benefits, which include unpriced environmental benefits.

References

Bailey, V.L. *et al.*, 2002. Fungal to bacterial ratios in soils investigated for enhanced carbon sequestration. *Soil Biology and Biochemistry*, 34: 997–1007.

Bardgett, R.D., and McAlister, E., 1999. The measurement of soil fungal bacterial biomass ratios as an indicator of ecosystem self regulation in temperate meadow grass lands. *Biology and Fertility of Soils*: 29: 282–290.

Bhodwal, S. and Singh, Roma, 2002. Carbon sequestration estimates for forestry options under different land use scenario in India. *Current Science*, 83(11).

Buyansky, G.A. and Wagner, G.H., 1998. Changing role of cultivated land in the global carbon cycle. *Biology and Fertility of Soils*, 27: 242–245.

Capoor, Karan *et al.*, 2006. *State and Trends of Carbon Market 2006*. International Emission Trading Association. The World Bank Washington D.C.

Cole, C.V. *et al.*, 1997. Global estimates of potential mitigation of greenhouse gas emission by agriculture. *Nutrient Cycling in Agro-ecosystems*, 49: 221–228.

Gebhart, D.L. *et al.*, 1994. The CRP increases soil carbon. *Journal of Soil and Water Conservation*, 49: 488–492.

Jenkinson, D.S., 1990. The turnover of organic carbon and nitrogen is soil-philosophical transactions. *Biological Sciences*, 329: 361–368.

Johl, S.S., 2006. Environmental degradation and its correctives in agriculture sector. *Indian Journal of Agril. Economics*, 61(4).

Lal, R. *et al.*, 1999. The Potential of U.S. Cropland to Sequester Carbon and Mitigate Greenhouse Effects. *Lewis Publishers*

Parton, W.J. *et al.*, 1995. Models to evaluate soil organic matter storage and dynamics. In: *Structure and Organic Matter Storage in Agricultural Soils*. CRC Press, New York.

Pattanayak, S., *et al.*, 2005. Water quality co-effects of hreenhouse gas mitigation in U.S. agriculture. *Climate Change*.

Postwn, Kwon K.C., 2000. Soil carbon sequestration and land use change: Process and potential. *Global Change Biology*, 6: 317–327.

Ravi Kumar, K.S., 2002. Carbon sequestration as greenhouse gas mitigation policy. *Current Science*.

Sathaye, J.A., *et al.*, 1998. Concerns about climate change mitigation projects: Summary of findings from case studies in Brazil, India, Mexico and South Africa. Eastern Orlando Lawrence Berkeley National Laboratory.

Stahal, P.D., 1999. Fungal presence in paired cultivated and uncultivated soils in Central Iowa, USA. *Biology and Fertility of Soils*, 29: 92–97.

Stern Review, 2006. The Economics of Climate Change. World Bank Report.

Wilfred, M. *et al.*, 2004. Enhancement of carbon sequestration in U.S. soils. *Political Science*, 54(10).

Ecosystem Diversity and Carbon Sequestration: Climate Change Challenges and a Way Out for Ushering in a Sustainable Future *Pages 309–319*
Editors: P.L. Gautam, Vir Singh and Uma Melkania
Published by: DAYA PUBLISHING HOUSE, 2009

Chapter 33

Climate Change and Kyoto Protocol: Global and Indian Concerns

Tirthankar Banerjee, Jyotsana Pathak and R.K. Srivastava

ABSTRACT

The biggest contributor of frequent incidence of extreme weather is mainly the release of greenhouse gases (GHGs) and aerosols which change the climate and one of the most important global environmental challenges with repercussion on agriculture production, natural ecosystem, hydrological cycle, energy balance and public health. As is now well known, anthropogenic sources of a number of gases and aerosols are affecting the greenhouse effect, leading us into a future of uncertain global climate. The UN Conference on Environment and Development (UNCED) in 1992 led to Framework Conventions on Climate Change (FCCC) with a commitment to "prevent dangerous anthropogenic interference with the climate system", is a climate treaty that lays groundwork for nations to stabilize concentrations of greenhouse gases (GHGs) at a level that would prevent dangerous interference with the climate system. In 1997, for stabilizing GHGs concentrations and adhering to sustainable development principles, Kyoto Protocol was drawn up, which implies that ratified countries must make efforts to reduce their GHGs emission or engage in emission trading if they are not able to bring down their emission to acceptable levels. To mitigate climate change by reducing emission of greenhouse gases and aerosols, meeting rural energy needs, protecting the environment and generating gainful employment, a shift is needed towards environmentally sustainable technologies and promotion of renewable energy, forest conservation, reforestation, water conservation etc. India, as a developing country with nearly two-thirds of population depending directly on climate sensitive sectors, the cost of not addressing climate change or not to promote any mitigation or adaptation strategies, may have very uncertain implications on its socio-economic development and early actions are therefore needed as a protective measures. Environmental conservations towards sustainability are not a luxury concept but must have to adopt in sacrifice with current economic growth for better future, particularly in context of developing countries.

Factors Contributing to Climate Change: GHGs and Aerosols

Periodical increase of global temperature is measured by different approaches like change of velocity of sound with the change of sea depth or by determining concentration of heavy isotope of oxygen in world's ice cover. Isotopic analysis of composition of air bubbles sealed off in glacial ice provides information about the concentration of different gases in atmosphere in different time periods. A continuous 2,20,000 years records of temperature variation was obtained by analysis of a 2.083 km ice core recovered from Soviets at Vostok in East Antarctica (Jouzel *et al.*, 1993). Since 1856, by actual measurement of global mean temperature, an increase of 0.3–0.6°C was found with about 0.2-0.3°C of that occurring over the last 40 years.

Outgoing radiation from earth is essentially greater than 3μm according to Wien's displacement rule and radiatively active gases that absorb wavelengths longer than 4 μm of radiation are greenhouse gases (GHGs), which are mainly carbon dioxide (accounted for 63 per cent of radiative forcing), methane (24 per cent), nitrous oxide (10 per cent), chlorofluorocarbons, hydroflurocarbons, sulfur hexafluoride and perfluorcarbons. Basic reasons of absorption of radiation by these gases are their vibrational and rotational energy. According to their varying geometrics and sizes, different molecules absorb radiation of different wavelengths. Apart from GHGs, aerosols also contribute to the climate change by absorbing and scattering solar radiation, changing microphysical properties of cloud.

Concept of radiative forcing is used to specify the potential of any gases to alter atmospheric temperature. Radiative forcing is the balance between incoming solar energy absorbed and outgoing radiant energy. Insolation interacts with atmosphere, change of insolation by the introduction of aerosol is aerosol forcing and interaction of GHGs with radiant energy is GHG forcing. Total radiative forcing from the sum of all human activities is a warming force of about +1.6 watts/m^2. Radiative forcing from carbon dioxide, methane, and nitrous oxide combined is very likely increasing more quickly during the current era (from 1750-2007) than at any other time in the last 10,000 years [Summary for Policymakers (SPM) Report, 2007].

Aerosol climate forcing is some what complex because of its heterogeneous distribution in space and time, short life span and regional impact but long-lived GHGs are more homogenously distributed and only have an impact on outgoing long wave radiation.

Emission Scenario

Carbon dioxide plays the most vital role in climate change, that's why it has received most attention than other gases and aerosols. Since 1850, fossil fuel combustion released about 270 giga tonnes of carbon (GtC) and change of land use pattern released about 136 GtC, of which 180 GtC ended in atmosphere, while 110 GtC absorbed by vegetation and 116 GtC dissolved and stored in ocean (Climate

Change, 2001), almost all of it in form of bicarbonate (HCO_3^-). In 1990, India was responsible for 200TgC (around 5 per cent of world emission) release, but due to rapid industrial and economic growth, by 1997-98, it increased to 800 Tg CO_2/yr (Van Aardenne *et al.*, 2001). According to an estimate, between 1990 and 1998, around the world, 6.3 GtC/yr was released with another 1.6 GtC/yr by burning of biomass. This was offset by absorption of 2.3 GtC/yr each by growing vegetation and oceans. Remaining 3.3 GtC/yr was left in atmosphere which is the basic cause of global temperature increase (Climate Change, 2001). There are some other estimates of loss of carbon from geologic and terrestrial pool to atmosphere. From the beginning of industrialization era up to recent time (1850-1998), 270±30 petagram (Pg) of carbon were emitted from the burning of fossil fuel and cement industries (Marland *et al.*, 1999), of which 176±55 Pg carbon were absorbed by atmosphere (Etheridge *et al.*, 1996). Combustion of fossil fuel accounts for over 80 per cent of CO_2 emissions. In 1990s, industrialized countries were responsible for 53 per cent of CO_2 emission, from burning of petroleum (Loske, 1996). According to another research, industrialized countries are responsible for 63 per cent of emission whereas developing countries accounted for only 37 per cent of emission. US, Canada and Europe emit 46 per cent of CO_2 where India and China are responsible for only 19 per cent of emission (World Resources 2000–2001). Per capita emission of carbon in US is 5.4T/yr (1992), where it is just 0.2 and 0.6T/yr for India and China, respectively (Asian Development Bank,

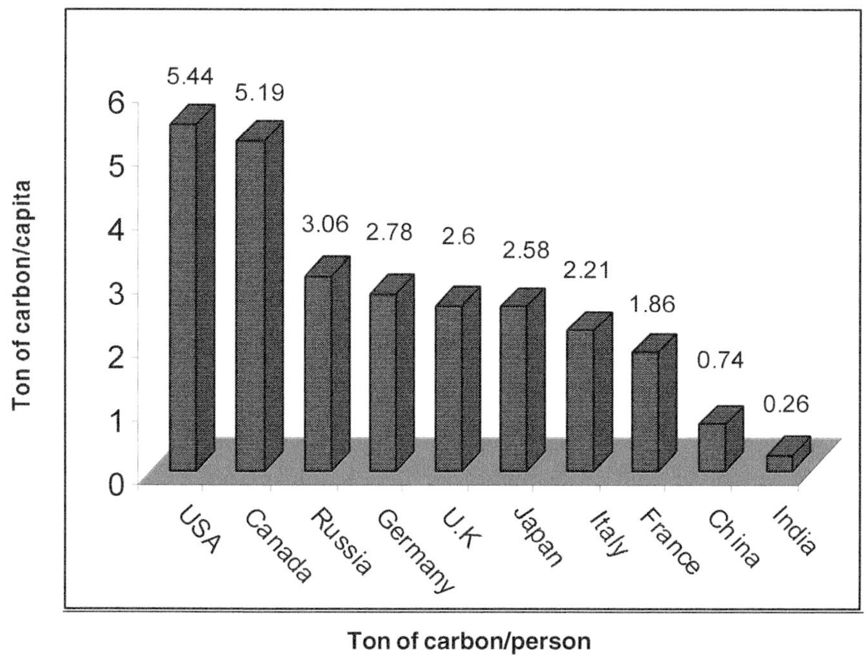

Figure 33.1: Per Capita Carbon Emissions from Energy for the Year 2003 (Source: International Energy Outlook, 2005)

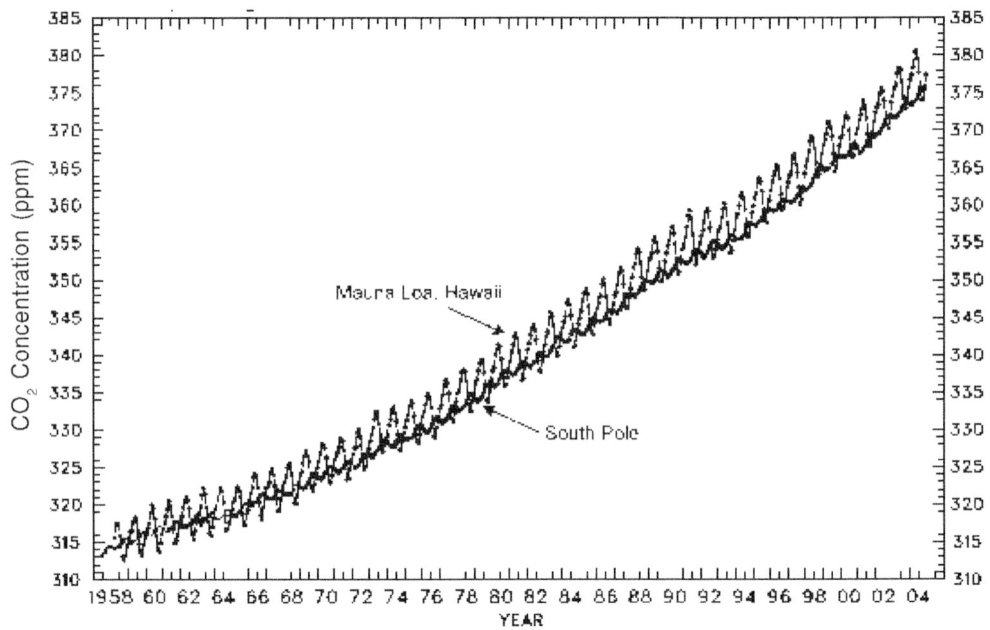

Figure 33.2: Carbon Dioxide Concentration from 1958 taken from South Pole and Mauna Loa (Inter-governmental Panel on Climate Change, 1995)

1998).

IPCC have developed a set of six emission scenarios referred as IS92a through IS92f ("IS" refer as "IPCC Scenarios" and 92 stands year). According to IS92a which assume an intermediate economic growth with medium growth projections of population, estimate an increase of 230 per cent of global CO_2 emission up to 2100 over 1990 emissions with maximum portion of emission from Central and South-East Asia (0.6 GtC/yr in 1990 to 4.2 GtC/yr in 2100). CO_2 concentration at atmosphere has increased by 30 per cent over pre-industrial era (280 ppm) and may rise to 700 ppm in 2100 with 1.53 ppm/yr growth rate [Intergovernmental Panel on Climate Change (IPCC), 1990]. The responsibility for increased level of CO_2 concentration in atmosphere lies with industrialized countries, but developing countries are likely to be primary emitter in near future (Sathaye *et al.*, 2006).

Among other GHGs, concentrations of CH_4 and N_2O increase to 1774ppb and 319ppb from their pre-industrial concentrations (320-790ppb and 270ppb, respectively) primarily due to agricultural operations and fossil fuel burning [Summery for Policymakers (SPM) Report, 2007].

Impacts of Climate Change: Global Perspective

The greatest uncertainty of the projected climate change is the issue of its impact on the natural resources, biodiversity and socio-economic system of the developing countries. By the year 2100, the range of estimates for realized, global mean temperature change is between 0.8° and 4.5°C. The middle estimate, corresponding to scenario IS92a, earth temperature may raise 2.4°C over current surface temperature (Intergovernmental Panel on Climate Change, 1996). According to IPCC fourth assessment report, 2.4°C is projected temperature rise with a likely range of 1.4 to 3.8°C and about 0.1°C per decade temperature increase is expected for the next two decades, even if greenhouse gas and aerosol concentrations were kept at year 2000 levels. The combined effect of GHGs and aerosols on regional climate is eleven of the twelve years in the period (1995-2006) rank among the top 12 warmest years in the instrumental record (since 1850) [Summary for Policymakers (SPM) Report, 2007].

Global mean surface warming will be accompanied by an increase in global precipitation by about 1 to 2 per cent per Kelvin temperature increase. The precipitation will enhanced due to increase in evaporation with increase in saturation vapor pressure with temperature. Due to thermal expansion and retreat in glacier, sea level rose at an average rate of about 1.8mm/year over 1961-2003 but average rate increased to 3.1mm/year over 1993-2003. By the year 2100, sea level rise is expected to a range of 20 to 45 cm [Summary for Policymakers (SPM) Report, 2007] which may cause displacement of population, loss of crop yield and increase flooding in coastal zones.

Indian Concern about Climate Change

The potential impacts associated with climate change lie in the interaction of several systems with many variables that must be collectively considered. With a hundred billion populations, the natural resources and climate-sensitive sectors of India are already under vulnerability. Food security will be diminished due to decrease in agricultural productivity for thermal and water stress; sea level rises, floods and droughts, and tropical cyclones would exacerbate the degradation process.

The global mean precipitation has been decreasing since the 1950s (Hulme *et al.*, 1998). This decrease is pronounced in the tropics between 25S and 30N, and a smaller increase in the extra-tropics. Climate models with GHGs and sulphate aerosols able to stimulate the negative trend observed in the tropics (Hulme *et al.*, 1998; Climate Change: Synthesis Report, 2001). The link between aerosol and reduced rainfall has also simulated by some climate models for tropical regions as north-south shift in East Asian monsoon during recent decades (Xu, 2001); and rainfall changes in the south and southwest Asian region (Chung *et al.*, 2001). In China, monsoon has moved southwards leading to drying in the northern part and flooding in the south (Xu, 2001).

Projected climate change also affects adversely soil moisture and temperature profile which may affect the species composition in the ecosystem. Due to projected increases in temperature, there will be general reduction in potential crop yield in most tropical and sub-tropical regions with the shifting of crop production belts. Theoretically, with 1°C temperature increase is equivalent to approximate pole ward shift of vegetation zones by 200Km (Ozenda and Borel, 1990). According to other research work, it is predicted that a 1°C temperature increase would cause an average altitudinal shift of vegetation belts by 150m (Bottner *et al.*, 1995). With 1°C temperature increase, climax vegetation shows a shift of 20km on the ground (Duckworth *et al.*, 2000).

The potential impacts of climate change on agriculture are highly uncertain. Potential future climate changes are also made more uncertain because of the recently recognized role of sulphate aerosols which may partly offset the warming. Research concluded that the severe yield losses in South, Southeast and East Asia for rice in many scenarios were due to a threshold temperature effect that caused spikelet sterility (Matthews *et al.*, 1994). Despite some projected increase in photosynthesis brought about by the increased concentrations in CO_2 (Ministry of Environment and Forests, 2004) the changes in temperature will have a far greater detrimental effect, resulting in a general trend of reduction in productivity. Agriculture will be negatively affected by shifts in the timing of the rainfall. For instance, over the last few years, the Chattisgarh region has received less than its share of pre-monsoon showers in May and June (Ramakrishna *et al.*). Coastal regions of Gujarat and Maharashtra, which are agriculturally fertile areas are vulnerable to inundation and salinisation (O'Brien and Leichenko, 2001). Climate change is also likely to impact on agricultural genetic resources. For example, impacts on the distributions of crop wild relatives are predicted to be significant, with 18 per cent of species in important crop gene pools likely to become extinct by 2050 (Lal *et al.*, 2001). Changes in climate have also affect on soil organic pool and the affect may be different in various regions as projected increase in temperature and decrease in rainfall may decrease net primary productivity in many tropical regions, but increases it in Boreal regions (White *et al.*, 1999).

In 1994, the IPCC warned that global warming could intensify and increase the variability of the monsoon. Global warming is intensifying the monsoon in Central India, according to a study that warns of increasing risk from heavier rains during the season. Indian summer monsoon rainfall may increase by 20 per cent (Long *et al.*, 2006), but winter rainfall may decrease up to 5 to 25 per cent with variability on onset of summer monsoon (Lane *et al.*). Changes to India's annual monsoon are expected to result in severe droughts and intense flooding in parts of India.

Kyoto Protocol and Emission Trading

The Kyoto Protocol is the approach to fortify the international responses to climate change. By assigning mandatory targets, it commits the developed countries,

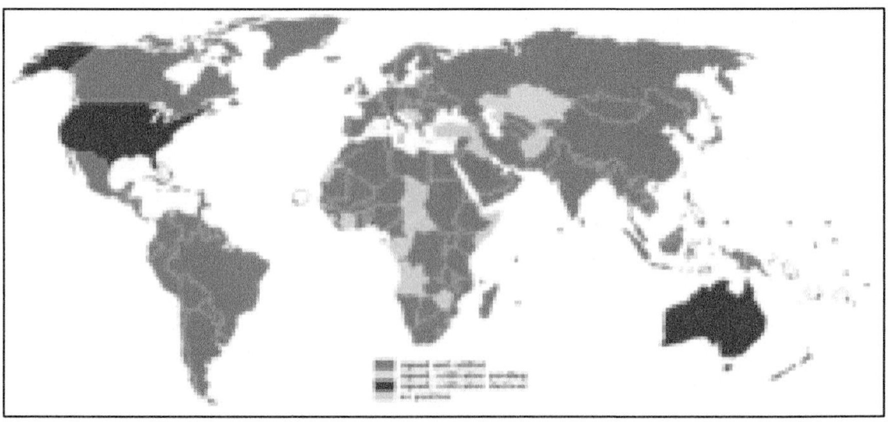

Figure 33.3: Countries Participation in Kyoto Protocol

including economics in transition to reduce emissions of GHGs by an average of 5.2 per cent below 1990 levels during the first commitment period (2008-2012). By arresting and reversing the upward trend in GHGs emissions, the Protocol promises to move the international community one step closer to achieve the ultimate intention of preventing dangerous anthropogenic interference with the climate system.

The treaty was negotiated in Kyoto, Japan in December 11 on 1997 and only after ratification of Iceland as 55th party of Annex-1 countries and Russia on 18 November 2004, satisfied the 55 per cent emission clause, brought the treaty into force on 16th Feb, 2005. As of December 2006, a total of 169 countries and other governmental entities have ratified the agreement representing over 61.6 per cent of emissions from Annex-1 countries and 55 per cent globally.

Kyoto includes "flexible mechanisms" for increasing global collaboration among the signatory countries, which allow Annex-1 economies to purchase Certified Emission Reductions (CERs) either by financial exchanges or from projects which reduce emissions in non-Annex-1 economies under the Clean Development Mechanism (CDM), or in other Annex-1 countries under the Joint Implementation (JI).

Table 33.1: Share in Global CO_2 Emissions (per cent)

	1990	*2003*
United States	23.04	23.06
China	10.41	14.07
Russia	9.67	6.38
Japan	4.54	4.79
India	2.63	4.07
Germany	4.24	3.35
Canada	2.19	2.39
United Kingdom	2.76	2.24
Italy	1.91	1.85
France	1.8	1.63
Rest of World	38.61	36.17

Source: International Energy Outlook (2005).

Participation in the Kyoto Protocol, where grey indicates countries that have signed and ratified the treaty and white indicates states that have signed and hope to ratify the treaty. Australia and the United States have signed the treaty but refuse to ratify it.

Indian Perspectives on Clean Development Mechanism

CDM seems to be an opportunity for the developing countries to profit in the emissions-trading market by selling emission reduction credits which provides a new source of funding for activities that will also protect biological diversity. Not only to achieve investment capital, it also provides an opportunity to implement innovative technical and institutional support to promote more energy efficient and sustainable projects throughout the country. For instance, Annex-1 countries provide institutional support to developing countries to replace kerosene lamps with solar light to obtain CERs.

India signed and ratified the Protocol in August, 2002. Since India is exempted from the framework of the treaty, it is expected to gain from the protocol in terms of transfer of technologies and related foreign investments. As a developing country, concepts of CDM was taken as an opportunity in India and a National CDM Authority was approved by the Cabinet in December 2003 and till 10th march 2006, it has approved 252 projects in diverse fields biomass based cogeneration. Mainly the projects are in the field of renewable energy, municipal solid waste, hydro-flurocarbons, small hydro and energy efficiency, etc.

Gujarat Fluorochemicals limited (GFL) near Godhra sold millions of certified carbon reduction units each year. This unit produce HCFC-22, generates by- product HFC-23 which is potent global warming gas. Projects are designed to incinerate the gas to reduce its emission. HFC-23 projects are the bulk of CERs sold from India.

Global Concern: Not Addressing Climate Change

The ultimate objective of the United Nations Framework Convention on Climate Change (UNFCCC) is to achieve "stabilisation of greenhouse gas concentrations in the atmosphere at a level that would prevent dangerous interference with the climate system." Even after greenhouse gas concentrations are stabilised, it will take a considerable amount of time for a balance, if greenhouse gas concentrations were stabilised today, which would require an immediate reduction in emissions of around 60 per cent, we would still expect the temperature to eventually rise by another 1°C due to the inertia of the climate system.

It is assumed that, it would be far cheaper to take strong measure now to cut emissions and stabilize GHGs at acceptable levels than to face the economic consequences of catastrophic climate change. The cost of not addressing climate change is very indecisive, but for India with ultimate dependence on climatic sectors, early actions on adaptation must be taken as a preventive measure. A wide variety of adaptive actions may be taken to minimize or overcome adverse effects. In agriculture sector, adjustments may include the introduction of late maturing crop varieties, switching cropping pattern, adjusting timing of agricultural operations, conserving soil moisture through appropriate tillage methods, and improving irrigation efficiency.

Conclusion

After the adoption of Agenda-21 and other conventions in the field of environment and natural resource conservation, concept of sustainable development has become part of different policies. Sustainability is the approach to obtain maximum profit form any sector with the maintaining stock of assets. It requires taking longer term perspectives, integrating local and regional effects of global change into the developmental process with the maximum utilization of best scientific knowledge. But the point of concern is whether it will be realistic approach to adopt environmental conservation and sustainable development concept in the sacrifice of current economic growth. But all the current trends of climate change, trends in declination and degradation of natural resources, projected impact on economy and sustainability of human beings by regional and global shift of climatic balance, clearly shows that the long term conservation of natural resources and simultaneously adoption of sustainable development concept is now not a luxury concept. The developmental concept should be constantly re-evaluated to ensure that resource utilization has reduced impacts on Earth system. Above all of these, better communication is required among scientists, decision makers and the general public.

References

Asia Least Cost Greenhouse Gas Abatement Strategy (ALGAS), Asian Development Bank, Manila, 1998.

Bottner, P., Couteaux, M.M. and Vallejo, V.R., 1995. Soil organic matter in Mediterranean-type ecosystems and global climate changes: A case study–the soils of the Mediterranean Basin. In: *Global Change and Mediterranean Type Ecosystems*, (Eds.) J.M. Moreno and W.C., Oechel. Springer-Verlag, New York, pp. 306–325.

Chung, C.E., Ramanathan, V. and Kiehl, J.T., 2001. Effects of the south-asian absorbing haze on the north-east monsoon and surface-air heat exchange. *Journal of Climate*, 15: 2462–2476.

Climate Change 2001. The Scientific Basis, Summary for Policy Makers and Technical Summary of the Working Group I Report, Intergovernmental Panel on Climate Change, Geneva, Switzerland.

Climate Change: Synthesis Report, 2001. Inter-governmental Panel on Climate Change, Cambridge University Press.

Duckworth, J.C., Bunce, R.G.H. and Malloch, A.J.C., 2000. Modelling the potential effects of climate change on calcareous grasslands in Atlantic. *European Journal of Biogeology*, 27: 347–358.

Etheridge, D.M., Steele, L.P., Langenfelds, R.L., Francey, R.J., Barnola, J.M. and Morgan, V.I., 1996. Natural and anthropogenic changes in atmospheric CO_2 over the last 1000 years from air in Antarctic ice and fern. *Journal of Geophysical Research*, 101: 4115–4128.

Hulme, M., Osborn, T. and Johns, T., 1998. Precipitation sensitivity to global warming: Comparison to observation with HadCM2 simulation. *Geophys. Res. Lett.*, 25(17): 17379–17382.

India's Initial National Communications to the United Nations Framework Convention on Climate Change, Ministry of Environment and Forests, New Delhi, 2004.

Intergovernmental Panel on Climate Change, 1990. *Climate Change: The IPCC Scientific Assessment*. Cambridge University Press, Cambridge, U.K.

Intergovernmental Panel on Climate Change, 1995. *Climate Change 1994: Radioactive Forcing of Climate Change*. Cambridge University Press, Cambridge, UK.

Intergovernmental Panel on Climate Change, 1996. *Climate Change 1995: The Science of Climate Change*. Cambridge University Press, Cambridge, UK.

International Energy Outlook, 2005. US Department of Energy, Energy Information Administration of Government of United States of America.

Jouzel, J., Barkov, N.I., Barnola, J.M., Bender, M., Chappellaz, J., Genthon, C., Kotlyakov, V.M., Lipenkov, V., Lorius, C., Petit, J.R., Raynaud, D., Raisbeck, G., Ritz, C., Sowers, T., Sitevenard, M., Yiou, F. and Yiou, P., 1993, Extending the Vostok ice-core record of paleoclimate to the penultimate glacial period. *Nature*, 364:407–412.

Lal, M., Nozawa, T., Emori, S., Harasawa, H., Takahashi, K., Kimoto, M., Abe-Ouchi, A., Nakajima, T., Takemura, T. and Numaguti, A., 2001. Future climate change: Implications for Indian summer monsoon and its variability. *Current Science*, 81: 1196–1207.

Lane, A., Jarvis, A., Hijmans, R. and Mukema, I. (In Prep.) Conservation and Use of Crop Wild Relatives in the Face of Climate Change.

Long P., Ainsworth, E.A., Leakey, A.D.B. and J. Ort D.R. Nösberger, 2006. Food for thought: Lower-than-expected crop yield stimulation with rising CO2 concentrations. *Science*, 312(5782): 1918–1921.

Loske, R., 1996. Scope of the report: Setting the stage. In: Climate Change and Sustainable Development, Third Assessment Report of Working Group III to Intergovernmental Panel on Climate Change.

Marland, G., Andres, R.G., Boden, T.A., Johnson, C. and Brenkert, A., 1999. Global, Regional and National CO_2 Emission Estimates from Fossil Fuel Burning, Cement Production and Gas Flaring, 1751–1996. Report NDP–030. Carbon Dioxide Information Analysis Center, Oakridge National Laboratory, Oakridge, TN, USA.

Matthews, R.B., Kropff, M.J., Bachelet, D. and Van Laar, H.H., 1994. The impact of global climate change on rice production in Asia: A simulation study. Report No. ERL–COR–821. US Environmental Protection Agency, Environmental Research Laboratory, Corvallis.

O'Brien, Karen and Leichenko, R., 2001. The dynamics of vulnerability to global change, retrieved from http://www.ihdp.uni–bonn.de/IHDPUpdate0102/article4.html, on June 28, 2002.

Ozenda, P. and Borel, J.L., 1990. The possible response of vegetation to a global climate change: Scenario for Western Europe with special reference to the Alps. In: *Landscape: Ecological Impact of Climate Change*, (Eds.) Boer, M., de Groot, R.S. Proc. European Conf., Luntern, The Netherlands. IOS Press, Washington, Amsterdam, pp. 221–249.

Ramakrishna, Y.S., *et al. Impacts of Climate Change Scenarios on Indian Agriculture: Evidences*. Central Research Institute for Dry land Agriculture, Hyderabad, pp. 6.

Sathaye, J., Shukla, P.R. and Ravindranath, N.H., 2006. Climate change, sustainable development and India: Global and national concerns. *Currernt Science*, 90(3): 314–325.

Summary for Policymakers (SPM) Report, 2007. The Physical Science Basis, Working Group I (WG1) Fourth Assessment Report (AR4), retrieved from http://www.ipcc.ch/SPM2feb07.pdf.

Van Aardenne, J.A., Dentener, F.J., Olivier, G.J., Goldewijk, C.G.M.K. and Lelieveld, J., 2001. A1 degrees x 1 degrees resolution data set of historical anthropogenic trace gas emissions for the period 1890–1990. *Global Biogeochem. Cycles*, 15(4):909–928.

White, A., Cannell, M.G.R. and Friend, A.D., 1999. Climate change impacts on ecosystems and the terrestrial carbon sink: A new assessment. *Global Environmental Change*, 9: S21–S30.

World Resources 2000–2001. *People and Ecosystems: The Fraying Web of Life*. World Resources Institute, Oxford University Press, New York.

Xu, Q., 2001. Abrupt change of the mid-summer climate in central east China by the influence of atmospheric pollution. *Atmos. Environ.*, 35: 5029–5040.

Ecosystem Diversity and Carbon Sequestration: Climate Change Challenges and a Way Out for Ushering in a Sustainable Future *Pages 320–333*
Editors: **P.L. Gautam, Vir Singh and Uma Melkania**
Published by: **DAYA PUBLISHING HOUSE, 2009**

Chapter 34

Carbon Sequestration, Global Climate and Laws: What Has Been Done and What Remains?

Rinku Verma

Removal of excess carbon from the atmosphere is "carbon sequestration" which has caused climate change and various factors pertaining to global warming especially with high levels of CO_2 rise (Greenhouse gas–GHG) due to human activity. Hurricanes impact carbon sequestration by forests and climate change will affect carbon sequestration in oceans. Delay in flowering and presence of leaves which persist are some of the indicators of climate change.

Various policies and legislations are now coming into picture with the various ill effects of carbon sequestration and climate change. Senators of the US have sent a write up to the Government of US to have legislation towards global warming (GW) which is on its way (Jan 2007). Intergovernmental Panel on Climate Change (IPCC) 4th assessment report released on 2nd Feb 2007 does not raise various issues related to the impact of GW on climate and vulnerability of people. Various Federal Global Warming Legislations (2005, 2006 and 2007) are in limelight. The 3rd Assessment Report (TAR) issued in 2001 (The Scientific Basis, Impacts, Adaptation and Vulnerability, Mitigation and Synthesis Report) and 4th Assessment Report does not alter the basic global warming scenario outlined. Kyoto Protocol (1997) to the United Nations Framework Convention on Climate Change is an amendment to the international treaty on climate change, assigning mandatory emission limitations for the reduction of greenhouse gas (GHG) emissions to signatory nations. This is the most important global initiatives to address climate change. The U.N. Framework Convention on Climate Change (UNFCCC) was established in 1992. A recent IPCC report has projected a global mean sea level rise of 0.59m by the end of the 21st century with increase in sea surface temperature increasing and speeding of cyclones particularly in the Bay of Bengal, using an RCM (Regional Climate Model) of the Hadley centre with emission scenarios of the IPCC. With storm surge heights which will be much greater with its disaster on the Indian coastline. RCM of British Hadley Center for Climate Prediction and Research (PRECIS) used by Indian Institute of Tropical Management, Pune, India, has outlined GHG emission scenarios with reference to TAR with increase in precipitation and Surface Air Temperature (SAT). India needs to address this issue with proper legislations with policy makers; scientists along with the governmental officials having proper carbon sequestration, GW policies and laws. With temperature rise and IPCC emission scenario, malaria persists in many parts of India and may shift to other parts. Coral reefs are at serious threat with respect to Global temperature rise, bleached corals have been reported. Mitigation and adaptation measures countries must take to minimize adverse effects have to be addressed.

Carbon Sequestration

Atmospheric levels of CO_2 have risen from pre-industrial levels of 280 parts per million (ppm) to present levels of 375 ppm. Evidence suggests this observed rise in atmospheric CO_2 levels is due primarily to expanding use of fossil fuels for energy. Use of carbon more efficiently and to increase use of low-carbon and carbon-free fuels and technologies (nuclear power and renewable sources such as solar energy, wind power, and biomass fuels). The third and newest way to manage carbon is through carbon sequestration.

Removal of excess carbon from the atmosphere is "Carbon sequestration" which has caused climate change and various factors pertaining to global warming especially with high levels of carbon dioxide (CO_2) rise [the principal Greenhouse gas (GHG)] due to human activity.

Carbon sequestration refers to the provision of long-term storage of carbon in the terrestrial biosphere, underground, or the oceans so that the buildup of CO_2 concentration in the atmosphere will reduce or slow. In some cases, this is accomplished by maintaining or enhancing natural processes; in other cases, novel techniques are developed to dispose of carbon.

To help mitigate global warming, a variety of means of artificially capturing and storing carbon–as well as of enhancing natural sequestration processes–are being explored.

- ☆ Sequestering Carbon in Underground Geologic Repositories;
- ☆ Enhancing the Natural Terrestrial Cycle: Carbon sequestration of the terrestrial biosphere through CO_2 removal from the atmosphere by vegetation and storage in biomass and soils;
- ☆ Ongoing projects: Carbon Sequestration in the Oceans: Enhancing the net oceanic uptake from the atmosphere by fertilization of phytoplanktons with nutrients, and injecting CO_2 to ocean depths greater than 1000 meters; and
- ☆ Sequencing Genomes of Micro-organism for Carbon Management: Sequencing the genomes of microbes that produce fuels such as methane and hydrogen or aid in carbon sequestration.

Natural Sinks and Enhancing Natural Sequestration

Forests

Forests are carbon stores, and they are CO_2 sinks when they are increasing in density or area. Tropical reforestation can mitigate global warming until all available land has been reforested with mature forest (Chester 200 and Gibbard *et al.*, 2005). In the United States in 2004 (the most recent year for which EPA statistics, Raven and Falkowski 1999 are available), forests sequestered 10.6 per cent (637 teragrams) (Takashaki *et al.*, 2002) of the CO_2 released in the United States by the combustion of fossil fuels (coal, oil and natural gas; 5988 tetragrams) (Swift 2001). A long-term sequestration of carbon from forests comes from the use of wood products such as "stick built" (*i.e.*, with lumber) scarp lumber from homebuilding which increases atmospheric carbon, predominant in the US.

Oceans

Oceans are natural CO_2 sinks, and represent the largest active carbon sink on Earth. This role as a sink for CO_2 is driven by two processes, the solubility pump and the biological pump (*Raven and Falkowski 1999)*. At the present time, approximately one third of anthropogenic emissions are estimated to be entering the ocean (Takahashi *et al.*, 2002)

Another proposed form of carbon sequestration in the ocean is direct injection. In this method, CO_2 is pumped directly into the water at depth, and expected to form "lakes" of liquid CO_2 at the bottom. Experiments carried out in moderate to deep waters (350–3600 meters) indicate that the liquid CO_2 reacts to form solid CO_2 clathrate hydrates, which gradually dissolve in the surrounding waters.

Damages Caused by Carbon Sequestration in Ocean Sequestration

This method, too, has potentially dangerous environmental consequences. The CO_2 does react with the water to form carbonic acid, H_2CO_3; however, most (as much as 99 per cent) remains as dissolved molecular CO_2. The equilibrium would no doubt be quite different under the high pressure conditions in the deep ocean. Even though life appears to be rather sparse in the deep ocean basins, energy and chemical effects in these deep basins could have far-reaching implications. Much more work is needed here to define the extent of the potential problems. Carbon storage in or under oceans may not be compatible with the London Convention (Convention on the Prevention of Marine Pollution by Dumping of Wastes and Other Matter) (Rosenbaum *et al.*, 2004).

The direct injection of unwanted carbon dioxide deep into the ocean is one suggested strategy to help control rising atmospheric carbon dioxide levels and mitigate the effects of global warming. But, like the problems associate with the long term storage of nuclear waste, finding a safe place to sequester the carbon may be more difficult than scientists first anticipated was stated by the University of Illinois, 2002). Indeed estimating the impact of carbon injection is complicated because of limited understanding of climate and oceanic carbon cycle feedback mechanisms.

CO_2 dumped in the oceans won't stay there forever. Eventually it will percolate to the surface and into the atmosphere.

Soils

Carbon as plant organic matter is sequestered in soils: Soils contain more carbon than is contained in vegetation and the atmosphere combined (Swift 2001).

Benefits of Carbon Sequestration in Soil

Farmers can benefit from carbon sequestration through the use of conservation tillage, crop rotation, the use of buffer strips, and permanent vegetation for highly eroded soils. These benefits include improved soil productivity, an improved environment due to less erosion and improved physical and biological properties of soil. Carbon credit is another benefit that has been explored recently by many different entities. Carbon credit is worth exploring but needs careful consideration. The issue of market value, policy, carbon monitoring procedures and management entities are among those that need to be addressed when considering carbon sequestration and credits.

The overall benefit of soil carbon sequestration needs to be viewed as an opportunity to improve soil quality as well as the environment (http://www.extension.iastate.edu/Publications/PM1871.pdf).

In the United States in 2004 (the most recent year for which EPA statistics are available), agricultural soils including pastureland sequestered 0.8 per cent (46 teragrams (Lal, 2004) as much carbon as was released in the United States by the combustion of fossil fuels (5988 teragrams (http://www.css.cornell.edu/faculty/lehmann/biochar/Biochar_home.htm). The annual amount of this sequestration has been gradually increasing since 1998 (http://www.architecture2030.or g/home.html).

Artificial Sequestration

For carbon to be sequestered artificially it must first be captured, or it must be significantly delayed or prevented from being re-released into the atmosphere (by combustion, decay, etc.) from an existing carbon-rich material, by being incorporated into an enduring usage (such as in construction). Thereafter it can be passively stored or remain productively utilized over time in a variety of ways. One ton of dry wood is equivalent to 1.8 tons of CO_2. Building construction and operation (electricity usage, heating, etc.) are estimated to contribute nearly half of the annual human-caused carbon additions to the atmosphere.

Natural-gas purification plants often already have to remove CO_2, either to avoid dry ice clogging gas tankers or to prevent CO_2 concentrations exceeding the 3 per cent maximum permitted on the natural-gas distribution grid.

Early applications of carbon capture is the capture of carbon dioxide from flue gases at power stations (in the case of coal, this is known as "clean coal"). A typical new 1000-MW coal-fired power station produces around 6 million tons of CO_2 annually. Adding carbon capture to existing plants can add significantly to the costs of energy production; scrubbing costs aside, a 1000-MW coal plant will require the storage of about 50 million barrels of CO_2 a year. However, scrubbing is relatively affordable when added to new plants based on coal gasification technology, where it is estimated to raise energy costs for households in the US using only coal-fired electricity sources from 10 cents per kWh to 12 (http://www.architecture2030.org/home.html).

Carbon Capture and Storage

Currently, capture of CO_2 is performed on a large scale by absorption of CO_2 onto various amine-based solvents. Other techniques are currently being investigated, such as pressure and temperature swing absorption, gas separation membranes, and cryogenics.

In coal-fired power stations, the main alternatives to retrofitting amine-based absorbers to existing power stations are two new technologies: coal gasification combined-cycle and oxyfuel combustion.

Another long-term option is carbon capture directly from the air using hydroxides. The air would literally be scrubbed of its CO_2 content. This idea offers an alternative to non-carbon-based fuels for the transportation sector.

Geological Sequestration

Also known as geo-sequestration or geological storage, this method involves injecting CO_2 directly into underground geological formations. Declining oil fields, saline aquifers, and unminable coal seams have been suggested as storage sites. Caverns and old mines that are commonly used to store natural gas are not considered, because of a lack of storage safety. CO_2 has been injected into declining oil fields for more than 30 years, to increase oil recovery. These geological barriers will also be sufficient as long-term barrier to contain the injected CO_2. Identified possible problems are the many 'leak' opportunities provided by old oil wells, the need for very high pressures (about 80 times environment pressure) and low temperatures (below about 20 degrees Celsius) to keep the CO_2 liquified (only practical very deep underneath the sea) and the conversion of CO_2 into acids which can damage the geological barrier. Other disadvantages of old oil fields are their geographic distribution and their limited capacity. A major research project examining the geological sequestration of carbon dioxide is currently being performed at an oil field at Weyburn in southeastern Saskatchewan.

Mineral Sequestration

Mineral sequestration aims to trap carbon by placing it in its thermodynamics ground state where it will be non-reactive. This occurs naturally and is responsible for much of the surface limestone. Acids are used to convert mineral silicates to mineral carbonates. Ongoing research aims to speed up the kinetics of the reactions. Serpentinite sequestration is favored because of the non-toxic and predictable nature of magnesium carbonate with favourable conditions.

Various Policies, Conferences and Legislations

Various policies and legislations before and now are coming into picture with the various ill effects of carbon sequestration and climate change.

By 1979 the first World Climate Conference in Geneva took place. The establishment of a world atmosphere fund to be paid for by a levy on fossil fuel consumption in developed countries came into being. This was the first mention of a "carbon tax". None of the targets adopted by the conference has yet been attained and of course they were not binding.

The conference called for 20 per cent cut of Global CO_2 emissions till 2005 was held in June 1988 in Toronto called "The Changing Atmosphere: Implications for Global Security (Paul 1998).

The U.N. Framework Convention on Climate Change (UNFCCC) was established in 1992 which was adopted at Earth Summit. All parties to this can sign or ratify the Kyoto Protocol (KP), while non parties to this cannot. The KP was adopted at the 3rd session of the Conference of Parties to UNFCCC in 1997 Kyoto, Japan. The UNFCCC agreed to a set of a common but differentiated responsibilities. The parties agreed that; The largest share of historical and current global emissions of GHG has originated in developed countries; Per capita emissions in developing countries are still relatively low; The share of global emissions originating in developing countries will grow to meet their social and development needs (Wigley, 2006).

The first collective step the nations of the world took to deal with the issue of luxury emissions and survival emissions that led to the principle of common but differentiated responsibilities and respective capabilities, which was agreed to by the nations of the world-the US included in this historic 1992 (UNFCCC). There is then the responsibility approach. In other words, polluters pay (Karunakaran, 2007).

In the Rio de Janeiro conference in June in 1992 objective and related legal instruments was the stabilization of greenhouse gas concentrations in the atmosphere at a level that would prevent dangerous anthropogenic interference with the climate system. Berlin's First Conference of the Parties (COP-1) which began in March and lasted until April 1995. Most likely course was the creation of a new legal agreement called protocol.

According to the Chancellor Kohl (Germany) said climate protection was a joint task aimed at safeguarding the future.

1. The industrialized countries have the responsibility to limit emissions permanently beyond the year 2000. This is a vital first step. We should stand by this goal
2. With a substantial negotiation mandate we must set the course here in Berlin for a noticeable reduction in climate-damaging greenhouse gases after year 2000.
3. Industrialized and developing countries should reach an agreement on the joint implementation of climate protection measures and thus make possible the necessary transfer of know-how and technology (Paul 1998).

In the ICPP III 1996, report scientist put it this way: A highly political question of how much the world cares about imposing risks upon some of the poorest and most vulnerable people.

If the success of international responses is measured in terms of aggregate impacts on global wealth, then their position will not feature. If it is judged by how well it protects the weakest and minimizes the most severe suffering an entirely different approach may be called for. The costs associated with climate change involve inherently difficult judgements.

Carbon Sinks and the Kyoto Protocol (KP)

Kyoto Protocol (1997) to the United Nations Framework Convention on Climate Change is an amendment to the international treaty on climate change, assessing mandatory emission limitations for the reduction of GHG emissions to signatory nations. This is the most important global initiatives to address climate change.

Countries that ratify this protocol commit to reduce their emissions of CO_2 and 5 other GHG or engage in emissions trading if they maintain or increase emissions of these gases. The following principles included: Kyoto is underwritten by governments and is governed by global legislation enacted under the UN's aegis., Governments are separated into two general categories: developed countries referred to as Annex I countries (who have accepted GHG emission reduction obligations and must submit an annual GHG inventory); and developing countries referred to as Non Annex I countries (who have no GHG emission reduction obligations but may participate in the Clean Development Mechanism). But when a GHG emission reduction project is implemented in these countries, that GHG Project will receive Carbon Credit which can be sold to Annex I buyers. All the Annex I economies have established Designated National Authorities to manage their GHG protfolious under Kyoto. Countries including Japan, Canada, Italy, The Netherlands, Germany, France, Spain and many more are actively promoting government carbon funds and supporting multilateral carbon funds intend on purchasing Carbon Credits from Non-Annex I countries. These government organization are working closely with their major utility, energy, oil and gas and chemicals conglomerates to try to acquire as many GHG certificates as cheaply as possible. Annex I entities want Carbon Credits as cheaply as possible, whilst Non Annex I entities want to maximise the value of Carbon Credits generated from their domestic GHG Projects.

Main contributors to the GHG emissions during the industrialization period that is believed to be causing today's climate change are the developed countries. India and China which have ratified the protocol are not required to reduce carbon emissions (as their contribution to this is meager) under the present agreement despite their relatively large population (http://en.wikipedia.org/wiki/Kyoto_Protocol). The present G8 summit (2007) insists both India and China must reduce their check on GHG. KP is an agreement under which industrialized countries will reduce their collective emissions of greenhouse gases by 5.2 per cent compared to the year 1990 (compared to the emissions levels that would be expected by 2010 without the Protocol, this limitation represents a 29 per cent cut). The goal is to lower overall emissions of six greenhouse gases- CO_2, methane, nitrous oxide, sulfur hexafluoride, HFCs, and PFCs-(perfluorocarbons) calculated as an average over the 5 year period of 2008-12.

However, critics of Kyoto argue that China, India and other developing countries will soon be the top contributors to greenhouse gases. Also, without Kyoto restrictions on these countries, industries in developed countries would be driven towards these non-restricted countries, thus there would be no net reduction in carbon.

There are series of linked markets, rather than a single carbon market. The common theme across most of them is the adoption of market-based mechanisms centered on Carbon credits that represent a reduction of CO_2 emissions. The fact that most of these initiatives have similar approaches to certify their credits makes it conceivable that Carbon credits in one market may in the long run be tradable in most other schemes. An obvious precondition, however, is a realignment of penalties and fines to similar levels, since these create an effective ceiling for each market.

In 2001, a continuation of the previous meeting was held in Bonn where the required decisions were adopted. After some concessions, the supporters of the protocol (led by the European Union) managed to get Japan and Russia in as well by allowing more use of CO_2 sinks. COP was held in 2001 in Marrakech to establish the final details of the protocol.

The first meeting of the Parties to the KP (MOPI) was held in Montreal from November 28 to December 9, 2005, along with 11th Conference of the Parties to the UNFCCC (COP11).

Enforcement

If the Enforcement Branch determines that an Annex I country is not in compliance with its emissions limitation, then that country is required to make up the difference plus an additional 30 per cent. In addition, that country will be suspended from making transfers under an emissions trading program (http://unfccc.int/essential_background/convention/background/items/1362.php).

India Stand

India signed and ratified the Protocol in August, 2002. Since India is exempted from the framework of the treaty, it is expected to gain from the protocol in terms of transfer of technology and related foreign investments. At the G-8 meeting in June 2005, Indian Prime Minister Manmohan Singh pointed out that the per-capita emission rates of the developing countries are a tiny fraction of those in the developed world. Following the principle of common but differentiated responsibility, India maintains that the major responsibility of curbing emission rests with the developed countries, which have accumulated emissions over a long period of time.

Opposition

The two major countries who have signed the Protocol but are not intending to ratify it are the US and Australia. Some public policy experts who are skeptical of global warming see Kyoto as a scheme to either slow the growth of world's industrial democracies or to transfer wealth of the third world in what they claim is a global socialism initiative (http://www.cnn.com/ALLPOLITICS/1997/12/11/kyoto/) they argue the protocol does not go far enough to curb greenhouse emissions.

As of year-end 2006, UK and Sweden were the only EU countries on pace to meet their Kyoto emissions commitments by 2010. While UN statistics indicate that, as a group, the 36 Kyoto signatory countries can meet the 5 per cent reduction target by 2012, most of the progress in GHG reduction has come form the stark decline in Eastern European countries emissions after the fall of communism in the 1990's.

Successor

In the non-binding Washington declaration agreed on Feb 16, 2007, Presidents/ Prime Minister from various countries including India agreed in principle on the outline of a successor to the Kyoto Protocol. They envisage a global cap and trade system that would apply to both industrialized nations and developing countries and hoped that this will be in place by 2009 (Hilsenrath 2001; Adams 2006).

Because growing vegetation absorbs CO_2, the KP allows countries that have large areas of forest (or other vegetation) to deduct a certain amount from their emissions, thus making it easier for them to achieve the desired net emission levels.

Some countries want to be able to trade in emission rights in carbon emission markets, to make it possible for one country to buy the benefit of CO_2 sinks in another country. If overall limits on greenhouse gas emission are put into place, such a "cap-and-trade" market mechanism will tend to find cost-effective ways to reduce emissions. (http://unfccc.int/kyoto_protocol/compliance/introduction/items/3024.php). There is as yet no carbon audit regime for all such markets globally, and none is specified in the KP. Each nation is on its own to verify actual carbon emission reductions and to account for carbon sequestration using some less formal method.

In the Clean Development Mechanism, only afforestation and reforestation are eligible to produce Certified Emission Reductions (CERs) in the first commitment period of the KP (2008–2012). Forest conservation activities or activities avoiding deforestation, which would result in emission reduction through the conservation of existing carbon stocks, are not eligible at this time (http://unfccc.int/files/essential_background/kyoto_protocol/application/pdf/kpstats.pdf). Also agricultural carbon sequestration is not possible yet (http://www.climnet.org/EUenergy/ratification/calendar.htm).

Senators of the US have sent a write up to the Government of US to have legislation towards Global warming (GW) which is on its way (Jan 2007). IPCC, 4th assessment report released on 2nd Feb 2007 does not raise various issues related to the impact of GW on climate and vulnerability of people. Various Federal Global Warming legislations (2005, 2006 and 2007) are in limelight. "The warming of the climate system is unequivocal, as is now evident from increases in global average air and ocean temperatures, widespread melting of snow and ice and rising global mean sea level, says the report IPCC (AR4) on climate change (Ramachandran 2007). The WGI report builds on past IPCC assessments and incorporates new findings since 2001 to provide, on the basis of climate science, the strongest statement so far on the extent and causes of climate change.

A recent IPCC report has projected a global mean sea level rise of 0.59 m by the end of the 21st century with increase in sea surface temperature increasing and speeding of cyclones particularly in the Bay of Bengal, using an RCM (Regional Climate Model) of the Hadley centre with emission scenarios of the IPCC. With storm surge heights which will be much greater with its disaster on the Indian coastline. RCM of British Hadley Center for Climate Prediction and Research (PRECIS) used by Indian Institute of Tropical Management, Pune, India, has outlined GHG emission scenarios with reference to Third Assessment Report with increase in precipitation and surface air temperature (SAT). The sea level rise itself, can inundate a large swath of predominately agricultural land on the coast, and the surviving coastline faces the threat of extreme storm surges. India is one of the 27 countries the UNEP has identified as most vulnerable to sea level rise (Ramachandran, 2007).

Climate models and observations indicate that rainfall extremes such as the Mumbai deluge of 2005 could become more frequent in India under the impact of climate change. Both 2005 and 2006 had spells of excessive rainfall that normally would have occurred once in a hundred years or so.

The all round warming seen in the mean is also reflected in the extreme temperatures, and both nights and days will get warmer in the future. Interestingly, night temperatures are seen to be increasing at a faster rate than day temperatures.

The latest report of IPCC fore casts that by the end of this century the probable temperature rise will be in the range of 1.8°C to 4.0°C if we go along business as usual (Karunakaran 2007).

The TAR issued in 2001 (The Scientific Basis, Impacts, Adaptation and Vulnerability, Mitigation and Synthesis Report) and 4th Assessment Report does not alter the basic global warming scenario outlined.

With reference to India, a subject on which substantial scientific literature has appeared since the TAR was released:

One of the direct effects of global warming is the altering of the heat budget and its regional variations in the atmosphere, the primary driver of weather systems around the world. The potential impact of alarming on the monsoon in India is of serious concern, particularly with respect to agriculture, much of which is rain fed.

There is pressure mounting on all countries to act decisively, including developing countries that have no targets to meet up to 2012 under Phase-I of the Kyoto Protocol. At the 12th conference of the Parties of UNFCCC held in Nov 2006, besides the reviewing of progress and the laying down of commitments beyond 2012 by developed countries, there were proposals mooted to encourage voluntary commitments by developing countries as well (Ramachandran 2007).

The CO_2 radiative forcing increased by 20 per cent during 1995-2005, the largest change for a decade in at least the last 200 years," points out R.K Pachauri, Chairman of the IPCC and Director-General of (TERI) based in Delhi.

Ministers from the Group of Eight rich industrialized countries (Germany, France, UK and Italy (G-8) plus 5 major developing countries (including India which is under G77) met in June 2007 at Hamburg, Germany. Important decisions on climate change including global negotiations are in the agenda (Ravindranath 2007). The main agenda is that all of these countries agree on settling binding targets to curb climate change by reducing greenhouse emissions. Carbon Trading–Curb climate change; limit global warming to two degrees Celsius this century. Lay ground work to strengthen Kyoto Protocol

Global Climate: Some Environmental Concerns

Delay in flowering and presence of leaves which persist are some of the indicators of climate change. Some other indicators are: Loosing of ice cores, Corel reefs bleaching, Precipitation trend; more rainfall extremes and side effects, Himalayan concern, Glaciers and polar ice caps melting faster, Cyclones and Storm surges/Storm warning, Food Security issues, unseasoned rain disrupts flowering cycle, senescence, Crop damage, Communication and Power hit, Vector diseases, Poverty, Natural disaster, Desertification, Forest loss, Loss of species by mid century, Paucity of drinking water and finally it is a pity we becomes Environmental refugees.

Temperature rise of 3°–5°C more would lead to extinction of perhaps a 4th generation of the worlds species to inundation of coasts and islands particularly crowded shores of Southern Asia and mega deltas, Ganga, Indus and Brahamaputra, including Arctic coral reefs. There is a threat to Gangotri glaciers, Sunderbans, (two islands have already submerged), and Low lying rice fields in Kerala as well.

Shanghai and Mumbai are likely to be affected by rising sea levels and stronger storm surges caused by global warming according to studies.

For a tenth of the world's population, live within 10 meters above the current sea level. "Low elevation coastal zone (LECZ) which the scientists define as coastal areas within an elevation of 10 meters (Journal of Environment and Urbanization) found that Asia, with its high population densities and countries having large populations was particularly at risk from increases in sea level. China with an estimated 144 million people, India with more than 63 million and Bangladesh with more that 62 million lead the ranking of countries with the largest populations in the zone. In coastal locations of Asian cities, in which 32 per cent of the population of cities, over 5 million resides in the LECZ, according to the journal.

Carbon Credits

Maximum carbon credits have been earned by the developed nations than the real poor of those countries who need it most.

Carbon Debt

China and India have contributed to < 8 per cent greenhouse gases since 1850, U.S on the top with 29 per cent, West Europe with 27 per cent, with Atmospheric space being reduced and still more in the process due to these GHG; the First World owes the Third World Climate dept. Let the developing countries get their returns!

Emission Sharing Among Nations

Several countries have a huge carbon divide.

If all the people of the world had the same living style as the average American, the holocaust predicted for the distant future would have already visited us (Karunakaran 2007).

People who most risk from climate change live in countries that, have contributed the least to atmosphere build up of CO_2 and GHG linked to recent warming of planet. Most vulnerable countries tend to be the poorest while the richest countries are well equipped. Large industrialized countries are more resilient partly because of geography they are mostly in mid latitude regions with Goldilocks climate (neither too hot nor too cold). Many enjoy gifts like thick, rich soil and generous growing seasons of the American Corn Belt or forgiving weather of France and New Zealand. Hence North America and northern Europe will see some benefits, at least in the short terms, from longer growing seasons and milder winters.

"If you drive your car into your neighbour's living room, don't you owe your neighbor something?" said Gleick (founder of the Pacific Institute for Studies in Development, Environment and Security in Berkeley, California). "On this planet, we are driving the climate car into our neighbours living room and they don't have insurance and we do."

With temperature rise and IPCC emission scenario, malaria persists in many parts of India and may shift to other parts. Coral reefs are at serious threat with respect to Global temperature rise, bleached corals have been reported.

Governments under the present world order can be expected to take the courageous step to foreground intergenerational as well as intra-generational equity the only solution that will work-in addressing the complex issue.

It is the ordinary people of different countries, collaborating with each other, who can ultimately bring about the social change needed to prevent effectively the environmental disaster that looms ahead added by Sir Nicholas Stern, the author of the Stern Review.

There are abundant examples of how wealth is already enabling some countries to grid against climatic and coastal risks, while poverty, geography and history place some of the world's most crowded vulnerable regions, directly in harms way.

Indian Scenario

Union environment ministry is taking up measures for conservation of rivers through pollution abatement works under the "National River Conservation Plan". Due to population growth per capita water availability has been declining over the years. Project for Repair, Renovation and Restoration of water bodies linked directly to agriculture were approved in 2005. Another interesting feature which figured in New Delhi in the early 2007 was "just change the yellow light emitting bulbs you use at your homes. Such a simple act can save the earth from loads of deadly CO_2 that triggers global warming. If these incandescent bulbs are altered in most India, it can reduce the country's CO_2 emission load by 4 per cent which is as good as the entire CO_2 emission of Denmark, says Green peace. The alternate is compact fluorescent lights which is more energy efficient and lasts longer 7000 hours as compared to incandescent lasts1000 hours. The gross per capita water availability in India will decline from 1820 cubic m per year in 2001 to 1140 cubic m per year in 2050. Decreases in crop yield, loss in biodiversity are direct probable consequence of climate change.

Future Directions

1. GHG are warming the globe and damaging our environment with long term repercussions.
2. If we and our future generations are to lead a normal life, right decisions ought to be taken and right policies have to be formulated and followed without any further delay. Strict penalties should be imposed on those violating the rules.
3. Government should fund research and encourage people to move towards renewable energy alternatives.
4. Issues related to Carbon sequestration should be dealt with utmost care after in-depth knowledge.
5. The clock is ticking and now is the time to realize, stand up and take the right decision. Tomorrow may be too late.

References

Adams, Cecil, 2006–04–30. *Is Global Warming for Real?*. Straight Dope, http://www.straightdope.com/columns/060407.html.

BBC, 2005. Trees only cool the planet if they are planted in the tropics. News Release: *Models show growing more forests in temperate regions could contribute to global warming*. December 5, 2005 NR–05–12–04, Website of Lawrence Livermore National Laboratory http://www.llnl.gov/pao/news/news_releases/2005/NR–05–12–04.html. S. Gibbard, K. Caldeira, G. Bala, T. J. Phillips, M. Wickett, "Climate effects of global land cover change", *Geophysical Research Letters*, Vol. 32, December.

Chester, Bronwyn April 20, 2000–Volume 32 Number 15, *McGill Reporter*.

Hilsenrath, Jon E. 2001. Environmental economists debate merit of U.S.'s kyoto withdrawal. *Wall Street Journal*, 2001–08–07.

http://en.wikipedia.org/wiki/Kyoto_Protocol.

http://unfccc.int/essential_background/convention/background/items/1362.php, UNFCCC.

http://www.architecture2030.org/home.html Climate Change, Global Warming, and the Built Environment–Architecture 2030

http://www.climnet.org/EUenergy/ratification/calendar.htm).

http://www.cnn.com/ALLPOLITICS/1997/12/11/kyoto/Clinton Hails Global Warming Pact", *All Politics*, CNN, 1997–12–11.

http://www.css.cornell.edu/faculty/lehmann/biochar/Biochar_home.htm

http://www.extension.iastate.edu/Publications/PM1871.pdf, Kaisi Mahdi Al, 2001, Impact of Tillage and Crop Rotation Systems on Soil Carbon Sequestration, Iowa State University, University Extension.

Karunakaran, 2007. *Frontline*, 24: 25–30.

Kyoto Protocol: Status of Ratification, 10 July 2006 (PDF). UNFCCC. http://unfccc.int/files/essential_background/kyoto_protocol/application/pdf/kpstats.pdf.

Lal, R. *et al.*, 2004. Managing soil carbon. *Science* 304: 393.

Paul Brown, 1998. *Global Warming, Can Civilization Survive*. Universities Press (India) Limited, pp. 235.

Ramachandran, 2007. *Frontline*, 24: 10–16.

Raven, J.A. and Falkowski, P.G., 1999. Oceanic sinks for atmospheric CO_2. *Plant Cell and Environment*, 22: 741–755.

Ravindranath, 2007, *Frontline*, 24: 20–24.

Rosenbaum, K.L., Schoene, D. and Mekouar, A., 2004. *Climate Change and the Forests Sector: Possible National and Sub-national Legislation*, Number 144 in *'FAO Forestry Papers'*, FAO. Available from: http://www.fao.org/docrep/007/y5647e/y5647e00.HTM

Swift, Roger S., 2001. Sequestration of carbon by soil. *Soil Science Abstract*, 166(11): 858–871.

Takahashi, T., Sutherland, S.C., Sweeney, C., Poisson, A., Metzl, N., Tilbrook, B., Bates, N., Wanninkhof, R., Feely, R.A., Sabine, C., Olafsson, J. and Nojiri, Y.C., 2002. Global sea-air CO_2 flux based on climatological surface ocean pCO_2, and seasonal biological and temperature effects. *Deep Sea Research II*, 49: 1601–1622.

Wigley, Tom. The Kyoto Protocol: CO_2, CH_4, and climate implications. *Geophys. Res. Lett.*, 25(13): 2285.

USHERING IN A SUSTAINABLE FUTURE

Eco-ethics, Eco-philosophy and Psychology as Core Elements Pivoting Conservation-oriented Transcendental Development

Ecosystem Diversity and Carbon Sequestration: Climate Change Challenges and a Way Out for Ushering in a Sustainable Future *Pages 337–341*
Editors: P.L. Gautam, Vir Singh and Uma Melkania
Published by: DAYA PUBLISHING HOUSE, 2009

Chapter 35

Conservation of Biodiversity for Sustainable Development: Eco-ethics as an Indispensable Element

A. Vanmathy and Abha Ahuja

Pure objective science has little use for such concepts as 'conservation', 'benefit to all', 'survival value', for these are categories outside the content of objective science, and they are outside the so-called 'cognitive content'. These categories are immensely important to us and our well-being in the short run and in the long run. Thus the entire premise of right conservation on the global scale is steeped in values. Simply, the conservation strategy is a value programme. We need to know in great detail what we are doing, but at the same time we must not forget for a moment why we are doing what we are doing. The scientific component of conservation strategies and the value component do not clash with each other, but complement each other.

– Henryk Skolimowski

Forests of the Earth are fast vanishing. And with that the Earth is gradually going into the worst trap of what is being referred to as 'global warming'. There was peace and harmony in the intricate complex of mutual relationships in nature's creation. But when man tried to out smart himself to design artificial environments it was the beginning towards his end. Vanishing of forests has created a very large gap between man and nature.

Forests are the major habitats for numerous kinds of life forms. A forest, we can say, is a mother of biodiversity. Our thought of conservation must focus on forests. They are absolutely needed for the conservation of terrestrial biodiversity. They are also of absolute importance for the conservation of human race, human cultures and human existence.

Biodiversity Conservation and Environmental Consciousness

The *Oxford Advanced Learners Dictionary* gives the meaning of biodiversity as the existence of a large number of different minds of animals and plants which make a balanced environment. Biodiversity refers to the variety and variability among living organisms, the genetic diversity they contain, the assemblages they form and the ecological complexes in which they occur. Man is a social and technical animal; his place in nature is peculiar because of the artificial environments he is able to create. Nature is auto-sustainable and self regulating if there are no man-made perturbations. The natural world is a far different place now than it was about ten thousand years ago. Different natural ecosystem on the planet have been affected and modified by the human populace. While it may be argued that the industrialized nations, having enjoyed the fruits of unthinking exploitation of nature, have now suddenly become environmentally conscious and are obstructing the developing nations in their march towards material progress. However for the long term interests of the earth and human species, environmental management must be taken seriously by one and all.

India faces environmental challenges on two fronts: Poverty as well as economic development. Poverty and population pressures clearly result in overuse of land, water, forest and other resources without thought of future-for who wants to think of the coming generations, when it is so difficult to survey today? Uncontrolled economic development, urbanization and industrialization are equally responsible for deforestation, overuse of groundwater systems and pollution of natural resources. It was only in the 1970s that the Indian government took the matter of environment seriously. In June 1972, the stockholder conference of Human Environment propounded the concept of "eco-development" implying a process of ecologically sound development for positive management of the environment for human benefit. It was the same year that a committee was set up in Indian under the initiative of the then Prime Minister, Mrs. Indira Gandhi to co-ordinate environmental issues.

The Call for Conservation

Conservation is a race between catastrophe and right utilization. In attempting to accelerate the evolutionary process wisely, we must be one with it and should serve it. The purpose of the laboratory of nature is to enhance what is, not to reduce it; to create diversity, not homogeneity. In that way, human psychology should give more importance to the social and cultural values to emphasis right conservation strategies for the benefit of all human kind. The weight of cultural responsibility is most clearly recognized in the social context. The community spirit in biodiversity conservation is obviously visible in the rich socio-cultural practices. The traditional life style is much intertwined with environment, which respects forest as the mother's home. The *Chipko Andulan movement* of hill women against the destruction of forest in the Himalayas has proved that they can protect, conserve and mange in environment. The *Maiti Movement* in Chamoli district of Uttarakhand region has made this region to flourish with trees. During the marriage ceremony of the Maiti group of girls, the bride presents sapling to her bridegroom, who plants it under the auspicious chanting of mantras. The idea of Rakish Sutra Movement, to save the nature wealth of Himalayas is totally a new thought that came in August 1994. The women of different villages tie *Rakish Sutra* to the trees of their forest and jointly raised their voice against forest business and felling of trees. All such movements are blended with the ongoing practices of the people, imbibed with their emotional sentiments and strong determination for the preservation of natural resources.

Biodiversity and Sustainable Development

There is a direct relationship between biodiversity and sustainable development. Sustainable development, in essence, is the one that ensures ecological integrity of a system, according to ecologist Prof. Vir Singh. Coining the phrase "ecological integrity", he says that no development can be sustainable unless it has ecological integrity of its own. Functional diversity, achievable through combining complementary plant and animal species in synergetic interactions, infuses sustainability into an ecosystem, according to Vir Singh.

Biodiversity causes, enhances, or otherwise contributes to ecological integrity of a system. Biodiversity contributes to enhance the vitality of an entire ecosystem which is best ensured when the soil is well managed and health of the plants, animals, and people is maintained through ecological (biological + environmental) processes, says Singh. According to him, "the natural ecosystems endowed with biodiversity are extremely resilient. Biodiversity creates barriers against any natural calamity (drought, for example) and pest epidemics. Some living organisms may be host to certain insects or diseases, some may be little more prone to weather extremes or other adverse conditions, but a complex mix of naturally occurring (in natural forest areas) organisms will do away any chances of widespread loss to the farmers. Simple, less complex ecosystems and monocultures are extremely prone to various kinds of damaging agents and hence threaten the very sustainability of ecosystems".

The Need for Reverential Thinking

Biodiversity conservation is discussed in many global forums since few decades back but still we could not land up in an effective technique to heal our earth. There is some lacuna in our approach towards nature. So, let us think different. Let us have a reverential thinking, which make us aware not only of the physical existence but goes beyond it to inherit the worth and beauty of nature. For such type of reverential thinking, the forerunner of the ecological consciousness is the humanistic psychology. It will be ludicrous to handle human psychology as a luddite. It should be handled against the temper of mechanistic age. I strongly believe that human psychology can best regulate the socio-cultural entities with the emotional respect on nature for its blessings bestowed upon us.

Conservation as a Function of Ethics

T.S. Eliot says, "A wrong attitude towards nature implies somewhere, a wrong attitude towards God, and the consequence is an inevitable doom". Conservation is an inevitable content of our responsibility. Responsibility is a moral category. Excessive specialization in science cannot be of much help in conserving nature in its intact form. Eco-philosopher and poet of our contemporary times, Henryk Skolimowski says, "All ethics have always been in the services of conservation and preservation.

All ethics wish to preserve and conserve the sanctity of human being, his wholeness, his integrity. The virtues that various systems of ethics advocate and wish us to uphold always serve some programme of conservation. Conservation should be seen as an ethical principle. For all ethic is conservation; and all conservation is ethic." The Skolimowskian eco-philosophy synthesizes the following theories in relationship to conservation and ethics:

1. Man did not weave a web of life, he is merely a strand in it. Whatever he does to the web, he does to himself.
2. Conservation work is about responsibility. We abrogate our responsibility if we know the truth and do not convey it to others.
3. Cost-benefit analysis is insufficient as the basis for long range conservation strategies, and is in the service of myopic values.
4. Every form of life is unique, warranting respect, regardless of its present worth to man.
5. Reverence for life is both an ecological value and an important principle of conservationist strategies, for without endorsing it, our defence of living habitats will hang in thin air.
6. Our evolutionary heritage is a value, and evolution can be used as the criterion as to which forms of life are more valuable than others: the more accomplished a creation of evolution, the more worthy it is of preservation.
7. Self-reliance of individuals, nations and culture is an ethical imperative so that the idea of responsibility (of individuals and nations) does not become meaningless.

Conclusion

The need of the hour is ecological reconstruction to assure a life of dignity and grace for future. The wounded nature calls for the ecological catharsis, so let us regenerate our ethical values to join her. The healing of our environment requires the healing form of thinking and the change in our attitude geared to healing, not to exploitation. The healing nature of ecological thinking has more power to fetch us a permanent solution to our entire struggle towards nature conservation, if it emphasizes on the humans' ethical approach united by social and cultural solidarity. This would seek us meaning beyond the mere survival for consumption and lead us towards a spiritual journey. So, let us join our hands for a noble cause, to pass on the green earth and the blue sky to our next generation as we inherited from our ancestors. It is an imperative of our eco-ethics.

References

Paria, D., 2006. Biodiversity: Its perspective and constraints. *Science and Culture*, 7–8: 217-219.

Pathak, A., Ashish, N., Neushiu, S. and Rao, G., 2006. Community conservation survival against all odds. *Environment and People*, pp. 21–24.

Singh, V., 2005. *Agrobiodiversity, Sustainability and Food Security in the Himalayan Mountains: An Uttaranchal Perspective.* Gorakhpur Environmental Action Group, Gorakhpur

Skolimowski, H., 1995. *Dancing Shiva in the Ecological Age.* Dr. Henryk Skolimowski International Center for Eco-philosophy and Indian Institute of Ecology of Environment, New Delhi.

Ecosystem Diversity and Carbon Sequestration: Climate Change Challenges and a Way Out for Ushering in a Sustainable Future *Pages 342–355*
Editors: **P.L. Gautam, Vir Singh and Uma Melkania**
Published by: **DAYA PUBLISHING HOUSE, 2009**

Chapter 36

Environmental Services Emanating from the Himalayan Mountains: Valuation Against the Backdrop of Eco-philosophy and Chasing the Goal of Global Happiness

Vir Singh

The concept of "ecosystem services" or "environmental services" or "natural services" is becoming increasingly popular since the last decade of the last Century. The term actually was coined by the economists and since the publication of the oft-quoted article of Costanza *et al.*, in 1997 on economic value of ecosystem services at global scale it is being heard all over and gets frequently encountered in documents generated by several national and international development agencies. Vocabularies of scientists, academicians, NGOs and development departments of public sectors, international organizations are now replete with the phrase "ecosystem services" or "environmental services" or "natural services", "environmental services" having become more popular.

Development of privatisation of natural areas and ecosystems reached its climax towards the close of the 20th Century with the conception of monetary values of the "services" rendered by natural systems. *Nature* magazine (Costanza *et al.*, 1997) and the book *Nature's Services* (Daily 1997) brought the concept to the fore. Ecosystem services emanating from the Earth's ecosystems have been tagged with price. For example, of late, Costanza *et al.* (1997) estimated the value of the Earth's ecosystem services at USD 33 trillion per year.

The emerging concept of ecosystem services, in actual sense, provides critical means of taking privatization to a new level–a means of privatizing many things that have as yet been unavailable for privatization: air, water, and all sorts of other ecological processes (GRAIN 2005). The environmental services are being largely accepted not only by private sectors but also by government agencies. The Kyoto Protocol has already created an environment through the provision of economic processes that puts a value on not emitting CO_2 and enables countries to trade carbon emissions (Singh 2006). Latin American nations, particularly Costa Rica, and Mexico, Ecuador and Brazil have been the pioneers in environmental services. Australia and the Philippines are the front-runners. The fields in which the greatest practical implementation has been made are the sequestration of atmospheric carbon, the capture and storage of water, and biodiversity, and landscape conservation (primarily for tourism) (GRAIN 2005).

In the Himalayan context, these invaluable services are not only of greater magnitude but are also of crucial implications for the region as well as for the large areas in the plains. Much of the appropriate environment conducive to food production in the plains is attributable to the ecosystem services rendered by the Himalayan mountains. This further increases the importance of mountain ecosystems in the mainstream context.

In the backdrop of current situation, it is necessary to go deeper into the many aspects of ecosystem functioning in the Himalayan mountains so that we could help ourselves develop a sound perspective compatible with our specific geo-ecological and socio-cultural framework. What is the ethical basis of the valuation of these ecosystem services? Can the Western system (rather mind) which is also reflected in the Kyoto Protocol suffice to all regions and cultures of the Earth? These are some of the issues that emerge out of the valuation system of ecosystem services. This piece is an attempt to value the ecosystem "services" against the backdrop of eco-philosophy.

Spinning Dollars through Environmental Services

Quite a few well-known NGOs, institutions and intergovernmental organizations are involved in environmental services and there are numerous projects and examples as presented in Table 36.1. These organizations are building up and environment for spinning dollars out of the environmental services. These are chalking out strategies, making partners, involving larger people/communities and nations, implementing projects and alluring masses towards harp on the environmental services

A variety of hosts (such as described in Table 36.1) descending into the arena of "environmental services" are developing new concepts, refining the old ones, redefining "values", designing new tactics, evolving new strategies, using seductive language and opening new fronts of global trades with alluring terms and conditions.

Concept of Natural Capital and Ethics of Living

The sunlight is the extra-terrestrial resource which is the first and the foremost source of life on Earth. This is the "service" rendered by the Cosmos that is primarily responsible to infuse life into planet Earth, evolve and design biomes and ecosystems and to create an enormous diversity in life forms. This is the first resource that makes the living Earth generate "environmental services". Can this resource be privatized?

The following concepts of natural capital developed by the World Bank in 1993 and further modified in 2003 (GRAIN 2005) would imply that this can be done. Let us have a perusal of the concept of "natural capital":

Table 36.1: World's Organisations Actively Involved in Environmental Services

OrganisationWebsite	*Activities in Environmental Services*
Worldwide Fund for Nature (WWF) www.wwfca.org/php/proyectos/agua/fondo02eng.php;www.panda.org/downloads/policy/shilling.pdf	The Water Fund model; Motagua-Polochic System, Guatemala; a novel environmental Payment Scheme with the Private Sector. WWF has also published a series of booklets and studies on payment of environmental services (PES). The first step in promoting stewardship of natural resources and the services they provide their true value. The next step is making environmental services more marketable. Creating environmental markets is the third step.
Conservation International (CI) http://conservation.org.br/programas/?id=98	CI-Brazil describes a partnership with various corporations, such as DuPont Brazil, and identifies PES as one of their activities.
World Conservation Union (IUCN) http://www.worldagroforestry.org/sea/Networks/RUPES/	IUCN supports or is part of programmes such as RUPES (Rewarding the Upland Poor for Environmental Services) in Sri Lanka, Philippines, Thailand, Vietnam, Indonesia, India, China and Nepal.
The Nature Conservancy (TNC) http://nature.org/aboutus/howwework/consevationmethods/conservationfunding/	The Nature Conservancy clearly identifies "market incentives for conservation", and PES as part of their central strategies. Examples of recent projects can be found in Chipas, Mexico; Logo de Yojoa, Honduras; Quito, Ecuador, and Sierra de la Minas, Guatemala. TNC is also part of the PES initiative in the Noel Kempf National Park in Bolivia.
International Institute for Environment and Development (IIED) http://www.iied.org/eep	IIED runs an Environmental Economics Programme within which runs a project named "Markets for Environmental Services". The aim of this project is to promote the provision and maintenance of environmental services in ways that reduce poverty and improve livelihoods. IIED aims to develop and test a general framework for analyzing the environmental and poverty impacts of market-based approaches to environmental protection.
World Resources Institute (WRI) http://www.maweb.org/en/Products.EHWB.aspx#downloads	WRI has presented policy proposals to make the marketing of environmental services more efficient and attractive. The WRI has also been involved, amongst others, with the Millennium Ecological Assessment Initiative (MA). This initiative aims to assess environmental services at a global scale, and its marketing is one of the lines of actions to be explored as part of "strategic recommendations".
World Bankhttp://www.fao.org///wairdocs/lead/x6154e/x6154e07.htm	The World Bank has a strong policy of promoting PES around the world.

Contd...

Table 36.1–Contd...

OrganisationWebsite	*Activities in Environmental Services*
United nations Food and Agriculture Organisation (FAO)http://www.fao.org/docrep/006/y5305b/y5305b01.htm	FAO has published several documents regarding PES. In one of them "Payment Schemes for Environmental Services in watersheds" can be read as the first line of its summary: "Payment Shemes for Environmental Services (PES) are flexible, direct and promising compensation mechanisms", and later on: "PES systems present a series of advantages and opportunities which make them a promising mechanism to improve the conditions of water resources in watersheds".
Global Environmental Facility (GEF)	The main funding mechanism through which the World Bank implements its environmental policy.
Tropical Agriculture Research and Higher Learning Centre (CATIE) http://catie.notlong.com	CATIE provides technical support to different projects that include PES. The institute has also created a "Group on the Socio-economy of Environmental Services" dedicated to research and teaching on this topic. The group identifies PES as one important area of work.

Source: Based on the information of GRAIN (2005).

"The capital of an economy is its stock of real goods, with the power to produce further goods (or utilities) in the future. This definition of capital would probably be acceptable to most economists. Viewed as such, capital would comprise land, which in classical economic thinking is considered a separate factor of production, for land would qualify as part of the stock of real goods, capable of producing further goods. It is but a short step to extend this definition to nature, both as a source of raw materials and as a receptor of wastes generated in the course of economic activity."

Further modified in 2003, the concept defines the natural capital like this (Hawken *et al.*, 2003):

"Natural capital includes all the familiar resources used by humankind: water, minerals, oil, trees, fish, soil, air, etc. But it also encompasses living systems, which include grasslands, savannas, wetlands, estuaries, oceans, coral reefs, riparian corridors, tundra, and rainforests."

The new concept of the "natural capital" would virtually include everything on Earth. It would also suffice to any element of nature that produces new goods. As the sun is the supreme source of producing natural goods through the vital process of photosynthesis and unabatedly generating "environmental services", it ought to be a natural capital. Everything imaginable on Earth, in fact, comes under the cover of "natural capital", even the non-terrestrial resources, like the sunlight, are not beyond the realm of this concept.

We have been living on Earth for millennia in perfect harmony with nature. The other life forms and the natural resources, now referred to as "natural capital" came into existence billions of year ago than the human species. Evolution and existence of human species might not have been possible without the natural state of harmony and equilibrium. We have been living on Earth in the Cosmos, not in any sphere comprised of "natural capital". Is the very concept of "natural capital" to which the

The Importance of the Word "Services"

"Services" is a vague economic term with broad applications, and includes goods and processes that are not strictly productive, but which are a working part of the economy: highways, communications, banking, advertising, and so on. In practices, it has been sufficient to call something a service for it to be considered such from an economic and legal point of view. The reason why we talk about "environmental services" today, rather than "environmental processes" or "environmental functions" is because the concept of services fits perfectly with possibilities of maximizing the earnings generated by obligatory consumption:

(*a*) Unlike a product that we buy and pay for just once, a service must be paid for each time it is used. Once again, as Hawken states: "An economy based on a service-and-flow model could also help stabilize the business cycle, because customers would be purchasing flow of services, which they need continuously, rather than durable equipment that is affordable only in good years". How long can you hold your breath?

(*b*) Environmental services have a captive market that is constant, endless and free of capital depreciation.

(*c*) The concept allows the claim to ownership of not only tremendous components of the planet Earth, but also of intangible elements such as the regulatory capability of ecosystems.

(*d*) Because they are intangible, services can be consolidated or broken down into separate parts freely and according to the criteria of the seller. For example, a business could sell "weather control", but it would be much more profitable to sell individual packages of "the right amount of rain", "equable temperatures", "the absence of floods", "the absence of extreme temperatures", "freedom from drought, "perfect summers", "beautiful spring times", "the absence of storms", "tolerable winds", and so on. The creation of "new services" is limited only by entrepreneurial imagination.

Source: GRAIN (2005)

concept of "environmental services" is inherently bound not very much against the very ethics of living? Does such state of affairs not violate the very sanctity of life?

Every life-infusing and life-supporting element in nature is being given distortion in the psyche, thinking and right kind of action of the society. For instance "atmospheric regulation" that emanates from the very concept of "natural capital" is defined as the ability to keep air quality at breathable levels and is considered today to be an environmental service. It implies that we are not breathing in the very atmosphere provided by the mother Earth, but we are "receiving an environmental service" for which we will have to pay should we opt for "air quality at breathable level". Doesn't it put a question mark on the very fundamental aspects of life? It also implies that our happiness, our joy, our recreation, our safety, and our feelings are all attributable to the "environmental services" for which we have to inevitably pay.

The twin concepts of "environmental service" and "natural capital", as GRAIN (2005) suggests, are about privatization and exploitation and, above all, making payments to those who have claimed property rights over that "capital". And payment is obligatory because we might deny ourselves the purchase of television or a hamburger, but we cannot deny ourselves the act of breathing." Hawken *et al.* (2003) refer to it as Natural Capitalism, which is aimed at creating the next industrial revolution. Had V.I. Lenin, the founder of erstwhile USSR and one of the greatest revolutionaries of the 20th Century, would have called it the 'highest stage of imperialism'. This kind of imperialism is definitely worse than that of the one based on colonialism.

Environmental Services from the Himalayan Mountains

There is much hullabaloo about the ecosystem services emanating from the Himalayan mountains for last few years. A number of academicians in the Universities as well as scientists with NGOs throughout the Hindu Kush-Himalayan Region are now 'devoted' to evaluate the environmental services generated by Himalayan ecosystems. Regional governments are also seriously working to articulate the services into their policies. Importance of the Himalayan region further increases on account of the fact that this has been a hot spot in the regional as well as in the global politics and directly and indirectly supports significant proportion of the world population. The youngest, the highest, the most fragile and vulnerable mountain chain of the Himalaya supports myriad life forms and human cultures. These Himalayan mountains also serve as "water towers" for the mankind. Further, critical role of the majestic mountains in purification of air and water, hydrological regulation, giving refuge to numerous rare and endemic species of potential economic value, carbon sequestration and moderation of climate, etc. allure the interested organizations and their partners to evaluate the economic value of every thing, every process hitherto imaginable and unimaginable. Valuation of the "environmental services" oozing out of the Himalayan ecosystems is being linked with human well-being.

It has been argued that in spite of the crucial ecological, cultural and economic importance of the environmental services, ecosystems are being continually deteriorating worldwide as the value of ecosystems to human welfare is still underestimated (Negi and Agrawal 2006). But is it really true? Or, even if it is true, why to lament on it? Because the trading agencies could not extract as much money as they could aspire? It does not require citing references to put forth the fact that natural resources, including Himalayan forest ecosystems, have been wantonly destroyed to satiate greed of some vested interests who knew the economic value of these resources. As the economic values of different resources have been worked out, the pace of ecosystem degradation has increased. So long as the forest ecosystems were not linked with the external markets, or, in other words, so long as they were utilized only by local communities (who did not put any price tag on any resource), the forest resources were blossoming and flourishing.

Can there be a foolproof method of evaluation of the environmental services? If one claims that these services amount to, for example, USD 0.5 trillion or so, should one be trusted? And what, if these services equate to exactly that figure? Contribution of the ecosystem services to the Himalayan communities and–through ecosystem linkages–to the foreland communities in the plains is not an easy job to evaluate in monitory terms. We can ponder over an attempt made for the ecosystem evaluation of the millennium in 2002. The main motive was to evaluate the costs of compensation for rural communities (http://www.prisma.org.sv/pubs/CES_RC_Es.pdf) quoted in *Seedling* by GRAIN (2005). These are categorized into supply, regulation or control, cultural and back-up, or supporting activities. On the basis of these, an attempt has been made to list the specific contributions of environmental services in the context of the Himalayan mountains (Table 36.2).

Table 36.2: Environmental Services in the Specific Contexts of the Himalayan Mountains

Ecosystem Services in Relation to	*Contributions the Mountain Ecosystem in Terms of*
Supply Goods produced or provided by ecosystem such as foods, water, fuel, fibre, biodiversity or natural medicines	☆ Variety of foods from the biodiversity (species and varieties) ☆ Fuel, fodder, fibre, fertilizers and water ☆ Milk, milk products, wool, meat, eggs, hides, furs, draught power, etc. ☆ Ethno medicines
Regulation, or Control Services obtained by regulating or controlling ecosystem processes, such as the quality of the air, the climate, water (distribution and quality), erosion, the causes of illness, the manipulation of biological processes, risk reduction and so on.	☆ Conserving very high degree of rare endemic species ☆ Managing agroforestry systems having a number of deep-rooted trees, shrubs and perennial grasses ☆ Maintaining forests/rangelands as the core components of mountain farming systems ☆ Increasing resilience of the agro-ecosystems ☆ Ecological integrity of the agro-ecosystems ☆ Improving/maintaining quality of environment and its products as would stem from the above points ☆ Preventing/controlling diseases by means of ethno medicines and traditional practices ☆ Managing high degree of agrobiodiversity for minimising or averting risks to the farming communities ☆ Maintaining flow of stream waters and regulation of water cycle ☆ Regulation of appropriate weather cycle and micro-climate
Cultural Aspects The non-material benefits that enrich the quality of life such as cultural diversity, religious or spiritual values, knowledge (traditional or formal), inspiration, aesthetic values, social relations, a feeling of place, the values of a group's cultural patrimony, recreation and ecotourism.	☆ Sanctity associated with forests, trees, seeds, cattle, etc. ☆ Aesthetic values associated with nature and nature's objects ☆ Cultural attributes of endemism ☆ Social cohesion in the community based systems ☆ Native plant varieties and animal breeds with unique and superb traits developed in the specific mountain environments ☆ Unique diversity in the cultures of livestock-dependent communities ☆ Recreation values emanating from natural bounties such as forest-capped landscapes, alpine meadows, festivals and rituals ☆ Ecotourism in mountain ecosystems

Contd...

Table 36.2–Contd...

Ecosystem Services in Relation to	*Contributions the Mountain Ecosystem in Terms of*
Back-up, or Supporting ActivitiesThe services required to produce the other services, including primary production, the formation and/or fixture of soil, oxygen production, pollination, habitat creation, nutrient recycling and so on	☆ *In-situ* or on-farm conservation of native germplasm ☆ Services to produce cereals, pulses, oilseeds, medicinal and aromatic plants, grasses, trees, fruits, vegetables, vegetable seed production, etc. ☆ Soil enrichment by means of traditional farming practices ☆ Pollination (by birds, native as well as migratory) and dispersal of seeds, wild animals and a variety of insects prospering in mountain habitats ☆ Niche-based socio-economic activities ☆ Nutrient flows from rangelands to croplands mediated by livestock ☆ Moisture regulation in the environment ☆ Enhancement in resilience and reduction in systems' vulnerability ☆ Sustenance of ecological integrity of the system

A comprehensive attempt of the evaluation of the ecosystem services in the context of the parts of the Himalayan mountains has been made by Singh (2005, 2006). All services, of course, cannot be measurable. And, if any economic value is attached to them, it, unlike in case of goods, would amount to deception of the self. An estimate of the monitory value of these ecosystem services is often made using conventional economic framework, but this would be extremely vague and irrelevant on several grounds. How would they measure the value in Dollars of water cycle, air circulation, photosynthesis, rate of decomposition by soil microbes, precipitation, nutrient cycling, climate regulation, and such vital natural processes which could be considered "environmental services"? Can it be possible to privatize these natural phenomena which are vitally responsible for the origin, blossoming and sustenance of life and life systems on Earth?

Are the scenic beauty, breath-taking panoramas, soothing climate, etc., which have been inseparably associated with the Himalayan mountains are useless without putting a price tag on them? And, is anybody holding a fool-proof formula to measure the economic value of the natural phenomena? Existence values, optional values, aesthetic values and several other non-consumptive values of nature are not only immeasurable in monitory terms, but doing an attempt in this direction amounts to make a measurable attempt to set out a condition to live in a fools' paradise.

Clinging to Eco-philosophy

Eco-philosophy, according to the Father of Eco-philosophy Prof. Henryk Skolimowski, is the meaningful fusion of ecology and philosophy (Skolimowski 1995). Ecology goes even beyond the conventional boundaries of "study of the relationship between living organisms and their physical environment" when it is analysed on the fertile ground of philosophy, *i.e.*, love for wisdom. Ecology links everything with everything else. So does philosophy. Eco-philosophy is not accustomed to look at the things and situations superficially. It tends to go deeper into the things and issues. Eco-philosophy does not analyse the things in their parts or bits but in a holistic fashion. Eco-philosophy loves to see the whole world as a unitary whole. It has a universal view point. The idea of the universe perpetuates in the womb of eco-philosophy which regards our universe as a single unified entity and the living planet Earth and her life systems as an attribute of the same. The message is "All is One".

Eco-philosophy guides us to see the world in its perfect unity amidst enormous diversity. Eco-philosophy, therefore, is a force to intertwine the diversity into perfect unity. It teaches us how ecosystem vibrate into biomes and biomes into a single biosphere. Eco-philosophy calls for uniting the world (unlike politics, which often divides). Roots of sustainability lie deeper into ecology, and eco-philosophy guides us to save and strengthen these roots.

Eco-philosophy conjures up sanctity of life in its fullness. It values frugality, sanity, austerity, culture, symbiosis, harmony and sustainability. No economy of the world can sustain itself without regenerating natural resources. Ecological wellbeing is economical wellbeing. Ecology is an indomitable source of economy. Ecology shrinks, economy shrinks proportionately. Ecology blossoms, economy is sure to boom. Ecological bankruptcy not only debases economy, but also leads to generating the conditions that do not favour further regeneration of resources as the means for survival. Global warming and Earth's increasingly becoming uninhabitable are the result of that. Eco-philosophy guides us to keep everything in balance. Eco-philosophy calls for preserving the balance of our Cosmos by cultivating life-enhancing philosophy absolutely necessary for saving and sustaining life on Earth.

Ecosystems, Environmental Services and Sanctity of Life

The notion of "environmental services", or "ecosystem services" or "natural services" would suffice that the services are to originate only from an ecosystem vibrating with the myriad life forms existing in complex physical environment with complete coherence, equilibrium, mutualism and symbiosis. Environmental services mostly are to harp on the physical elements, but these too evolve out of a living system, that is, ecosystem. Biological and physical entities in an ecosystem are intertwined together and cannot be separated.

Sanctity of life emanates from the philosophy that values life in its wholeness, that takes care of the elements and complex processes and functions that support and enhance life, that holds reverential attitude towards nature and mother Earth, that strives to make commitments to and works for the enhancement and sustainability of life, that helps us enjoy amidst the plentiful beauty of the planet and of Cosmos, and that regards life as a celebration of the Universe.

"Environmental services" (*i.e.*, the economic notion to environmental functions) are the product of ecosystem and, in turn, also play critical role in the sustenance of the ecosystems and thus ensure sustainability and enhancement of life and life systems. Putting price tag on each service is neither desirable nor in tune with the fundamental philosophy brooding over the sanctity of life.

Evaluation of environmental services, in fact, is not necessitated for halting or reversing the processes of ecosystem degradation or attaining the state of sustainability. The social welfare is not the goal of the evaluation. For the constant flow of services from an ecosystem, ecological integrity of the system is the must. Then what is the need of monitory measurement of this aspect of the situation? Should philosophical gratification be an indispensable ingredient of the human intellect, it can be secured merely by knowing and understanding the elements, processes and complexity that make up a congenial environment to constantly generate and flow "services" vital for life.

All processes going on in the ecosystems are not regarded to be services for human beings. Some of the services generated out of these ecosystem processes are of potential use for human welfare. The valuation of the services, indeed, is in relation to human species only. Ecosystem services for the welfare and sustenance of other innumerable species on the planet are not in question. These species, seemingly existing on the mercy of human species, are themselves valuated for utilization in the human economic system virtually for "human welfare". Symbiosis between human species and the rest of the life forms is not the crux of the thinking behind ecosystem analysis. Ecosystem services criterion merely for human species is opposed to the very concept of symbiosis amidst the ecosystem, species and varieties of species.

The concept of ecosystem valuation promotes the preponderance of one species on Earth, *i.e.*, the human species. The valuation is selective too. It makes gradation in the services to originate from complex ecosystem processes and thus naturally emphasizes what is economically more important for human species. In essence, this is opposed to heterogeneity, which is the rule of life in totality. What is the most pronounced danger for humanity on Earth is not the religious fundamentalism, terrorism, or nuclear catastrophe, but the turning of Earth into human monocultures (Singh 2005). Monocultures introduce vulnerability in the systems that support them and are ultimately liable to destruction and elimination. Their care, their raising, their protection and short-term existence requires heavy cost to pay. Heterogeneity, on the other hand, depicts considerably high degree of resilience, stability and sustainability.

One thing inherent in the concept of economic valuation is the identification of providers and receivers. These two categories are of the people who should pay the prices and those who should be the gainer. Amongst all sorts of the stakeholders, they might be the communities inhabiting in different areas, such as ecologically more flourishing ones and those facing a state of ecological deprivation. The latter are naturally the receivers of the services and the former the providers of the services. In terms of the economic gains through ecosystem services the reverse would be the case. This pattern, however, is neither ethical nor practical and, in the setting of Indian cultural ethos, unacceptable also.

More pronounced beneficiaries, however, are not the communities, but the ones ruling over the economic models developed with the eventual aim of globalisation. The rapid processes of globalisation, in fact, are setting the terms and conditions for the maximum possible exploitation of all the natural resources–products as well as services–that are life-producing, life-supporting, life-enhancing, and life-sustaining. The concept of ecosystem services' evaluation emanates from the very basic ideology of globalisation. We must pose the questions: Globalization for what? Whose globalisation? Globalisation for whom?

Jodha (1998) identifies three elements which individually or jointly strengthen the ecosystem-social links and contribute to the natural resource-friendly traditional management systems: (a) a total dependence-driven stake in the protection of natural resources; (b) close proximity and a functional knowledge-driven approach to resource use; and (c) local control-determined sanctions and facilities governing resource use. We can derive some modalities for the evaluation of ecosystem services from the ecosystem-social system links as elaborated by Jodha.

Ecosystem services, from the eco-philosophical angle, are meant for all the life partners on Earth, *i.e.*, all species and all varieties of the species, not merely for human species. Human welfare and posterity are inevitably linked with the blossoming of all life forms on Earth. Again, the ecosystem services should be linked with the welfare of entire humanity and they should not be monopolised in the interest of the multinational corporate or in the interest of elite sections of the people. Commoditization of vital ecosystem services is tied with the reductionist thinking, not with the progressive thinking or ecological philosophy or sanctity of life.

Therefore, there should be some ethical basis of the valuation of ecosystem services; a basis that must ensure conservation and equitable utilization of resources and promote symbiosis and life-enhancing values. It must embrace reverential attitude towards nature and all processes and services that lead to the affluence of nature and sustainable economies. Ecosystem services' evaluation concept, instead of linking up with global GNP, should articulate in Gross Global Happiness.

Mountain-Lowland Linkages for the Flow of Global Happiness

Because of their geological characteristics, particularly the verticality, the mountain ecosystems inevitably have very strong linkages with the low-lying areas. Mountains provide origin to rivers that take the form of crucial river systems bringing affluence in agricultural production in the vast low-lying areas. Silt carried through mountain rivers adds to soil fertility in the plains. Mountains regulate hydrological and climatic cycles which eventually have significant influence in the plains.

Mountains thus send happiness to the vast lowland areas, which is owing to the natural highland-lowland interaction. This happiness is reflected in the comparatively better economies in the plains. The land-based economies of the plains are likely to collapse if the highland ecosystems are paralysed. Ecological affluence in the mountains has enormous bearing on the socio-economic boom in the plains.

It is the functioning of the mountain ecosystems that is converted into happiness, not only within the mountains but also in the low-lying areas. The ecosystem functions (referred to as "services" for the pure purpose of economic gains) are directly proportional to the ecological state of the ecosystems. More vibrant the ecosystem the greater and more effective the functions they perform and greater the flow of happiness within the mountains as well as in the lowlands.

The natural highland-lowland linkages cannot be de-linked. Economic evaluation of the "ecosystem services" emanating from the Himalayan mountains also implies de-linking the upland ecosystems from the lowlands. Mountain cultures throughout the Himalayan Region are comprised of the people filled with the ideas of self-respect, freedom and they are known throughout for their uncompromising honesty. They are not only lovers of nature, but, as eco-philosopher Prof. Henryk Skolimowski rightly says, average man and woman of the Himalayas is creating eco-philosophy. They would not yield to the parochial ideological efforts like 'compensating' native communities for centuries of preserving the earth's ecosystems. Accepting privatization of the planet and spinning money out of every ecosystem function (which, in fact, is the highest stage of privatization, or rather of imperialism of the globalization) is unacceptable to the majority of the sons and daughters of the Earth. It is also clearly against the ethos of Himalayan cultures. Idea of 'compensation' to the native people for their resource preservation (which has been a sacred work for ensuring safe and secure future for the next generations) sounds as if a child when grows up into an adult person attempts his or her mother to 'compensate' for his or her birth and rearing.

Mountain dwellers have been conserving their resources as part of their vital responsibility because they knew–perhaps more than other communities–that conservation is the foundation for eternal happiness. It is thanks to their philosophically oriented kind efforts that despite much degradation owing to external intervention their ecosystem are still functioning and are contributing to the reverberation of life far and wide. Mountain communities are rich repository of the knowledge through which they handle the processes that lead to and ensure sustainability. Their knowledge systems have not been static. They have transcendentally improved their systems and have been learning from lessons they experienced during the course of evolution. They have learnt from circumstances and have changed and refined their strategies at appropriate point of time. Conservation-oriented approaches have always been at the heart of the management of natural resources they developed. The world knows the unquestionable importance of conservation today. Natives of the mountains knew this and are following this for millennia. The world has learnt a lot from the lifestyles and strategies of mountain dwellers. Chipko Andolan (Hug-the-Trees Movement aimed at conservation of the Himalayan forests), for instance, stirred the conscience of people across the globe during 1970s and many a governments were prompted to formulate policies and constituted laws for the protection of natural resources.

Can any sort of compensation, then, suffice to the state-of-the-art role and unique achievements of the mountain communities as a result of their conscious efforts towards developing most efficient ways of the management of natural resources? Is not a talk of 'compensation' in the name of "environmental services" an insult to the mountain communities? Compensation would only undermine reputation and dignity and self-respect mountain communities have gained over millennia.

Mountain communities' efforts of conserving resources have been largely responsible for the constant flow of happiness to the mainstream economies in the plains. For this, the natives should be appreciated and their efforts be duly recognized. Mountain people are amongst the poorest in the world. It is owing to the fact that they have been marginalized by the mainstream areas and economies. It's the responsibility of the mainstream economy to provide opportunities for the mountain people. What mountain people need is their share in the mainstream economies and not compensation. They need be rewarded respectable jobs. Further, they should be paid remunerative prices against their produce, which is often rare and of very high value.

Ecosystems on earth interact in synergy. The highland-lowland linkages, however, are largely in favour of the lowland ecosystems. The role of highland ecosystems, naturally is to send happiness in terms of contributions to maintenance of appropriate conditions in the lower ecosystems. The ideology of "ecosystem services" would only undermine the specific and unique functioning of these ecosystems. Wellbeing and sustainability of the world's economies is largely proportional to the wellbeing of the mountains. Mountains wellbeing can be ensured by ensuring the socio-economic wellbeing of mountain dwellers, through implementing policies and programmes that articulate in the amelioration of mountain ecology and socioeconomic wellbeing of mountain dwellers. Mountain ecosystems' most efficient functioning thus would ensure sustainable flow of what should be regarded the most important indicator of human development and progress: global happiness. And sustainable flow of global happiness from the mountains can be ensured when functioning of these ecosystems is valued against their life-giving, life-enhancing and life-sustaining attributes.

Acknowledgements

Original idea which became basis for the comprehensive analysis of the situation in this article was picked up from the GRAIN's perspective published in the *Seedling*, April 2005.

References

Costanza, R., d'Arye, R., deGroot, R., Farber, S., Grasso, M., Hannon, B., Ruskin, R.G., Sutton, P. and van den Belt, M., 1997. The Value of the World's Ecosystem Services and Natural Capital. *Nature*, 387: 253–260.

Daily, G. (ed.). 1997. *Nature's Services: Societal Dependence on Natural Ecosystems*. Island Press.

GRAIN, 2005. Air, don't sell yourself. *Seedling*, p. 34–40.

Hawken, P., Lovins, A. and Lovins, H., 2003. *Natural Capitalism: Creating the Next Industrial Revolution*. Rocky Mountain Institute.

Hobson, P.N., 1988. *The Rumen Microbial Ecosystem*. Elsevier Applied Science, New York.

http://www.prisma.org.sv/pubs.

Jodha, N.S., 1998. Reviving the social system: Ecosystem links in the Himalayas. In: *Linking Social and Ecological Systems*, (Eds.) F. Berkes, C. Folke, and J. Colding. Cambridge University Press, Cambridge.

Negi, G.C.S. and Agrawal, 2006. Measuring and valuing ecosystem services: Himalayan mountain context. *Current Science*, 91(5): 573–575.

Singh, S.P., 2005. Ecosystem services. In: *Keynote Address to the Workshop on Integrated Management of Natural Resources in Mountains* at Pantnagar, 28 Nov–2 Dec.

Singh, S.P., 2006. Incorporating values of ecosystem services of ecosystem services of Uttaranchal and other Himalayan states in national accounting systems. In: *Resource Appraisal, Technology Application and Environmental Challenges in Central Himalaya*, (Ed.) M.S.S. Rawat. Department of Geography, Srinagar (Garhwal), 1: 1–14.

Singh, V., 2005. Eco-philosophy, equity and sustainable development: Potential tools for world peace. In: *Paper presented at the National Conference on Role of India in Promoting World Peace* organized by UNESCO and World Peace Centre, Pune on February 3–4.

Skolimowski, H., 1995. *Dancing Shiva in the Ecological Age*. Dr. Henryk Skolimowski International Centre for Eco-philosophy, New Delhi.

Ecosystem Diversity and Carbon Sequestration: Climate Change Challenges and a Way Out for Ushering in a Sustainable Future *Pages 356–361*
Editors: **P.L. Gautam, Vir Singh and Uma Melkania**
Published by: **DAYA PUBLISHING HOUSE, 2009**

Chapter 37

Ecosystem Conservation for Carbon Sequestration: Let it be in the Popular Psyche of India

Subaran Singh

Carbon sequestration is an inevitable function of an ecosystem harbouring green plants. Putting it in different words, an ecosystem–terrestrial as well as aquatic–that comprises photosynthetic organisms, inevitably performs sequestration of carbon-dioxide. Ecosystem structure determines the rate of carbon sequestration in nature. It is therefore necessary to improve the health of the diverse ecosystems if we are aimed at enhancing carbon sequestration and, consequently, mitigating global warming leading to ominous climate change. This paper attempts to describe some general and frequently talked about things relating to carbon sequestration vis-à-vis ecosystem conservation. Ecosystem conservation and carbon sequestration are everybody's business. These should be at the heart of our thinking, our philosophy, our plans and our policies. In fact, these should be the issues deeply linked with our secure future, our wellbeing, and even with our existence.

The Ecosystem

A biological community together with its associated physical environment is termed an ecosystem. An Ecosystem includes all the organisms interacting themselves and with the physical environment so that a flow of energy leads to a clearly defined trophic structure, biotic diversity and material cycle within system. Ecosystem diversity denotes the diversity of habitats, communities and ecological processes within the biosphere. Ecosystem is a natural functional unit. Ecosystem is also a dynamic function unit.

Soil is built up out of parent rock material and decaying organic matter. Photosynthetic plants absorb light energy, and this energy is released back into the atmosphere as heat during the lives of plants and animals, as well as after they die and decompose. Plants absorb carbon dioxide and release oxygen during respiration, while animals and fungi absorb oxygen and release carbon dioxide. Each species utilizes a unique set of resources that constitute its niche.

The greatest biological diversity is in the tropical regions of the world, with large concentrations in tropical rain forest, coral reefs, tropical lakes, and the deep sea (tropical rain forests occupy only 7 per cent of the Earth's land area)

About 1.5 million species have been described and at least twice that number of species remain to be described, leading to a conservative estimated that there are about 5 million species worldwide.

Cause of Increasing Carbon Level

- Industrial smoke
- Refrigerator gases (CFC, N_2O)
- Deposition of heavy polymers (Polythene)
- Coal gases (CO_2)

Problems of Increasing Carbon

- Global warming
- Melting of glaciers
- Effect on crops' canopy
- Flooding in lowland areas

Managing Increased Levels of Carbon

- To manage carbon is to use energy more efficient to reduce our need for major energy and carbon source-fossils fuel combustion.
- To increase our use of low carbon and carbon free fuels (bio fuels) and technologies (renewal resources; solar energy, wind power, nuclear power).
- The newest way to mange carbon is through carbon sequestration

Carbon Sequestration

Carbon sequestration is the term describing processes that remove carbon from the atmosphere. To help mitigate global warming, a variety of means of artificially capturing and storing carbon–as well as of enhancing natural sequestration processes–are being explored.

It refers to provision of long term storage of carbon in the terrestrial biosphere, under ground or the oceans to reduce greenhouse gas concentration in the atmosphere.

Carbon sequestration is a way to reduce greenhouse gas emissions. It complements two other major approaches for greenhouse gas reduction, namely improving energy efficiency and increasing use of non-carbon energy sources. Interest has been increasing in the carbon sequestration option because it is very compatible with the large energy production and delivery infrastructure now in place that is the stabilization of greenhouse gas concentrations in the atmosphere at a level that would prevent dangerous anthropogenic interference with the climate system.

The Sleipner project in Norway's North Sea is the world's first commercial carbon dioxide capture and storage project. Started in 1996, it sequesters about one million metric tons of carbon-dioxide each year.

Carbon Capture and Storage (CCS)

It is an approach to mitigating climate change by capturing carbon dioxide (CO_2) from large point sources such as power plants and subsequently storing it away safely instead of releasing it into the atmosphere. CCS applied to a modern conventional power plant could reduce CO_2 emissions to the atmosphere by approximately 80-90 per cent compared to a plant without CCS. Capturing and compressing CO_2 requires much energy and would increase the fuel needs of a plant with CCS by about 10-40 per cent. These and other system costs are estimated to increase the cost of energy from a power plant with CCS by 30-60 per cent depending on the specific circumstances.

Storage of the CO_2 is envisaged either in deep geological formations, deep oceans, or in the form of mineral carbonates. The major merit of CCS systems is the reduction of CO_2 emissions by up to 90 per cent, depending on plant type. Generally, environmental impacts from use of CCS arise during power production, CO_2 transport and CO_2 storage. Capturing CO_2 can be applied to large point sources, such as large fossil fuel or biomass energy facilities, major CO_2 emitting industries, natural gas production, synthetic fuel plants and fossil fuel-based hydrogen production plants. Broadly, three different types of technologies exist:

- Post-combustion
- Pre-combustion
- Oxyfuel combustion

These are storage in various deep geological formations (including saline formations and exhausted gas fields), ocean storage, and reaction of CO_2 with metal oxides to produce stable carbonates.

Geological Storage

Also known as geo-sequestration, this method involves injecting carbon-dioxide directly into underground geological formations. Oil fields, gas fields, saline formations, unminable coal seams, and saline-filled basalt formations have been suggested as storage sites. Current research shows that several *trapping mechanisms* immobilize the CO_2 underground, reducing the risk of leakage.

For well-selected, designed and managed geological storage sites, IPCC estimates that CO_2 could be trapped for millions of years, and the sites are likely to retain over 99 per cent of the injected CO_2 over 1,000 years.

Ocean Storage

Carbon storage in the oceans is plentiful. Two main concepts exist. The 'dissolution' type injects CO_2 by ship or pipeline into the water column at depths of 1000 m or more, and the CO_2 subsequently dissolves.

The 'lake' type deposits CO_2 directly onto the sea floor at depths greater than 3000 m, where CO_2 is denser than water and is expected to form a 'lake' that would delay dissolution of CO_2 into the environment. A third concept is to convert the CO_2 to bicarbonates (using limestone) or hydrates. The bicarbonate approach would reduce the pH effects and enhance the retention of CO_2 in the ocean, but this would also increase the costs and other environmental impacts.

An additional method of long term ocean based sequestration is to gather crop residue such as corn stalks or excess hay into large weighted bales of biomass and deposit it in the alluvial fan areas of the deep ocean basin. Dropping these residues in alluvial fans would cause the residues to be quickly buried in silt on the sea floor, sequestering the biomass for very long-time spans. Alluvial fans exist in all of the world's oceans and seas where river deltas fall off the edge of the continental shelf such as the Mississippi alluvial fan in the gulf of Mexico and the Nile alluvial fan in the Mediterranean Sea.

Mineral Storage

Mineral storage aims to trap carbon in stable minerals, and CO_2 would be forever trapped. In this process, CO_2 is reacted with (abundantly available) metal oxides which produces stable carbonates. This process occurs naturally and is responsible for much of the surface limestone.

Carbon-dioxide Sink

A carbon-dioxide (CO_2) sink is a carbon reservoir that is increasing in size, and is the opposite of a carbon "source". The main natural sinks are (1) the oceans and (2) plants and other organisms that use photosynthesis to remove carbon from the atmosphere by incorporating it into biomass. This concept of CO_2 sinks has become more widely known because the Kyoto Protocol allows the use of carbon dioxide sinks as a form of carbon offset.

Sequestration activities can help prevent global climate change by enhancing carbon storage in trees and soils, preserving existing tree and soil carbon, and by reducing emissions of CO_2, methane (CH_4) and nitrous oxide (N_2O). For more information on the science, emissions, and reduction opportunities for these and other non-CO_2 gases, please concern our non-CO_2 gases page.

Natural Sinks

Forests' Carbon is incorporated into forests and forest soils by trees and other plants. Through photosynthesis, plants absorb carbon dioxide from the atmosphere, store the carbon in sugars, starch and cellulose, and release the oxygen into the atmosphere. Oceans are natural CO_2 sinks, and represent the largest active carbon sink on Earth. This role as a sink for CO_2 is driven by two processes, the solubility pump and the pump. The former is primarily a function of differential CO_2 solubility in seawater and the thermohaline circulation,

Soils carbon as plant organic matter is sequestered in soils: Soils contain more carbon than is contained in vegetation and the atmosphere combined.

Artificial Sequestration

it must first be captured, or it must be significantly delayed or prevented from being re-released into the atmosphere (by combustion, decay, etc.) from an existing carbon-rich material, by being incorporated into an enduring usage (such as in construction).

Natural-gas purification plants often already have to remove carbon-dioxide, either to avoid dry ice clogging gas tankers or to prevent carbon-dioxide concentrations exceeding the 3 per cent maximum permitted on the natural-gas distribution grid.

Carbon Capture

Currently, capture of carbon-dioxide is performed on a large scale by absorption of carbon-dioxide onto various amine-based solvents. Other techniques are currently being investigated, such as pressure and temperature swing absorption, gas separation membranes, and cryogenics. Another long-term option is carbon capture directly from the air using hydroxides.

Oceans

Another proposed form of carbon sequestration in the ocean is direct injection. In this method, carbon-dioxide is pumped directly into the water at depth, and expected to form "lakes" of liquid CO_2 at the bottom. Experiments carried out in moderate to deep waters (350-3600 meters) indicate that the liquid CO_2 reacts to form solid CO_2 clathrate hydrates, which gradually dissolve in the surrounding waters.

Mineral Sequestration

Mineral sequestration aims to trap carbon by placing it in its thermodynamics ground state where it will be non-reactive. This occurs naturally and is responsible for much of the surface limestone. Acids are used to convert mineral silicates to mineral carbonates. Ongoing research aims to speed up the kinetics of the reactions.

Serpentinite sequestration is favored because of the non-toxic and predictable nature of magnesium carbonate.

Serpentinite Reactions

Reaction

Mg-Olivine + Water + Carbon-dioxide → Serpentine + Magnesite + Silica

Reaction

Fe-Olivine + Water + Carbonic acid → Serpentine + Magnetite + Magnesite + Silica

Reaction

Serpentine + carbon dioxide → Magnesite + silica + water

Carbon Sinks and the Kyoto Protocol

Kyoto Protocol allows countries that have large areas of forest (or other vegetation) to deduct a certain amount from their emissions, thus making it easier for them to achieve the desired net emission levels.

In the Clean Development Mechanism, only afforestation and reforestation are eligible to produce CERs in the first commitment period of the Kyoto Protocol (2008–2012). Forest conservation activities or activities avoiding deforestation, which would result in emission reduction through the conservation of existing carbon stocks, are not eligible at this time.

Primacy of Ecosystem Conservation for Carbon Sequestration

The first and the foremost thing we should delve into is ecosystem conservation. Carbon-dioxide concentration in atmosphere is increasing thanks to the deplorable condition of the Earth's ecosystems. Ecosystem conservation would be a precondition to contain global warming and associated climate changes. We have a wonderful diversity of ecosystems in India. These ecosystems should be given due protection. Natural forest cover should be expanded, diversity and density of plants should be increased in the degraded forests and an environment of ecological regeneration should be provided to these degraded ecosystems. Agroforestry systems need be developed and be made more efficient from the carbon sequestration point of view. Aquatic ecosystems should be kept free from pollution.

Increasing awareness amongst masses rather than mere legislative measures would be instrumental for ecosystem conservation, which would automatically manifest into enhanced CO_2 sequestration and revert the processes leading to global warming and climate change. The vital contemporary issues relating to ecosystem conservation and carbon sequestration should remain in the popular psyche of India. That would be instrumental in mitigating the most serious environmental problems in the history of humankind.

Ecosystem Diversity and Carbon Sequestration: Climate Change Challenges and a Way Out for Ushering in a Sustainable Future *Pages 362–366*
Editors: **P.L. Gautam, Vir Singh and Uma Melkania**
Published by: **DAYA PUBLISHING HOUSE, 2009**

Chapter 38

Socio-cultural Values Promoting Conservation of Nature's Biodiversity: Heal the Earth for Enhancing Carbon Sequestration

A. Vanmathy and Abha Ahuja

We are divided by nationalities;
We are divided by languages;
We are divided by ideologies;
We are divided by cultures;
Yet, we have in common–
The common heritage of life
Our mother earth!

Mother earth has delivered all the present lives, right from the single-celled to the complex structures contributing to the biodiversity. Biodiversity is defined as the variety of life forms, measured in terms of biomes, ecosystems, species, genetic varieties and the interaction between them. She has a unique identity from the rest of the universe, to gift us with the accentuation of quality life, to share and thrive in harmony with nature. Though we have diversified biological heritage, we still stand united with our common ecological heritage–our mother earth.

We have of course benefited from her blessings, but what shall we leave for our forth comers? This remains as the biggest question mark in the universal context. Ecological concerns are repeatedly hitting the front pages of leading dailies, not because we are unaware of the ecological problems, but because we are unconcerned about future in striving for the present. The selfishness of humanity has handled capitalism as a destructive tool to exploit nature. This has intrusively harmed our relationship with nature that affects indifferently both the benefited and the innocents. Now the need of the hour is ecological reconstruction to assure a life of dignity and grace for future. The wounded nature calls for the ecological catharsis. Let us rejuvenate ourselves to join her!

The Relationship Between Culture and Environment

Plants, animals and human beings constitute

An integrated whole, it is this specific culture

That makes sense by ordering natural and social

Phenomena into a structured world

It is an amalgamation of nature and culture!

— Seeland (1997)

Man is identified by his society that hallmarks a unique culture. Man is a social animal; he craves for individuality and independence but his stature as a human being is shaped and enhanced by the social context. The relationship of culture to environment is one of the oldest matrixes in the science of anthropology. The concept of environment determination is developed based on the ways in which cultures are adjusted to the unique features of their local environment. The view that "Culture" can endow life with meaning and value by promoting truth, beauty and goodness is therefore no airy abstract, but on the contrary, it arises directly from everything that men and women are doing to fashion and direct their lives to improve their lot on earth.

Man did not weave a web of life,

He is merely a strand in it!

What ever he does to the web,

He hoes to himself!

— Chief Seatile

When humanity draws up its balance sheet of losses and gains, those who have rendered signal services in the cause of culture come into their own because it is seen that without their contribution, the human heritage of meanings and values would be impoverished. The responsibility for promoting cultural values in conservation of nature's biodiversity is widely shared between individuals, families and neighborhood groups as well as society at large scale.

Can Social and Cultural Values Save Our Earth?

Culture is that, which enhances the quality

Of life with meaning and value by making

Possible the formulation, progressive realization,
Appreciation and achievements of truth, beauty and
Moral value for betterment of humanity.

— F.R. Cowell

Can social and cultural values save our earth? The answer to this question is obviously "yes". In many societies with fine cultural values, there is a deep understanding that right thinking leads to right action. The idea of right thinking is a specific domain, be it farming or forest conservation. The right conservationist strategies emphasise on benefit to all humankind. The weight of social responsibility is most clearly recognized in the cultural context. If we talk of natural resource management from global perspective, whom do we find in the fore front of the race for protection and preservation of resources. The answer comes very naturally, it is the local communities. They possess social and cultural values that make them properly exploit their "local knowledge", to obtain forest products. It is an autopoietic system, which has functional efficiency of ecosystem and orienting livelihood towards environmental integrity. The community spirit in biodiversity conservation is obviously visible in the rich socio-cultural values they posses. There are many tribal traditions and practices which are in harmony with the ecological well-being of nature. They have spiritual and emotional relationship with the environment in which they live.

Socio-Cultural Values Promoting Biodiversity Conservation

Biodiversity management is an ancient phenomenon. The origins of biological conservation can be traced to philosophical and religious beliefs about man as a full part of nature. Vedas contain hymns in praise of animals. Sanatana dharma has linked some animals with specific Gods and Goddess as the best way of conservation of biodiversity; for example: Python has been associated with God *Vishnu*, Snake with God Shiva, Swan with Goddess Saraswati, Lion with Goddess Durga, rendering the animal pious and protected. In the *Mahabharata*, Rishis and Munis have been indicated to conserve wildlife fauna such as deer and birds around ashramas. In the *Arthshastra*, Chanakya had imposed severe penalties for killing, entrapping and molesting birds, fishes and deer etc. in protected areas.

The *Torah* or *Old Testament* prohibits the destruction of fruit bearing trees and this Commandment has been extended to encompass all manner of wastefulness. Islam recognizes each species as its own "Nation" and an obligation of man to "Khalifa" or "stewardship" of the earth. Specific conservation mechanism such as *harma* and *hima* zones and the origins of the idea of carrying capacity was a product of Islamic civilization. As early as 450 BC, according to Grove (1992), Artaxerxes 1 attempted to restrict cutting Lebanese timber. Plato, writing in 4th century BC, noted that the removal of trees in Attica produced social erosion "and what remained is like the skeleton of a body wasted by disease" (Glacken 1965).

Similarly, the traditional life style of local communities is much intertwined with their environment. For them, forest is the mother's home. We can understand the profound and unique relationship of local community with the forest. It can also be understood by the following example: A woman in Chomoli district of Uttarakhand gave birth to girl baby while collecting fodder in the forest. So the baby was named as Boni Devi, Goddess of forest. This shows their emotional relationship with forest.

Numerous initiatives taken by the local communities for biodiversity conservation and generating sustainable livelihood emerge as a strong ray of hope. The history of community forest protection and management in the state of Orissa dates back to the pre-independence days. Lapan village in the Sambalpur district which began protection as early as 1936 is a marvelous example. The world famous *Chipko Andulan* (Hug-the-Tress Movement) of the hill women against the destruction of forest in the Himalayas has proved that they can protect, conserve and manage the environment. The *Mahila Mangal Dal* has received the "Indira Priyadarshini Vriksha Mitra Award" from the late Prime Minister Shri Rajiv Gandhi in 1984 for their initiatives. The Maiti Movement in Chamoli district of Uttarakhand region has made this region flourish with trees. During the marriage ceremony of the Maiti group of girls, the bride presents sapling to her bridegroom, who plants it under the auspicious chanting of mantras.

The idea of *Raksha Sutra* Movement to save the nature wealth of Himalayas is totally anew though that came of in august 1994. It is similar to *Raksha Banadhan*, an Indian festival to celebrate the brother-sister relationship. The women of different villages tie *Raksha Sutra* to the trees of their forest. They jointly raise their voice against forest business and felling of trees.

All such movements are blended with the ongoing cultural practices of the people. People imbibe with their emotional sentiments and strong determination for the preservation of natural resources. It clearly demonstrates that conservation and protection of natural wealth always incorporate the traditional knowledge, practical experiences, religious sentiments, code of social conduct, cultural practices with community participation, and all such elements of a value system.

Another best example to illustrate the traditional conservation strategy is the Indigenous Biodiversity Conservation Areas (IBCAs) of Ghana forest in Africa. It is as old as traditional Ghanaian society itself (backing to 8th century). They follow regulatory strategies often enshrined with religious or cultural beliefs and superstitious and enforced by taboos. In fact Ghana is said to hold ancient keys to a meaningful model of forest conservation.

Biodiversity Conservation and Carbon Sequestration

Traditional natural resource conservation adopted by communities is based of the ethics they have been cherishing for millennia. Climate change is–in fact–rooted in the erosion of social values based on ethics. Social justice and ecological justice hardly have any value in the modern economic models that depend on excessive extraction of natural elements to flow into economic circle. The desperate economic growth thus steadily leads to bankruptcy of natural systems on which the very sustainability of livelihoods depends and which are the very source of weather and climate regulations.

It is on account of this state of ecological paucity or bankruptcy of natural systems that we are bound to face the worst ever situation of global warming and ominous climatic changes threatening our very survival and existence.

Blossoming of biodiversity in the natural environments and enhanced carbon sequestration go hand in hand. The natural factors and processes responsible for enhancement of carbon sequestration can be restored only through ecological regeneration of the earth's natural resources. Conservation and enhancement of biodiversity in natural ecosystems are the vital means to restore the ecological systems that imbibe with life and serve as potent human tactics for the healing of our mother earth.

Conclusion

We can't make the Earth sustainable; it is sustainable–but whether with us, or without us, is our choice.

Biodiversity conservation is discussed in many global forums since few decades back but still we could not land up in an effective technique to heal our earth. There is the lacuna in our approach towards nature. So let us think different. Let us have reverential thinking which makes us aware not only of the physical existence but goes beyond it to inherit the worth and beauty of nature. Such type of reverential thinking is practiced in many socio-cultural entities, under different names. We must strongly believe that the highly funded natural resource management projects of government can succeed only if they incorporate the rich values of eco-cultural practices, which have not only proved to be possible but have been practices in large scale.

Professor Henry Skolimowski, the highly acclaimed eco-philosopher of the world, guides us to eco-justice. Eco-justice is a value that makes us responsible for conservation and enhancement of all life forms on earth. It is the value that teaches us to live in symbiosis with other species and in coherence with nature. The healing nature of ecological thinking has more power to fetch us a permanent solution to our entire struggle towards nature conservation as it emphasizes on the participatory approach united by social and cultural solidarity. Only when we understand and promote the socio-cultural values in healing the nature, we can actively participate in the ecological catharsis.

So come on, let us continue, chanting the mantra of "conservation", by adding the right fuel of "socio-cultural values" to strengthen the yojna of "biodiversity".

Ecosystem Diversity and Carbon Sequestration: Climate Change Challenges and a Way Out for Ushering in a Sustainable Future *Pages 367–376*
Editors: P.L. Gautam, Vir Singh and Uma Melkania
Published by: DAYA PUBLISHING HOUSE, 2009

Chapter 39

Environmental Psychology in Landscaping: A Dimension of Sustainability Operations

Govind Singh Kushwaha and Vir Singh

Introduction

Operationalisation of sustainability in our agricultural production systems is a complex phenomenon. It is not dependent on linear sort of concept and approach to the systems, but on a complex of circumstances, type and environment of biophysical resources and methodologies applied. While Singh (2005) has presented a set of sustainability indicators, such as ecological stability, viable economy, energetic efficiency, resilience, and social justice, what is missing is the dimension of environmental psychology, without which the course of sustainability would be somewhat unsustainable.

Landscaping is one of the essential outcomes of a land-based development scenario. A landscape fulfils basic requirements of life. At the same time, it also contributes to intellectual, physical, ethical, aesthetic and psychological development of human beings. A type of landscape would determine the material flows and amount of human contentment. Landscaping concepts and development approaches are instrumental in deciding whether the outcome would have a dimension of sustainability or not. Development models are rejected or finalized based on landscapes evolved through the follow up of certain strategies.

Sustainability debate has come of age, but psychological factors have not been addressed adequately. Without psychological elements the debate should not be regarded as pregnant. This paper attempts to delve into several issues of environmental psychology that can be articulated in our concepts of and approaches to the development of a landscape fulfilling myriad purposes.

Psychology of Sustainability

Organizational development takes place on a foundation of four levels of infrastructure, *viz.*, physical, social, mental and psychological (IGNOU 2005). All these four types of resources contribute to development, of which only the most materials are inherently limited in nature. The productivity of resources increases enormously as the level of organization and input of knowledge rises.

In sustainable development processes the human resource is recognized as the driving force and the primary determinant of development. Human psychology is one of the key determinants of the courses of our development.

Psychology therefore ought to contribute enormously to the sustainability issues. Sustainability concepts and approaches inherently have elements of psychology. Psychology of sustainability was first of all used by Alice John (1996) who described it as follows:

I call it loosely "the psychology of sustainability", and it has to do with the very personal, individual, and sometimes conflicting experience of talking the talk of sustainability and then trying to walk the walk. My take on this is that we can't have a meaningful discussion about a national sustainability policy, or even a community sustainability policy without explicitly considering the very personal nature of what is implied.

Although the concept of sustainable development has been criticized for being too vague, anyway there is a considerable agreement among different scientific fields that it should remain the main aim to be pursued in the management of natural and human resources. Now sustainability is a trans-disciplinary concept which calls into question issues that are central in several social and human sciences and disciplines, ranging from economics, legal sciences, philosophy, psychology: within social and environmental psychology, in particular, some authors have recently proposed the term "psychology of sustainability" or "new ecological psychology" (Bonnes and Bonaiuto 2001).

These terms attempt to identify those theoretical and empirical contributions aiming at better understanding of the psychological processes involved in the development of a positive environmental awareness and concern in people's use of natural resources. A considerable effort in the direction of supporting the above-mentioned "integrated perspective" in ecological science, as well the concept of sustainable development, particularly for what it concerns the institution and managements of natural protected areas, came also from the UNESCO Program on Man and Biosphere (MAB).

Potentialities of Environmental Psychology

Environmental psychology, a branch of applied psychology, has enormous potential to operationalise sustainability. It is concerned both with psychological effects of the physical environment and with effects of human action on the socio-physical environment. Eenvironmental psychology attempts to examine the inter-relationship between environments and human behaviour. The field defines the term environment very largely, including all that is natural on the earth as well as social settings, learning environment and informational environments.

Environmental psychology can be instrumental in synchronizing the things on right track. When trying to solve problems involving human-environment interactions–global or local–one must try to deeply understand human nature that predicts the environmental conditions under which humans will behave in a civilized and creative manner. With such an understanding one can design models that might contribute to manage, protect, rejuvenate and restore environments that could help enhance reasonable behavior, predict what the likely outcome would be when these conditions are not met, and detect problem situations.

Environmental psychology contributes to develop such a model of human nature while retaining a broad and inherently multidisciplinary focus. It explores such dissimilar issues as common property resource (CPR) management, way finding in complex settings, the effect of environmental stress on human performance, the characteristics of regenerative and restorative environments, human information processing, and the promotion of durable conservation behaviour. Using theories and practical aspects of related disciplines–*e.g.*, psychology, sociology, anthropology, biology, and ecology–the field of environmental psychology seeks to recognize the need to be problem-oriented. Thus this attempts to address the problems in a holistic manner.

Insights of Environmental Psychology

Environmental psychology is rooted in three insights, namely:

1. There is a deeply bonded and reciprocal relationship between humans and nature. Environmental psychology has presented two metaphors for this relationship: (a) nature as home and family (including siblings, mother, and father) and (b) nature as self, in which self-identifications are broadened.
2. The illusion of a separation of humans and nature leads to suffering both for the environment (as ecological devastation) and for humans (as grief, despair, and alienation).
3. Realizing the connection between humans and nature is curative for both. This reconnection includes the healing potential of contact with nature, work on grief and despair about environmental destruction, eco-therapy, psycho-emotional bonding with the world as a source of environmental action, and sustainable lifestyles.

Perception is an important feature of behavior to understand human behavior. It starts with understanding how people perceive the different stimuli present in environment. This includes at least two kinds of stimuli: those that unwillingly, even distractingly, command human to perceive, as well as those places, things or ideas to which humans must willingly, and with some effort, direct their awareness. Restoring and enhancing people's capacity to willingly direct their attention is a major factor in maintaining human effectiveness.

Psychology in the Formation of Cognitive Maps

One of the interesting things in environmental psychology since the beginning is to know how people image the natural and built environment. Information is stored in the brain as spatial networks called cognitive maps. These structures link one's recall of experiences with perception of present events, ideas and emotions. It is through these neural networks that humans know and think about the environment, plan and carry out their plans. Interestingly, what humans know about an environment is both more than external reality in that they perceive with prior knowledge and expectations, and less than external reality in that they record only a portion of the entire visual frame yet recall it as complete and continuous.

Shaping Environments in Tune with Holistic Needs

People tend to seek out places where they feel capable and confident, places where they can make sense of the environment while also being engaged with it. Research has expanded the notion of preference to include coherence, that is a sense that things in the environment hang together and legibility (the inference that one can explore an environment without becoming lost) as contributors to environmental comprehension. Being involved and wanting to explore an environment requires that it has complexity (containing enough variety to make it worth learning about) and mystery (the prospect of gaining more information about an environment).

Preserving, restoring and creating a preferred environment are thought to increase sense of well being and behavioral effectiveness in humans. Human society does only depend on environment for just fulfilling needs for survival, but it is a part and parcel of the same. There is no other option than to depend on the environment. An environment ensures survival as well as fulfills myriad needs for holistic human development and a sustainable future. Environmental psychology does not merely attempt to capture a single dimension of development, but is aimed at its orientation ensuring fulfilling holistic needs keeping pace with its ability to regenerate.

Environmental psychology is committed to enhancing people's involvement in environmental design, management and restoration efforts. It is concerned not only with promoting citizen comprehension of environmental issues but with ensuring their early and genuine participation in the designs, modifications and management of environments.

Conservation Behavior

Ecological sustainability is the prime indicator of sustainable development (Singh 2005). Environmental psychology therefore is set to strengthen dimension of sustainable development of resources. Environmental psychology has also played a major role in bringing psychological knowledge to bear upon the issue of developing an ecologically sustainable society. It explores environmental attitudes, perceptions and values as well as devises intervention techniques for promoting environmentally appropriate behavior.

Environmental Psychology in Landscape Management

Landscaping is one of the basic approaches to acquiring sustainability. A landscape is always hanging in our mind and imagination. Sometimes used to designate land in the immediate environs of a town or city, not just natural scenery, a landscape tends to satiate our imagination. A landscape can also be used in terms of art, in which it designates the area of a religious, cultural, historical, natural painting that forms the setting for the central drama and its protagonists.

A landscape reflects our deep feelings and behaviour. The environment in which we live, work and involved in deep creative activities can influence the way we behave and feel. For example, in one landscape we might feel a sense of comfort and well-being, but in another, apprehension or fear (Kaplan and Ryan 1986). These feelings develop quickly and automatically from an evaluation of the visual cues available.

The reasons why we respond to our environment in this way are partly based on the information immediately available and partly based on the past experience of the individual observer. Some theorists argue that there are evolutionary reasons for such reactions (Appleton 1975). No wonder then, that planners, landscape designers, and architects want to know how we respond to particular features in our environment.

Research in environmental psychology has explored people's preferences for landscapes in order to identify the variables that make up the landscapes that people prefer. The studies are usually based on photographs that have been placed in *a-priori* categories. Many categories have been tested, examples of which include information-based and perception-based variables.

The information-based variables denote coherence (how orderly the landscape is), complexity (the heterogeneity, number of elements and 'activity' in the landscape), and mystery (the likelihood that there will be more in the landscape than meets the eye). The perception-based variables, on the other hand, include openness (the degree of visible space in the landscape), smoothness (uniformity and height of the elements on the surfaces and ground), ease of locomotion (how easy and effortlessly observers could move around), and age (how old the elements in the landscape appear to be) (Herzog 1992, 1995a, 1995b; Strumse 1994, 1996).

Counter intuitiveness of the variables sometimes is clearly visible. People's likings and enjoyment about landscape is quite diverse. For example, Kaplan and Ryan's (1989) review of a number of studies showed that, using the simple dichotomy of 'natural' versus 'built' environments, led to the finding that people preferred 'manicured' and planned parkland over natural forest or pasture. On the contrary, as authors' experience would reveal, inhabitants in Garhwal part of the Indian Central Himalayas prefer natural and undisturbed landscapes, for they also contribute to geo-ecological stability. Tourists pouring in the Himalayan mountains would love to be amidst wild beauty.

There is still a new and quite recognizable variety of hardcore ecologists, for whom absolute conservation-oriented natural landscapes should not be compromised with the elements of management by human species. Landscapes with high degree of biodiversity are more resilient and type of more sustainable ecosystem and a large number of ecologists, sociologists, and local inhabitants would overwhelmingly support maintenance of such landscapes (Shiva *et al.*, 2005).

Diverse issues relating to people's choice make the sustainability approaches more complex and entangling. Therefore when studying landscape type to be developed or maintained researchers need to undertake social studies and design experimental studies which can provide simple answers that can contribute to our understanding of complex issues.

Attributes of Landscaping

Landscaping is one of the essential conditions of human surroundings. Panorama presented by a landscape imparts soothing effect on human mind and working environment. It also has aesthetic, recreational and optional values. A landscape type fulfils the requirements of a production system while another type would satiate psychological requirements and even would contribute to psychological development of the people or a society. Building up of an environments of food sovereignty for a society that is thanks to landscaping management providing a culture for biodiversity conservation (*e.g.*, a forest or a grassland), for agro-biodiversity cultivation and utilization (*e.g.*, cultivated croplands).

The Wall Street Journal reported that landscape investments are recovered fully, and sometimes doubled, by the increased home value. An enchanting landscape helps get the environment crime-free. Landscaped areas are relatively graffiti-free, while open, non-landscaped areas are graffiti targets. Well planned and maintained landscapes are seen as safer than non-maintained plantings.

Some other attributes of a landscape, especially relating to ecological, economic and human development are summarized in Table 39.1. Many attributes are after Stokols and Altman (1987), Garling and Golledge (1993) and Herzog (1987, 1995a, 1995b).

Landscaping for Mitigating Global Environmental Problems

Carbon sequestration, global warming and climate changes are the latest issues that are occupying central stage in global politics and are also fast pervading human psychology. Global warming leading to climate changes is striking a sort of terror amongst the global communities. Such ominous changes will cost the whole life on Earth. Gradually and steadily culminating into serious situation throughout the globe, the global warming is all set to threaten the very existence of human race in the coming centuries. The governments of the world nations now appear to be totally concerned with adopting the measures that could help mitigate the most critical environmental problem of our times.

Societies' concerns of landscape management now no more confined to just the objective of fulfilling materialistic needs. Controlling, containing, arresting, curbing, and/or reversing the factors responsible for the ominous situation of global warming

Table 39.1: Attributes Vital for Environmental, Economic and Human Development Emanating from Landscape Management

Environmental Services	*Promoting Economic Development*	*Improving Human Health*
Biodiversity, soil and water conservation	**Property market value**	**Gardening**
Natural landscapes provide the most valuable service of biodiversity conservation. Thick cover of vegetation conserves soil and water in an environment. Dense cover of plants in a landscape also helps keeping sediment out of lakes, streams, storm drains, and roads, thus reducing flooding, mudslides, and dust storms. Natural forests are regarded to be soil-builders and the store houses of water. The conserved water is released in the form of springs, streams or rivers. Proper landscaping reduces nitrate leaching from the soil into the water supply. Plants also reduce surface water runoff, keeping phosphorus and other pollutants out of our water ways and preventing septic system overload.	An attractive landscape increases the value of a home by an average of 7.5 per cent and reduces the time on the market by five to six weeks.	Gardening brings lot of enhancement to life. Routine gardening tasks such as shoveling, rot tilling, and even mowing grass with a push-type, reel lawn mower can measure up to the exertion rates of jogging, bicycling, or aerobics. Studies have shown that one hour of weeding bums 300 calories - the same as walking or bicycling at a moderate pace.
Improved air quality	**Community appeal**	**Gardens produce healthy food**
Plants control pollution and keep the atmospheric composition at desirable level. One tree can remove 26 pounds of carbon-dioxide from the atmosphere annually, equaling 11,000 miles of car emissions. Landscape plants, including shrubs and turf, remove smoke, dust, and other pollutants from the air. One acre of trees has the ability to remove 13 tons of particles and gases annually.	Parks and street trees have been found to be second only to education in residents' perceived value of municipal services offered.	People are getting increasingly conscious of their health. Health-friendly organic foods are in demand everywhere. Fresh organic food from the garden can have up to three times as many vitamins and minerals as canned or frozen food. Community garden plots have become a valuable means of providing food for the homeless.
Lowering of summer air temperatures	**Job satisfaction**	**Horticulture is therapeutic**
Atmospheric temperature values during summer season are frequently becoming intolerable. Landscapes laden with mixed forests vegetation play a role to	Employees with an outside view of plants experience less job pressure and greater job satisfaction than workers viewing man-made objects or having no outside view. They also report	Horticultural therapy is a treatment for a variety of diagnoses. Working with and around plants improves quality of life through psychological and physical changes. Nurturing a

Contd...

Table 39.1–Contd...

Environmental Services	*Promoting Economic Development*	*Improving Human Health*
lower atmospheric temperature and make the environment more inhabitable. Urban forests reduce urban air temperatures significantly by shading heat sinks such as buildings and concrete and returning humidity to the air through evaporative cooling. Trees shading homes can reduce attic temperatures as much as 40°C. Properly placed deciduous trees reduce house temperatures in the summer, allowing air conditioning units to run 2 to 4 per cent more efficiently, but allow the sun to warm the house in the winter. Homes sheltered by evergreen windbreaks can reduce winter heat loss and are generally warmer than homes without such protection. By using trees to modify temperatures and protect against wind, the amount of fossil fuels used for cooling and heating is reduced.	fewer headaches and other ailments than workers without the view.	plant into maturity from seed is rewarding and builds self-confidence. Various horticulture-related tasks such as carrying plants, planting trees, or arranging flowers are used to improve co-ordination and motor control of injured or disabled individuals.
Screening out of busy streets	**Worker productivity**	**Landscapes heal**
Well-placed plantings offer privacy and tranquility by screening out busy streets noises and reducing glare from headlights.	Psychologists have found that plants and green spaces provide a sense of rest that allows workers with access to plants and nature to be more productive.	Restorative gardens offer an environment for people who are sick, injured, and under stress to recover and regain confidence in them. Such landscapes are also currently used by hospices in treatment of Alzheimer and AIDS patients. Roger Ulrich showed through a study of hospital patients that those whose rooms witnessed vegetation recovered faster and required less pain medication than did patients without a view of nature.
	Income from tourism	
	Landscapes are the prime objects to nourish tourist industry. Enchanting places charmed by captivating landscapes with variety of ornamental trees, flowering plants.	

now form the basic concerns associated with human psychology. This objective now appears to be a part of our approaches to landscape management.

Someone has rightly said, "We can't make the Earth sustainable; it is sustainable– but whether with us, or without us, is our choice." Nature has healing power but we don't have. Nature's healing will be possible without us. Ours survival however won't be possible without nature.

Landscaping for the Future

Landscaping is an integral part of social habitats and plays an important role in the quality of our environment, affecting our economic well being and our physical and psychological health. A landscape nurtures our multiple needs and also determines to a great extent our behavior in society as well as our attitude towards managing natural resources.

If we are to keep our communities strong and prosperous, we must take responsibility for the conservation and well-being of our environment. Environmental responsibility is a step beyond awareness, developed only through experience. Through our gardens and landscapes, we acquire a personal awareness and responsibility for the environment while we relieve the tensions and frustrations of everyday life.

Landscaping offers many opportunities for the encouragement and education of responsible, productive citizens. School grounds represent the world environment of a child and should be designed and integrated into the curriculum to instill responsibility, knowledge, and experience in caring for the environments, while teaching the math, science, and art associated with the cultivation of plants.

Public and commercial landscapes have a major influence on our environment, and on people's minds, actions and attitudes. Sustainable landscape maintenance techniques can be used to protect the environment while enhancing economic development and improving workers' productivity.

Fear of nature's worst fury is steadily stalking human psychology. This is set to articulate into the processes that are shaping as major agenda of natural resource management. Creating natural carbon sinks on significant scale would help us mitigate global warming and climate change problems.

Landscaping is one of the most cost effective tools for improving and sustaining the quality of life, whether in the city, the suburbs, or the country. With holistic approach–involving environmental psychology–to landscape development, we would be able to enhance sustainability of our natural resources, livelihoods and habitats. Incorporation of environmental psychology issues in operationalising sustainability in landscaping and/or other production systems, in essence, would be indispensable for our future strategies.

References

Alice Jones, 1996. The psychology of sustainability: What planners can learn from attitude research? *Journal of Planning Education and Research*, 16(1): 56–65 (aljones@magnus.acs.ohio–state.edu).

Appleton, J., 1975. *The Experience of Landscape*. John Wiley and Sons, New York.

Bonnes, M. and Bonaiuto, M., 2001. Environmental Psychology: from spatial physical environment to sustainable development. In: *The New Handbook of Environmental Psychology*, (Eds.) R. Bechtel, A. Churchman. Wiley & Sons, New York.

Garling, T. and Golledge, R. (eds.), 1993. *Behavior and Environment: Psychological and Geographical Approaches*. North Holland, Amsterdam.

Herzog, T.R., 1984. A cognitive analysis of preference for field- and forest environments. *Landscape Research*, 9(1): 10–16.

Herzog, T.R., 1987. A cognitive analysis of preference for natural environments: Mountains, canyons, and deserts. *Landscape Journal*, 6(2): 140–152.

Herzog, T.R., 1995a. A cognitive analysis of preference for urban nature. In: *Landscape Perception*, (Ed.) A. Sinha. Academic Press Inc., San Diego, CA.

Herzog, T.R., 1995b. A cognitive analysis of preference for waterscapes. In: *Landscape Perception*, (Ed.) A. Sinha. Academic Press Inc., San Diego, CA.

Herzog, T.R., 1992. A cognitive analysis of preference for urban nature. *Journal of Environmental Psychology*, 9: 27–43.

IGNOU, 2005. *Sociology of Development*. Indira Gandhi National Open University, New Delhi.

Kaplan, R. and Kaplan, S., 1989. *The Experience of Nature: A Psychological Perspective*. Cambridge University Press, Cambridge.

Kaplan, S. and Kaplan, R., 1982. *Cognition and Environment*. Praeger, NY.

Singh, V., 2005. *Agrobiodiversity, Food Security and Sustainability in the Himalayan Mountains: An Uttaranchal Perspective*. Gorakhpur Environmental Action, Gorakhpur.

Stokols, D. and Altman, I., (Eds.), 1987. *Handbook of Environmental Psychology*. Wiley, New York.

Strumse, E., 1994. Environmental attributes and the prediction of visual preferences for agrarian landscapes in western Norway. *Journal of Environmental Psychology*, 14: 293–303.

Strumse, E., 1996. Demographic differences in the visual preferences for agrarian landscapes in Western Norway. *Journal of Environmental Psychology*, 16: 17–31.

Index